PERSPECTIVES IN BEHAVIOR GENETICS

PERSPECTIVES IN BEHAVIOR GENETICS

Edited by

John L. Fuller
State University of New York
Binghamton

Edward C. Simmel
Miami University
Oxford, Ohio

LAWRENCE ERLBAUM ASSOCIATES, PUBLISHERS
1986 Hillsdale, New Jersey London

Lawrence Erlbaum Associates, Inc., Publishers
365 Broadway
Hillsdale, New Jersey 07642

Library of Congress Cataloging-in-Publication Data

Perspectives in behavior genetics.

Bibliography: p.
Includes index.
1. Behavior genetics. I. Fuller, John L.
II. Simmel, Edward C.
QH457.P47 1986 591.5′1 86-16613
ISBN 0-89859-869-9

Printed in the United States of America
10 9 8 7 6 5 4 3 2 1

Contents

List of Contributors

John C. DeFries, Institute for Behavioral Genetics, University of Colorado, Boulder.

John L. Fuller, Department of Psychology, State University of New York at Binghamton.

Kenneth R. Henry, Department of Psychology, University of California, Davis.

Jerry Hirsch, Departments of Psychology and Ecology, Ethology, and Evolution, University of Illinois at Urbana-Champaign.

Mark J. Holliday, Department of Psychology, University of Illinois at Urbana-Champaign.

Joseph K. Kovach, Research Department, The Menninger Foundation, Topeka, KS.

Michele C. LaBuda, Institute for Behavioral Genetics, University of Colorado, Boulder.

Jeffry P. Ricker, Department of Psychology, University of Illinois at Urbana-Champaign.

Edward C. Simmel, Department of Psychology, Miami University, Oxford, OH.

James R. Stabenau, Department of Psychiatry, University of Connecticut School of Medicine, Farmington.

Mark A. Vargo, Biology Department, Brandeis University, Waltham, MA.

George P. Vogler, Institute for Behavioral Genetics, University of Colorado, Boulder.

Preface

We live in a world in which the number of publications in behavior genetics has reached a point where it is difficult, even for those who teach the subject, to keep up with the literature. The editors of *Perspectives in Behavior Genetics* believe that there is a need for people who have planned and executed long-term research programs to summarize and comment on their results. We hope that the present volume will help to meet this need. The authors were given free choice of subject and format. The result is a variety of topics that have been researched mainly over the past decade. Chapter 1, by Fuller and Simmel, is an exception. We have simply looked at the work of others in behavior genetics over a quarter-century and tried to detect trends in the types of research done in the field. In many ways the objectives of today's researchers have been shaped by persons such as Dobzhansky and Thompson, whose ideas have influenced the directions of research since behavior genetics has been recognized as a discipline. Perhaps the most important advance over the past 25 years is the recognition of the need for methods of analyzing complex data banks.

In Chapter 2, DeFries, Vogler, and LaBuda describes a long-term project dealing with the heritability of differences in reading ability. The problem of dyslexia is serious, and it is difficult to separate the importance of biological and environmental factors. Their work is ongoing, but already important data has been collected.

Chapter 3 is concerned with the genetics of audiogenic seizures in inbred mice. It is well known that genes have an important influence on susceptibility to high-pitched sounds. Henry has investigated the genetic factors responsible for pathological features in the nervous system.

Kovach, in chapter 4, summarizes a long program of studying color preferences in newly hatched Japanese quail. Most tantalizing is his discovery that such preferences appear to be mainly innate although some learning also occurs. His research is a fine example of methodology for dealing with the heredity–environment question.

In chapter 5, Ricker, Hirsch, Holiday, and Vargo consider the learning ability of the blow-fly (*Phormia*) and the fruit-fly (*Drosophila*). The problems deal with distinguishing between learning in its usual sense and transient changes in responsivity.

Stabenau, in chapter 6, deals with the genetic aspects of human alcoholism. He discusses both biological and environmental factors involved in the pathology of this costly plague. It becomes clear that progress against this disease will require more multidisciplinary research than has been carried out up to this time.

As editors, we have learned a great deal from our authors. We hope that our readers will benefit as much as we.

1 Trends in Behavior Genetics: 1960–1985

John L. Fuller
State University of New York at Binghamton

Edward C. Simmel
Miami University, Oxford, OH

INTRODUCTION: A VIEW FROM 1957

Our contribution to this volume is a study of trends in the genetics of behavior over the past 25 years, 1961–1985. Before doing so, we are presenting brief excerpts from three papers delivered at the 1956 conference of the Milbank Memorial Fund entitled "The Nature and Transmission of the Genetic and Cultural Characteristics of Human Populations." Each of the authors eventually served a term as president of the Behavior Genetics Association (BGA). All three directed their remarks to general issues that are still relevant to this area of science.

Dobzhansky (1957) entitled his presentation "The Biological Concept of Heredity as Applied to Man." He noted at the start that "the traditional dichotomy of hereditary versus environmental traits is invalid" (p. 11). He emphasized the importance of genes by noting that "one has to be human to learn any human language, hence the learning process presupposes a human genotype" (p. 11). "Your genes . . . have determined your intelligence, but only in the sense that a person with a different genotype might have developed differently if his life experiences were approximately like yours (p. 13). We inherit genes—not characters or traits. . . . Genes determine processes, not states" (p. 14).

Continuing, he stated, "It is . . . legitimate to inquire what part of the observed variance in a given trait is due to the diversity of human genotypes, and what to the environments in which men develop. . . . No single or simple answer to this question (the relative weights of genotypes and environmental variables in the causation of observed human differences) is possible because the weights are different for different traits. . . . The demonstration that a given trait is conditioned by heredity does not . . . exclude the possibility that variation is controlled also by environmental influences" (p. 15). And finally: "The observed degree of heritability of a given character difference may be valid only for that time, place, and material studied" (p. 16). These precepts have become a guide for researchers in behavior genetics, and we believe that for the most part they are widely observed. Dobzhansky and other pioneers set a code of procedures that made behavior genetics the study of nature *and* nurture rather than of nature *or* nurture.

Thompson (1957) was particularly interested in personality and intelligence differences in populations. He noted that "Human groups must be demonstrated not only by physical, but also by psychological variables" (p. 38). "So long as there is assortative mating with respect to . . . these characters, we can expect to find genetically determined group differences" (p. 39). Thompson was not referring to "ethnic differences," but to groups of any kind that differ in modal behavior. On genic versus environmental explanations of behavioral differences, he stated that "within the limits of scientific facts we can feel free to proceed in either direction without embarrassment" (p. 40). The conciliation he predicted is not universal, but we believe that the majority of biologists and psychologists accept an interactive gene-environment model of behavioral development. Universal agreement on particular cases has not and may never be achieved. Thompson's view was that "The problem of separating genetic from environmental determination is difficult, but it is not insoluble" (p. 49). He considered culture-free tests to be an unsatisfactory approach, and favored multiple variance analysis. As a final way of disposing of the nature-nurture problem he suggested measuring phenotypic transmission of psychological characters without attempting to measure the relative importance of heredity and environment. We suspect this was written with tongue in cheek, but it is a convenient way to challenge die-hard environmentalists.

A third presentation at the Milbank conference dealt with pathways between genes and behavioral characteristics excluding defects in physical development or metabolism involving chromosomes that produce gross defects in physical or mental characteristics (Fuller, 1957). Behavior genetics asks questions regarding correlations between behavioral phenotypes and genotypes. Evidence for such correlations in animals comes from various sources: (1) behavioral effects of single-locus differences such as the reduction of male mating drive in *Drosophila* by the yellow gene; (2) successful selection for behavioral characters in insects and mammals; (3) examples of behavioral variation among inbred strains, and reduced variation within such strains.

With respect to the analysis of human behavior Fuller wrote, "Human populations are not good material for evaluating the effects of single genes on behavior unless the gene produces abnormal development" (p. 103). He suggested that, because we cannot carry out selection experiments on our own species, we should investigate the effects of assortative mating. On the question of ethnic differences in behavior he noted that differences exist in characteristics such as blood groups, pigmentation, hair distribution, and other somatic characters. Why should there not be a genetic contribution to ethnic differences in behavioral characteristics? He believed, however, that they would be of minor significance.

Fuller also advocated a physiological approach to psychogenetics, ranging from studies on cellular enzymes and hormones to the structure and functions of the somatic and autonomic nervous systems. Progress in this area should go beyond analysis of single-gene effects on mental deficiency. For example, it would be interesting to study the role of genes on variation of social behavior within a species. As an example he cited studies of social interactions among breeds of dogs (Scott & Fuller, 1965). He also proposed that humans were not selected for uniformity of behavior, but for diversity within an organized society. The contrary hypothesis that any healthy individual could fill any niche in society seemed to have little empirical support.

Clearly, by the late 1950s the framework of a new discipline was taking form. Although it was not based on eugenics, it would deal with a wide range of human problems that had interested the eugenicists. The difference was that it emphasized that both environments and genes should be considered when scientists tried to better the human condition through biology and social science. What we now call behavior genetics (alternatively behavioral genetics) has its roots in comparative and differential psychology, the behavioral element within zoology, and a variety of medical specialities that dealt with psychopathology and mental deficiency. Behavior genetics today is a synthesis of elements from these disciplines. Its openness to contributions from a variety of sources is a source of strength. It could not have survived without cooperation among scientists with different skills and common objectives.

NOTES ON BEHAVIOR GENETICS IN 1960

Recognition of Behavior Genetics as a designated discipline might be said to begin with Fuller and Thompson's *Behavior Genetics* published in 1960. The book did not propose a new orientation combining biology and psychology. It was an effort to summarize the research of many scientists, mainly from the United States and Europe. Its predecessors were scattered articles in various biological and psychological journals. Actually, the beginnings of behavior genetics precede written records. The humans who domesticated horses, cattle, dogs, and ducks altered their behavior by genetic selection using procedures similar to those of today's scientists who breed mice and fruit flies in their laboratories.

In the 1940s and 1950s considerable behavior genetic research with both animals and human subjects was carried on by psychologists and biologists. Hall

(1951) reviewed experimental animal studies (mostly with mice and rats) of selective breeding for maze learning, aggression, audiogenic seizures, and other behaviors that could be quantitatively measured. He outlined the basic genetics required for research on laboratory animals, and described the characteristics of inbred strains. McClearn (1963) traced the history of ideas regarding the heritability of behavior from about 8,000 B.C. to 1960. *Drosophila*, mice, rats, and humans were the species most studied, and they are still predominant. For human data, family correlations, twin studies, and adoption studies were the most used techniques. (They are now, but the statistical designs of present-day researchers are more sophisticated.) Animal research dealt with inbred and selected lines, and single-gene effects.

Scientists as a group are curious regarding relationships between phenomena, and behavioral geneticists are no exception. Regardless of immediate economic application, they study genetic and environmental factors (physical and social) that affect behavior. They seem to have divided their attention between topics that are relevant to human welfare (for example, psychopathology) and topics related to general issues dealing with intraspecific variation in individual and group behavior in mice and insects. It is an error to conclude that behavior geneticists neglect environmental contributions to behavioral development. Major areas of their research include (1) the effects of selection on behavior, (2) the interaction of genotypes with environmental factors, (3) the biochemical and anatomical factors associated with specific behavioral phenotypes, (4) identification of the genes involved with specific behavioral syndromes, and (5) the role of genes in the ontogeny of behavior. Some behavior geneticists concentrate on problems related to human behavioral disorders or deficiencies, mainly working directly with human subjects but sometimes with animal models. Others are concerned with understanding the nature of gene-environment interactions on behavior without considering their direct application to human or animal welfare. We come from a variety of backgrounds and have different skills and interests.

BEHAVIOR GENETICS IN ACADEMIA

Our major emphasis in this chapter is the "producers" of behavior genetics and their "products": research and the dissemination of knowledge through conferences, books, chapters, and journals. Equally important, however, are the "consumers" of the data obtained by the scientists and their reaction to the general principles espoused by the producers. Later in this chapter we consider some of the conflicts among scientists from a number of disciplines—essentially a continuation of the rather hoary nature-or-nurture argument. The fact is, however, that behavior genetics has consumers in many disciplines, among them medicine, physiology, education, evolutionary biology, and gerontology. However, the discipline with the closest and most direct ties to behavior genetics is almost certainly psychology. We believe that the relationship between genes and behavior should be part of a general education.

Two recent chapters by Fuller have dealt with the problems of a hybrid science. In "Psychology and Genetics: A Happy Marriage?" (1982) he predicted that disagreements between the disciplines will continue, but the results will be beneficial to both if reason wins over ideology. In "Psychology and Genetics" (1984) he considered relationships among psychology, genetics, and sociobiology. In this section we emphasize the influence of psychology on behavior genetics. It seems to us that further progress of the discipline depends as much (perhaps more) on new behavioral inputs as on progress in genetics. The cutting edge of genetics is at the molecular level, rather than at the level of genetics most directly related to behavior.

Psychology's contribution to behavior genetics includes research methodology, measuring instruments, and other technologies used by researchers. It has also provided a "home" for instruction. Most behavior genetics courses are taught within departments of psychology. Most important, psychology has contributed people—the majority of individuals who consider themselves to be behavior geneticists have received their training in psychology. However, many individuals with medical or zoological backgrounds have made substantial contributions to the field. Workers in a hybrid research area must educate themselves in areas outside their formal education.

What impact has behavior genetics had on psychology? It may be too early, considering the youth of behavior genetics, to attempt an overall assessment of its impact on a major discipline such as psychology. We will, however, attempt to give a rough idea of its influence on psychological thinking. One index of an organized system of beliefs, methods, and findings is its incorporation into the introductory textbooks of a discipline. We can gain a general idea of the acceptance of behavior genetics in psychology through examining contemporary textbooks.

If a person who is totally naive with respect to the subject matter of modern psychology were to select certain recent books, such as Darley, Glucksberg, Kamin, and Kinchla (1984) at the introductory level, or Sarbin and Mancuso (1980) at the graduate-professional level, this person might be led to conclude that genetic approaches to normal and abnormal behavior are viewed by psychologists as highly controversial if not downright wrong. Other books by psychologists and biologists (e.g. Lewontin, Rose, & Kamin, 1984) deprecate the significance of genes for any variation in personality or cognitive ability and would reinforce this view. (It is not clear if these authors would accept evidence for the heritability of behavioral characteristics in nonhumans.) The impression would be left that not only is the Nature *versus* Nurture issue a major topic of great controversy, but that the Nature side is contrary to the prevailing psychological evidence and may even be immoral.

Such a conclusion would be a distorted picture of the true state of affairs in current psychological thinking. A broad sampling of current introductory texts indicates quite clearly that most of them show a rather reasonable and balanced picture of the topics of behavior genetics, with at least one current book (Gleitman, 1983) presenting a most accurate and sophisticated incorporation of several issues and findings of behavior genetics with other subject matter. Some other authors do not cover the area as well, but they do discuss the genetic basis of various behavioral phenotypes; controversial issues are not overplayed; and, in general, behavior genetics seems to have been incorporated into modern psychology. Some faculty might criticize the tendency of an author to treat genetics as a separate topic rather than tying it specifically into various subject matters. Similarly, a physiological psychologist might criticize a general textbook treatment of brain function. Overall, behavior genetics seems to have been adopted into the mainstream of modern psychology. It receives reasonable treatment within all but a small minority of the current and recent crop of introductory psychology textbooks.

Is the acceptance of genetic concepts into contemporary psychology texts a recent development? Might it be the offshoot of the institutionalization of a new discipline with a society and a journal? Let us go back 25 years to the early and mid-1960s. At that time there was neither a society nor a journal. The only textbook devoted exclusively to the topic was simply entitled *Behavior Genetics* (Fuller & Thompson, 1960). The Behavior Genetics Association was nearly a decade away. Not surprisingly, few psychologists considered behavior genetics to be a major research interest, but a small number of pioneers from psychology and biology were finding common interests, sensing needs for collaborative research, and beginning to get together informally. Only a small handful of universities offered even one course devoted to behavior genetics, although the subject matter was sometimes taught under titles such as "Differential Psychology."

Given this contextual background, one would expect that the prevailing introductory textbooks of that era would be nearly devoid of behavior genetics content—save perhaps for a brief mention of Galton, or the Jukes and Kallikaks, buried somewhere in a chapter on the history of psychology. Look, then, at a sampling of seven of the most widely used introductory texts of the 1960s. Included for comparison are the current editions of two of these books.

Table 1 shows the pages on which material is given for any of the following topics: *genetics, heredity, hereditary, inherited, behavior genetics*, or close synonyms. The selected textbooks were all "best sellers," and were probably the books from which most students of the 1960s were introduced to psychology. Note that the topics related to behavior genetics were not only covered in the eclectic books (e.g., Morgan or Hilgard & Atkinson), but also in those having a strong behavioristic emphasis (e.g., Kendler, and Kimble & Germezy). Surprisingly, we found that behavior genetics received rather extensive coverage, although the institutionalization of the discipline via journals, formal courses, and professional societies was still in the future. Comparing the *amount* of coverage in these textbooks of two decades ago with their counterparts of today we found less change than we had expected.

TABLE 1.1

Authors	*Edition*	*Yr. Pub.*	*Pages*
Morgan	2	1961	30–35, 36–43, 101, 599, 636
Kimbel & Garmezy	2	1963	406–408
Hebb	2	1966	139–146, 155, 160, 162, 193, 195
Hutt, Isaacson, & Blum	1	1966	32–63, 78, 434, 739–740
Munn	5	1966	89–101, 139–146
Hilgard & Atkinson	4	1967	8, 444–460, 540–541
Kendler	2	1968	86–89, 489–491, 516–518, 623–632
Atkinson, Atkinson, & Hilgard	8	1983	53–57, 62–63, 373–378, 647–675
Kimble, Garmezy, & Zigler	6	1984	31–67

If introductory textbooks say anything about the current state of a discipline, it appears as though psychology was waiting for behavior genetics to join physiology as a link with the biological sciences. This is not to say that behavior genetics has been incorporated into psychology to the same extent as, say, information processing, or that all psychology students (and their teachers) are knowledgeable and sophisticated in its techniques. One can be an outstanding psychologist without understanding the nuances of interpreting data from twin studies, or the use of recombinant inbred strains to detect the location on a chromosome of a gene that is associated with an unusual form of behavior. But if the content of introductory textbooks tells us even a little about the *Zeitgeist* of psychology, behavior genetics has become a part of the mainstream. In fact, it seems to have stronger ties with contemporary psychology's emphasis on individuality and developmental processes than it does with such fields as molecular genetics.

Most college students will probably obtain their ideas of behavior genetics from a chapter in their psychology of animal behavior text. To go further, it turns out that there are fewer texts suitable for an undergraduate course than books at a professional/advanced level. The number of books intended as undergraduate texts is sparse. The first of these, by Fuller and Thompson, was published in 1960 and was the only book at this level for the following 13 years. It was a comprehensive text covering the findings and methods of both animal and human behavior genetics.

Fuller and Thompson was followed in 1973 by a text authored by McClearn and DeFries, also balanced between human and animal findings, and with an emphasis on quantitative genetics assuming a fair degree of biological and quantitative sophistication on the part of the student (and the instructor). Three years later, in 1976, a third text entered the scene. This book, by Ehrman and

Parsons, emphasized animal behavior genetics and evolution, with a considerable amount of coverage of research on invertebrates, especially *Drosophila*.

Reflecting the continuing and growing involvement of behavior genetics within the undergraduate curriculum (especially in psychology departments), each of these pioneering textbooks has undergone a second edition. Fuller and Thompson (1978) was reorganized and doubled in size over the 1960 version. McClearn and DeFries (1973) became Plomin, DeFries, and McClearn (1980), updated, extensively rewritten, and more comprehensible to the less biologically sophisticated student. In 1981, Ehrman and Parsons was not only brought up to date, but was renamed *Behavior Genetics and Evolution* to reflect its emphasis.

In 1985, Fuller and Thompson was out of print, but the remaining two were joined by a newcomer, a comprehensive textbook by David Hay. There is now a good sample of beginning-level books in behavior genetics for the instructor of an undergraduate course to select from. These texts would also be useful for any person interested in securing a background in the subject. New editions and perhaps new texts will probably come on the market in the future. As one peruses the older and newer volumes there is evidence of a maturation of the discipline during the past quarter of the century. It has been a steady growth, with general agreement about the direction of research. Although the emphases of the texts differ, there is no evidence of widely different paradigms.

Behavior Genetics in Europe and Australia. We sent a number of letters to members of the Behavior Genetics Association outside of North America asking about research activity and academic courses in behavior genetics in their countries. We received only four responses and will summarize them briefly.

Professor van Abeelen from the Netherlands provided information on six Institutes in departments of zoology, comparative and physiological psychology, genetics, population and evolutionary biology, zoology, and psychology. All of these centers provide undergraduate courses in Behavior Genetics. Graduate and postdoctoral studies are available to a limited number. Five of the institutes specialize in animal studies. The sixth is conducting twin studies on stress and cardiac responses.

K. Lagerspetz reported from Finland. The work of Kalervo Eriksson on selection for high and low consumption in alcohol is well known. Her well-known work with aggression in mice is continuing, but there are few others working with animals. Twin studies are being conducted on the same problems that interest the American continent, personality, intelligence, and schizophrenia.

From Belgrade in the eastern zone of Europe V. Kekic described activity in behavior genetics in his country (Jugoslavia) and Hungary. Since 1978 he has taught university courses with a substantial behavior genetics content. Medical schools have some interest in genetics, particularly in cytogenetic studies of Down's syndrome and the possible influence of genes on alcoholism. In Hungary the Laboratory of Behavior Genetics was founded in 1973 at the University of

Budapest. Topics of study include ethological studies of fish, neuro-behavioral studies with recombinant inbred strains, and reproductive behaviors in *Drosophila*.

P. A. Parsons reported that in Australia behavior genetics is mainly centered at La Trobe University. *Drosophila* studies are a central theme, but David Hay also works with humans, particularly with twins and others with mental retardation syndromes. No formal courses in the subject are given, but the area is included within general genetic and psychology courses.

REVIEWS OF BEHAVIOR GENETICS: 1960–1985

In this section we turn to seven reviews (with 15 authors) of behavior genetics published in the *Annual Review of Psychology*. The earlier reviews provide an overall account of contemporary research. Recent reviews are more specialized, probably because of the increase in the number of publications and increased specialization within the discipline. All authors identified major topics of interest, commented on their significance, and suggested opportunities for future research.

The first review was contributed by the co-author of this chapter (Fuller, 1960). In 30 pages he listed 130 articles or books.The number of research reports by taxonomic groups were rodents (58), humans (29), other vertebrates (17), *Drosophila* (6), and other invertebrates (3). Seventeen references dealt with general topics. Clearly, comparative psychology was of great interest. In space allotment, however, human studies beat out all other species by a significant margin. Fuller noted high interest in the identification of genes and chromosomes that could be matched with specified behaviors. He endorsed selection programs for behavioral phenotypes, and the use of inbred lines to ensure consistent biological characteristics in research with animals. Finally, he advised that experimenters should use both genetic and environmental variables in their research designs.

In the second review McClearn and Meredith (1966) listed nearly twice as many contributions (254) as Fuller. Space allocations for human and animal studies were approximately the same, but the number of citations for animal studies outnumbered those of humans by a wide margin. Emotionality, social behavior, dominance, mating behavior, learning, alcohol consumption, and audiogenic seizures were featured in animals. Studies on the behavioral correlates of gene mutations affecting pigment and locomotion had some popularity, but the results were judged to have little general interest. Reports on rodents outnumbered other species. Broadhurst's work on emotionality in rats was positively reviewed. In *Drosophila* research Hirsch's selection for geotaxis and phototaxis attracted attention. Manning in Britain and Parsons in Australia studied the genetics of mating behavior in *Drosophila*, thus bringing behavior genetics into the arena of evolutionary theory. In the human area, 6.5 pages were devoted to chromosomal anomalies and deleterious genes. Approximately the same space was devoted to twin study methodology, intelligence, personality, and psychopathology.

The third review (Lindzey, Loehlin, Manosevitz, & Thiessen, 1971) noted three debuts: a journal, *Behavior Genetics*; a society, *The Behavior Genetics Association*; and graduate programs in the Universities of Colorado (Boulder) and Texas (Austin). Particularly noted was progress in the quantitative analysis of behavioral phenotypes. About 430 citations were reviewed or listed, more than three times the number noted by Fuller 11 years earlier. Steady progress in animal behavior genetic research was described, but no outstanding breakthroughs were noted. Unlike McClearn, these reviewers gave high credit to single-gene techniques in animal research, particularly in the analysis of behavioral development. The genetic approach to psychopathology and normal personality was reported to be "on the upswing." Reviews of schizophrenia and the affective psychoses were detailed and balanced. Somewhat less attention was given to cognition, but relationships between behavior genetics and evolutionary processes were given more attention than in previous reviews. By now a new discipline, sociobiology, was attracting positive and negative comment and was briefly reviewed in a balanced way. Sociobiology and behavior genetics have some objectives in common, but for the most part they go their own ways. (For a recent discussion of their relationship see Fuller, 1983.)

The fourth review, with 283 citations, was authored by three eminent behavior geneticists from the United Kingdom (Broadhurst, Fulker, & Wilcock, 1974). They noted important differences between early Mendelian approach to behavior genetics and newer biometric methodologies that take both genetic and environmental factors into consideration. Animal research occupied 6 pages divided among major gene effects, and selection for open-field activity, aggression, avoidance learning, and alcohol consumption. Human research occupied 10.5 pages divided among cognition (4), mental retardation (1), normal personality (1), abnormal personality (3), and methodology (1.5). The author's view of the future of behavior genetics is shown in this quotation: "We foresee a period of concentration on gathering empirical data which can then be consolidated into a comprehensive structure of knowledge" (p. 406).

Review 5 (DeFries & Plomin, 1978) begins on a plaintive note. "If a successful area is one in which the rate of publication is so great that no one person could keep up with the current literature, and couldn't understand it if he tried, then behavioral genetics is a rousing success" (p. 473). Their citation list made a new record with 334 entries.

On the animal side the areas previously noted continued to attract attention, but some new topics were identified: recombinant inbred strains (6 citations); neurogenetics, mostly *Drosophila* (21 citations); psychopharmacogenetics (10 citations). Although large-scale adoption studies are expensive, they were applauded because of some advantages over twin designs. Psychopathology is "the most active area in behavioral genetics"; and "the time has passed when behavior can be assumed innocent of genetic influence until proven guilty." For the first time in this series of reviews there is recognition of attacks on the validity of behavioral genetics, particularly when humans are subjects. The authors deplored intimidation of researchers, as well as the polemics and politicizing that have been employed by critics of human behavioral genetics. Although the vigor of these attacks (they are doctrinaire preachings rather than honest debates on the interpretation of experiments) has declined, they still smolder in a few places.

The sixth reviewer (Henderson, 1982) broke the pattern of covering the entire field of behavior genetics in one review, and concentrated on human studies. Apparently he agreed with DeFries and Plomin's remark that no individual could possibly review and evaluate all the active areas. Still, he found 216 citations of interest. Henderson noted three trends in the genetics of human behavior. First, there was an increase of complexity in research designs with larger samples, better control of environments, and more multivariate analysis. Second, greater caution in the interpretation of data was apparent; multiple estimates of parameters based on alternative hypotheses were common. Third, statistical models for decomposition of genetic and environmental variance into components were in wide usage.

Other notes of interest were as follows: (1) There is a decline of support for the hypothesis that an X-linked gene has major impact on spatial perception. (2) A new technique, the twin-family design, is available for studying the genetics of cognitive abilities (Rose, Miller, Dumont-Driscoll, & Evans, 1979). (3) Compared with earlier estimates, recent studies on the heritability of intelligence have decreased to a range between .3 and .6 with lower values in more recent studies. Evidence for differences in the proportions of additive genetic (AG) and common environment (CE) for general intelligence, specific abilities, and school achievement is inconsistent. He concluded that most abilities are influenced by the same sets of genes and environments. We interpret this as an assertion that "personality genes" are also "intelligence genes," and vice versa. (4) On personality and temperament he commented, "Although a large group of personality traits . . . may show more or less similar heritabilities, other traits . . . involving other attitudes and beliefs . . . show varying degree of G and CE influences" (p. 423). A plea was made for longitudinal studies of sequential cohorts in order to detect interrelationships among genotypes, maturation, and changing environments. Such studies are costly and hard to finance, but are judged to be important and cost-effective in the long run. The authors of this chapter agree. (5) Psychopathology is still an important part of behavior genetics. Currently there are increasing doubts of a real genetic distinction between bipolar and unipolar affective disorders, and also of the hypothesis that an X-linked dominant gene transmits these disorders. Genetic studies suggest that schizoaffective disorders are related more closely to affective disorders than to schizophrenia. And after many years of research on the genetics of schizophrenia there is still no firm choice between a major locus and a polygenic hypothesis. Clearly in the eighties there are still issues that have not been resolved.

For the future, Henderson urged increased use of comparisons between studies. Large numbers of subjects and agreement on techniques are needed to test the generality of results gathered in different populations. He also recommended more complete reporting of experimental data so that they can be pooled and analyzed by other investigators. Full information on environmental conditions as well as the results of genetic analyses is essential. Finally he challenged investigators who are primarily interested in the environmental influences on behavior to become acquainted with the biobehavioral approach to development and individuality.

The most recent review in this series by Wimer and Wimer (1985) is entitled "Animal Behavior Genetics: A Search for the Biological Foundations of Behavior." The first part of the chapter deals with animal models of human diseases; the second with genetic variations in "normal" animals, and their psychological and neurological correlates. A total of 342 citations set a new record for this series. Among the models discussed are audiogenic seizures (with similarities to some epilepsies); alcohol dependency in rodents; emotionality in rats and pointer dogs; and autoimmune disorders in humans that have been associated with dyslexia and left-handedness. The animal model approach has been widely used in medicine for disorders such as cancer and hypertension. Judging the relation of behavioral phenotypes in different species is often difficult. Audiogenic seizures in mice are very different from typical human epilepsy. The Wimers suggest that for behavior geneticists the similarities of motor patterns (the behaviors that interest psychologists) are less important than similarities and differences in the brain structures that are a part of neuroscience. Current research on learning in animals is concentrated on taste conditioning in blowflies, and shock avoidance in rodents. Mate selection in mice has been reported associated with a major histocompatibility complex.

Behavior genetics is now associated with neurobiology. Heritable differences in testosterone concentration produce structural variation in the central nervous system. Gene mosaics in *Drosophila* are associated with localized neural functions and behavior. Two lines of rats selected for high and low shock avoidance differ significantly in the mossy fiber patterns of their intra- and infra-pyramidal areas. Differences in shuttle box avoidance have also been found in two inbred lines of mice and their hybrids. Correlated with the behavioral differences are variations in granule cell numbers and density. It appears that genic influences on brain development do have significant effects on behavior. The authors conclude: "Taken together the results of studies of mossy fiber patterning and granule cell density lead to two major conclusions. 1. The mouse brain represents a genetically differentiated system with various morphological characters under separate control. 2. Morphological variation can have very substantial behavioral associations" (p. 203).

What will be the major topics of future reviews of behavior genetics? We are not making predictions, but we have expectations. Many of the topics of the present will still be pursued in order to improve our understanding of basic psychobiology. We will continue to develop animal models of the effects of various stressors on humans. Research on variation in the effects of ethanol in selected lines of mice is an example of such topics. If it proves to be useful in understanding human alcoholism, it is likely that similar programs on other forms of substance abuse will proliferate. However, behavior genetics will not become an applied science. It will become a parent of a collection of hyphenated disciplines that deal with heredity, neuroscience, and psychology. Academic behavior geneticists will be found in a variety of departments, though psychology may still take on the task of formal education in the field, from textbooks to *Annual Reviews*.

INVITED CHAPTERS, SYMPOSIA, WORKSHOPS, READINGS

The years 1960 to 1984 saw a number of volumes devoted to behavior genetics. We have selected 20 of these for comment. Our choices were in part a matter of convenience, but we believe that they are representative of the period covered. Although all were written in English and the majority edited by Americans, there is also a substantial contribution from Europeans. Unfortunately, our lack of skill in the Russian language made it impossible to evaluate a 1975 publication of the Academy of Sciences of the USSR.

We have placed these volumes in four categories: *invited chapters* were solicited by editors from individuals; *selected readings* were reprinted from journals; *symposia* were the products of meetings that were open to an audience; and *workshops* were the output of meetings of leading scientists held over several days or even weeks. The distinctions between these classes are somewhat blurred, but all contribute to the history of behavior genetics. We consider our examples in order of appearance. Comments are brief and citations of individual papers sparse because of space limits. For further information readers may refer to the collections themselves as noted in the bibliography.

Roots of Behavior (Bliss, 1962) was the product of a symposium sponsored by the American Association for the Advancement of Science. Six of the chapters dealt with the genetics of animal behavior. Topics included geotaxis and phototaxis in *Drosophila*; neurological aspects of behavior in other insects; surveys of learning in insects, fish, birds, and mammals; and strain differences in alcohol consumption among inbred strains of mice. *Roots* also dealt with biological differences in behavior among species and phyla. Behavior genetics today concentrates on variation within species. Hirsch, one of the contributors, stated that few variations in behavior can be explained by simple Mendelian principles. The point of his statement was to correct the idea that there is a gene for every behavior. This does not apply universally, and medical genetics deals with genes and chromosomes that are associated with physical and behavioral deficiencies. The symposium demonstrated that behavior genetics was combining comparative psychology, ethology, Mendelian, and quantitative genetics.

Vandenberg (1965) hosted a workshop that produced *Methods and Goals in Human Genetics* with 18 chapters. Half of them dealt with twin studies ranging from statistical aspects of data analysis to a survey of intelligence, personality, and psychopathology. Family studies, ethnic comparisons, and even comparisons of human and animal techniques in behavior genetics were covered. Some contributors described results of studies; others submitted prospectuses for the future. The presentations and discussions are still of interest 20 years later.

In 1965 the Eugenics Society of Great Britain sponsored the first of three symposia (Meade & Parkes, 1965). The book, *Biological Aspects of Social Problems*, dealt mainly with the nature-nurture relationship and genetic disorders. Thoday considered geneticism and environmentalism as programs and gave an elementary exposition of the new discipline, behavior genetics. A second volume (Meade & Parkes, 1966) dealt with the genetics of intelligence and mental deficiency. A third volume (Thoday & Parkes, 1968) was mainly concerned with genetic factors in personality and social behavior. The sixties were the decade in which eugenics was being replaced by social biology on one hand, and by behavior genetics on the other. Eugenics was still a movement as well as a study, and along with reports of research were proposals for applying biological and psychological findings to educational and social problems.

Behavior-Genetic Analysis (Hirsch, 1967) is the product of two workshops held in 1961 and 1962. This was the first multiauthored volume dealing with a broad range of topics in human and animal behavior genetics. Evolutionary aspects of behavior were treated by Hirsch, Caspari, Washburn, and Manning. The "big three," humans, mice and *Drosophila*, were well represented, along with single chapters on primates and social insects. Roberts and DeFries contributed useful accounts of quantitative genetics. The title implies that behavior genetics must embrace the techniques and principles of its two parent disciplines. It is still a good source of ideas.

Spuhler's (1967) *Genetic Diversity and Human Behavior* is also the product of a workshop. Concepts of race, ethnic differences, mating patterns, individuality, and genetic variability among humans were discussed. Anthropology played a substantial part in the discussions. Evolutionary concepts were stressed by Dobzhansky and Spuhler. The concepts discussed were similar in some ways to those of E. O. Wilson's (1975) yet to come *Sociobiology*, although they are more specifically directed at human behavior.

Biology and Behavior: Genetics (Glass, 1968) is the outcome of a symposium supported by the Russell Sage Foundation, the Social Science Research Council, and the Rockefeller University. Many of the chapters deal with humans on topics such as intelligence, ethnicity, personality, and psychopathology. Comments were also made on eugenics, social Darwinism, and the perennial nature-nurture dichotomy. Included with reports on *Drosophila* and *Mus*, the usual representatives of nonhumans, were studies of domestic dogs, wolves, and red grouse. McClearn surveyed the social implications of behavioral genetics and speculated on the future of the human species. Bressler commented on the conflicting viewpoints of sociology and biology, and Haller reviewed the histories of social science and genetics. The issues of 1968 were not greatly different from those of 1985, but there are now signs of a reduction in the sharpness of the nature-nurture dichotomy.

Vandenberg (1968) conducted a second workshop whose proceedings were entitled *Progress in Human Behavior Genetics*. Although several articles dealt with methods of statistical analysis, a majority were related to behavioral phenotypes of current interest. Examples were phenylketonuria, Down's syndrome, Turner's syndrome, individual sensory differences, schizophrenia, cognitive abilities, and autonomic responses to stress. A majority of the studies used twins as subjects, but some were based on family groups. Environmental factors related to these disorders were also emphasized. Kringlen investigated clinical variability of schizophrenia in twin pairs. Erlenmeyer-Kimling studied the sibs of schizophrenic probands, and Vandenberg and Johnson reported on the relation between age of separation and similiarity of tested IQ in pairs of separated monozygotic pairs. For the human geneticist interested in behavior the two Vandenberg volumes are still valuable nearly 20 years later.

A quite different purpose was served by *Behavioral Genetics: Method and Research* (Manosevitz, Lindzey, & Thiessen, 1969). Its objective was to provide additional readings for classes in behavior genetics, and perhaps also for professors to use as material for lectures. In the first section of the readings, five veterans of behavior genetics introduced their objectives and philosophies, and six others described important research methods. Examples of contemporary research were grouped under seven categories: (1) sensory processes and perception; (2) learning; (3) intelligence and abilities; (4) mental retardation; (5) temperament; (6) personality; and (7) psychopathology. From the viewpoint of 1985, some of these articles have flaw but, for the beginning student and all who are interested in the history of behavior genetics, there is still much to be learned.

The laboratory mouse was the star of *Contributions to Behavior-Genetic Analysis*, subtitled *The Mouse as a Prototype* (Lindzey & Thiessen, 1969). The editors invited a number of individuals to contribute chapters. McClearn and his associates introduced the theme of the book by a chapter on the use of isogenic and heterogenic mouse stocks in experiments. Most authors concentrated on a particular phenotype such as open-field behavior, mating choice, hoarding, audiogenic seizures, and laterality. Bruell broke away from experiments with laboratory strains and studied a population of wild *Mus musculus*. Hawkins commented on the objectives and techniques of single-gene substitutions in experiments. Rodgers donated a critique of research strategies with these inbred and outbred small animals that were bred for biological research, and are now found to have new uses in behavioral studies.

Genetics, Environment, and Behavior is the product of another workshop with invited discussants (Ehrman, Omenn, & Caspari, 1972). Individuals were assigned to introduce topics such as quantitative aspects of genetics and environment in the determination of behavior. Others considered the qualitative aspects of behavioral differences, and compared on the technique of behavior-genetic research in animals and humans. The most interesting feature was the variety of ideas promulgated and the variability of comments made in the discussions. In spite of vigorous debates, there was little evidence of rancor or deprecation of a personal nature. A developing discipline needs continuing debate if it is to avoid orthodoxy.

In 1973, Claridge, Canter, and Hume of the Department of Psychological Medicine in Glasgow authored a collection of articles on twins with the title *Personality Differences and Biological Variations*. They discussed the problem of twin studies emphasizing that only large-scale, carefully monitored programs could produce useful results. In addition to cognitive functions and personality, physiological and pharmacological characteristics were examined. It is a useful addition to the twin literature.

Behavioral Genetics: Simple Systems (J. R. Wilson, 1973) is at the opposite pole of the human behavior studies. Instead of the complexities of cognitive and personality characteristics of humans, this small book is the product of a workshop in the Institute for Behavioral Genetics at the University of Colorado. The simple systems are animals such as *aplysia, phycomyces,* planarians, and spiders. The phenotypes available for genetic study are fewer than those commonly used by behavior geneticists, and they are more stereotyped. We mention this volume because it extends the scope of subjects that can be investigated. Behavior genetics has in general restricted its scope to species with readily observed individual differences, and ease in procurement and rearing. Eventually it may be important to extend behavior genetic analysis to a broader range of subjects.

The Genetics of Behavior is a collection of invited chapters from European researchers (van Abeelen, 1974). Emphasis is placed on animals: mainly inbred stains and selected lines of mice with other reports on rats, fish, and *Drosophila*. In mice, three chapters deal with psychopharmacology, three with aggression, and one each on exploration, and genes affecting behavior through their effect on the inner ear. Broadhurst's chapter on the concept of genetic architecture and Fulker's on applications of biometrical genetics to human behavior are summaries of their earlier work. With the emphasis on animal studies it is a bit surprising that no entry from the Finnish studies on the genetics of alcoholism in animals was included. The one human study was Bekker's chapter on personality development in Turner's syndrome. The individual entries are worth reading, but the scope of European research was not adequately covered.

The human side in Europe is represented by *Genetics, Environment and Psychopathology* (Mednick, Schulsinger, Higgins, & Bell, 1974). The contributors were mainly from Denmark (where most of the data were obtained) and the United States. The heritability of schizophrenia was the main objective, and particular attention was given to family studies and long-term observations on "high-risk" individuals (relatives of index cases). Considerable attention was also given to the autonomic nervous system and to adopted-away offspring of schizophrenic parents. In general, the conclusions were that genetic factors were more important than environment as causes of schizophrenia. Higgins, for example, reported that offspring of schizophrenic mothers had the same risk of the disorder whether reared by their own mother or an adopted mother.

Continuing with the psychopathological aspect of behavior genetics is *Genetic Research in Psychiatry* (Fieve, Rosenthal, & Brill, 1975). The book is based on the proceedings of an annual meeting of the American Psychopathological Association. Major topics are biochemistry (particularly neurochemistry), cytogenetics (XXY and XYY karyotypes), and polygenic versus major locus models of the inheritance of schizophrenia. Affective disorders were linked with a dominant X-linked factor and the X_g blood group. Many of these hypotheses have been questioned over the past decade, but there is still strong support for familial transmission of the depressive syndromes. One chapter deals with the transmission of alcoholism and concludes that it is heritable with a multifactorial genetic base.

Developmental Human Behavior Genetics is the product of a workshop organized and edited by Schaie, Anderson, McClearn, and Money (1975). As indicated by the title, the group was concerned with the influence of genetic heritage at various stages of the life span, with emphasis on early stages. On the phenotypic side, intelligence, personality, and rates of development were key topics. Issues such as the effects of adoption, and the effects of a common environment on the similarity of monozygotic twins were debated. Considerable space was also devoted to ethical issues and the merits of genetic counseling. The comments by members of the workshop add to the value of this useful volume on human behavior genetics. Developmental stages should be included with genotype and environment as necessary parameters for research in behavior genetics.

Sperber and Jarvik's (1976) thin volume *Psychiatry and Genetics: Psychological, Ethical and Legal Considerations* differs from others in its concentration on social issues. In addition to brief chapters on the standard topics of psychiatry (schizophrenia, affective disorders, etc.) are those on ethical and legal issues, genetic counseling, neurosis and personality disorders, gender disorders, and mental retardation. Some of these phenotypes have been matched with genotypes;

others have not. It is a volume to make a reader think about issues that are increasingly important in our society.

Genetics, Environment and Intelligence is a compilation of 21 chapters from the United States, Italy, Switzerland, France, Australia, the Federal Republic of Germany, Canada, and the United Kingdom (Oliverio, 1977). There is no single theme. Humans, primates, rats, mice, fish, *Drosophila*, and *Paramecia* share space. Some of the chapters deal separately with environment and genetics as factors affecting behavior, but Caspari balances this with an introductory review of basic principles. Oliverio's authors provide a good sample of behavior genetics research in the mid-1970s. In so large a field it was impossible to cover all aspects of interest, but it is a useful source of information.

Theoretical Advances in Behavior Genetics is the product of a workshop held in Banff, Canada (Royce & Mos, 1979). Like most other workshops, each of the main presentations elicited comments that were included in the text. The theoretical theme evoked papers dealing with statistical models, evolution, ethology, development, sociobiology, and critiques of the concept of heritability. The authors and their commentators often disagreed on issues, but this is commonplace in the world of science. Basically the workshop dealt with what we call the nature-nurture problem. In reality it is not a matter of assigning one or the other as the "cause" of differences in behavior. It is a continuing debate over the relative importance of the many factors that shape behavioral development. Since both genes and environments are essential, the best approach is to determine the constraints that each places on the other.

Our final volume is a collection of previously published papers, *Behavior-Genetic Analysis*, selected and edited by Hirsch and McGuire (1982). Its themes are based on an article by C. S. Hall (1951), published 9 years before our survey of trends begins. Their collection of articles is divided into sections such as demonstrations that variation in a behavior can be attributed to genes; identification of the genetic system responsible for a behavioral difference, particularly whether it can be explained by Mendelian principles or polygenic systems; and finally descriptions of the pathways that must exist between genotypes and behavioral phenotypes. Its objectives are similar to those of Manosevitz et al. (1969), but it concentrates much more on animal studies with emphasis on methodology and basic principles of research in the field.

Subjects and Topics in Selected Volumes. Our sample of selected volumes included 363 contributions. The subjects of these included humans (151), mice (67), *Drosophila* (16), rats (8), plus scattered references to other vertebrates and invertebrates. Twelve entries dealt mainly with statistical analyses. From the 151 articles on humans, 63 involved twins. Major topics were personality (22); intelligence (20); and schizophrenia (8). Adoption studies were less numerous. Seventeen entries were divided among intelligence (5); personality (4);

schizophrenia (3); and single articles on criminality, hyperactivity, and affective disorders. Eleven of the 151 samples gave unusually strong emphasis to environmental factors.

Mice were in second place for popularity with 67 entries. Sixteen emphasized comparisons of behavior among inbred strains with observations on hybrids. Almost always it was demonstrated that the behaviors of F1 and F2 hybrids and reciprocal backcrosses were intermediate to that of the parent strains. The degree of dominance varied. Among the phenotypes studied were activity, aggression, response to alcohol and drugs, audiogenic seizures, emotionality, exploration, learning, and vision.

The mouse sample also included seven reports on selective breeding starting with a heterogeneous stock. The behavioral phenotypes were similar to those studied by the Mendelian approach. In every case selection was effective. The authors were particularly interested in genotype-environment interactions. An interesting feature of selection was the action of certain "major genes" (often affecting pigmentation) on behavioral phenotypes. For example, DeFries and Hegmann (1970) found that selection for low open-field activity resulted in a sharp increase in the frequency of the albino gene (*c*). Selection for high activity was correlated with a decrease in *c*. This finding is not unique, and DeFries suggests that a combination of quantitative and major gene analyses is the best procedure for explaining the genetics of behavior. It seems likely that there are few, if any, genes specialized for characters such as open-field activity. Activity involves muscular and neurological characters that are the products of many genes. The albino gene has specific actions at the biochemical level, but its influence on behavioral phenotypes is indirect and probably fortuitous.

Many of the problems that interested researchers in the early 1960s were still being studied in the 1970s and 1980s, and with the same species. Workshops that include critiques of individual papers are of special worth in evaluating the directions that appear most promising. For the most part differences were argued with vigor but without rancor. It is a healthy sign that behavior genetics has not become doctrinaire, and that different views are tolerated.

An important advance in technique is the wide adoption of multivariate analysis, particularly with human data. It is now recognized that large samples of twins or adoptees are required for significant results. Inbred mice are still useful in behavioral experiments, but the demonstration that their hybrids are intermediate to their parents is by itself of little interest. Efforts to match specific genes with a particular animal behavior have not been as fruitful as hoped. For example, a survey of 10 studies on the mode of inheritance of audiogenic seizure susceptibility in mice came up with four diagnoses of a single- or two-locus model; five diagnoses of a polygenic system, and one of both, depending on whether the first or second trial was counted (Fuller & Thompson, 1978. p. 102). We believe that differences in origin of the subjects, details of rearing, differences in test procedures and the like can result in different hypotheses. Each may be plausible under the circumstances, but should not be accepted as *the* genetic system.

Workshops, symposia, and collections of invited chapters are an important source of information for students, teachers, and researchers alike. Workshops that make available in print the give and take of debates over undecided issues are of particular value to those who could not attend. Compendia of selected articles from the literature provide an overview of the variety of topics that are of particular interest to students who are considering doing research in the field.

BEHAVIOR GENETICS: THE SOCIETY'S JOURNAL

The literature of behavior genetics is widely dispersed among a variety of journals ranging from general periodicals such as *Science* and *Nature* to numerous periodicals with special interests such as alcoholism, ethology, and psychopathology. Many of these were scanned and evaluated by the authors of the seven reviews described in an earlier section of this chapter. Our present concerns are trends in the topics and research directions of behavior genetics as reflected in the contents of *Behavior Genetics* from the relatively short period from its beginning in 1970 until 1984.

Table 1.2 shows the proportion of articles dealing with humans and three animal genera over the journal's 15-year history. The total output was divided into three periods, 1970–1975, 1976–1980, and 1981–1984. Overall, about one third of the total entries dealt with humans, though the figure was only 24% for volumes 1–5. The proportions of mouse and *Drosophila* studies were approximately equal for the 15-year period as a whole, but the high representation of mouse studies (40%, in volumes 1–5) had slumped to 21% in volumes 6–14. This does not mean that behavioral studies with genetically defined mice were decreasing in number. They were being published in a variety of other journals

TABLE 1.2
Proportion of Subjects in *Behavior Genetics* Articles (1970–1984).

Subjects	*Vol. 1–5*	*Vol. 6–10*	*Vol. 11–14*	*Vol. 1–14*
Humans	.24	.40	.34	.32
Mouse	.40	.21	.20	.26
Drosophila	.16	.29	.20	.26
Rat	.08	.03	.08	.06
Methods	.12	.03	.07	.07
Others	.00	.04	.11	.03
No. Articles	146	196	169	511

and listed in a series of bibliographical articles in *Behavior Genetics* (Sprott & Staats, 1975, 1978, 1979, 1980, 1981). The variability in methods-articles was due to special issues of conference proceedings. Although analysis of a wider data base might produce different results, we expect that for the immediate future the big three (*Homo, Mus,* and *Drosophila*) will continue to be favorite material for behavior geneticists for many years.

The reasons are clear. The genetics of *Mus musculus* is the best understood of all mammals whose behavior is complex enough to be of interest for comparison with humans. There are many inbred strains and selected lines to be compared. Heterogeneous stocks are available (or can be synthesized) that can provide a base for selection programs. Mice serve as models for research on topics such as alcoholism and obesity that are relevant to humans. Rats, because of larger size, have some advantages over mice for behavior-genetic analyses of physiological functions. They have also been the choice of comparative psychologists for research on learning and emotional behavior. However, they will not displace mice for behavior-genetics research. In addition to their genetic advantage, these small rodents mature more rapidly and cost less to maintain.

Drosophila cannot serve as models for human behavior, but their genotypes are the best known of multicellular animals. Their advantages for behavior genetics are in the areas of evolutionary theory, specifically on topics such as mating choice and the process of selection. It is relatively easy to select *Drosophila* for behavioral traits, but when selection is discontinued the lines tend to regress to their original status. The same is true of other species, but *Drosophila* are ideal for behavior-genetic studies involving evolutionary theory.

The emphasis on human behavior genetics is not based on the advantages of *Homo sapiens* for research. It is solely due to self-centered interest in our own species. The nearest thing we have to an experimental program is to invite family members, twins, and adopted children to participate in tests. We have identified chromosomes and genes that produce behavioral syndromes, but have not done so for differences in the normal range of intelligence, personality, and emotional level. It is unlikely, and probably impossible, that we ever will. The best we can do is to use the knowledge that has been gained to improve environments, counsel prospective parents, and design remedial procedures based on experience. An example of the application of such an approach is chapter 2 in this volume entitled "Colorado Family Reading Study."

Statistics of Major Topics in "Behavior Genetics." Counts were made in *Behavior Genetics* of the major behavioral topics and methods of research for the big three. For humans in our sample the results were intelligence (38), personality (26), spatial ability (22), mate selection (20), and psychopathology (8). The sample of research techniques included 48 twin, 31 family, and 13 adoption studies. Twelve papers concentrated on chromosomal effects on spatial and other behavioral syndromes. For mice, the favored behavioral phenotypes were activity (18), learning (18), aggression (17), responses to alcohol (12), audiogenic seizures (9), and mating behavior (5). The majority of these papers dealt with inbred strains and their hybrids (72), but there was a sizable contingent of reports based on selected lines (28). *Drosophila* papers dealt mainly with mating behavior (54), and phototaxis or geotaxis (20). Both selected lines and inbred strains were well represented, and considerable attention was given to evolutionary issues.

We made an effort to determine, by dividing the total number of articles in *Behavior Genetics* into three sections, if there were significant differences in the attention given to major topics over a 15-year period. Table 1.3 summarizes our findings. There are irregularities, but for the most part we believe that they are not significant. Our samples are small, and *Behavior Genetics* is not representative of the whole discipline. Most of the studies of such topics as psychopathogenetics, psychopharmacogenetics, and neurogenetics are published in specialized journals. We also sense that publications with possible direct applications to social problems are less likely to appear in *Behavior Genetics* than are those that deal with evolutionary and methodological contributions. An exception to this hypothesis is the increase in mouse studies of responses to ethanol. Certainly those who work with alcoholic mice hope that they will lead to an understanding of the factors that influence consumption and behavioral effects of ethanol in humans.

TABLE 1.3
Most Numerous Topics in *Behavior Genetics*: (1970–1984)

Subjects	*Topics*	*Vol. 1–5*	*Vol. 6–10*	*Vol. 11–12*
Humans	Mate choice	3	8	9
	Twin intelligence	6	3	0
	Adoptee intelligence	1	2	0
	Family intelligence	1	8	2
	Twin personality	1	8	2
	Adoptee personality	0	0	2
	Family personality	1	0	5
	Spatial ability	5	13	4
	Methods	20	8	10
Mice	Aggression	5	8	4
	Learning	12	2	4
	Seizures	8	4	1
	Activity	7	5	6
	Alcohol	1	5	6
	Mate choice	5	2	0
Drosophila	Taxes	8	8	4
	Mate choice	13	26	17

Looking Ahead. There are signs of a trend for more interaction between behavior genetics and psychology. A special issue on "Psychology and Children: Current Research and Practice" (*American Psychologist*, 1979, *24*, 10) contained 41 articles, 3 of which mentioned genetic differences. In contrast, however, a special issue of the same journal devoted to alcoholism (American Psychologist, 1983, *38*, 10) made no mention of the possibility that differences of the efficacy of controlled drinking regimes might have a genetic base. The absence of any genetic input is surprising in view of the amount of research in the field. Perhaps the best example of a behavior genetic special issue is *Child Development* (1983, *54*, 2), which published the best collection of papers on development behavioral genetics extant. It is an outstanding example of bringing genetic and developmental techniques together. We hope that such articles are signs of truly interdisciplinary research, and that other issues of this type will appear in the future.

One important function of *Behavior Genetics* is to serve as a forum. Letters of criticism and responses in kind have been a tradition. For the most part the controversies are debated with strong rhetoric, but without personal vindictiveness. There are exceptions, but the most heated battles are now fought elsewhere rather than in the pages of *Behavior Genetics*. We had considered a critique of the anti-behavior-genetics literature, but decided that little or nothing would be gained. For the same reason we have not reviewed controversies regarding a possible genetic basis for differences between ethnic groups on intelligence tests. In our opinion the best treatment of this subject is still a review by Loehlin, Lindzey, & Spuhler (1975). They conclude their text with these words:

> We do not believe that the lack of a definitive answer to the questions with which we began is either disastrous or disappointing. Moral and political questions never have had scientific answers. . . . It is part of our own fundamental conviction as social scientists that on the whole better and wiser decisions are made with knowledge than without. (p. 258)

SOME FINAL THOUGHTS

A quarter-century is a short period for a scientific discipline to acquire a name, literature, society, and journal. But before these existed, behavior genetics' future interests and objectives had been defined in the Milbank conference of 1957. Its naming in 1960 was a convenience to note its emergence as a field of research that involved both genetics and psychology. Although there has been a substantial increase in the literature and new areas of research, many of the topics treated in Fuller and Thompson (1960) are still subjects of research. However, its roots have spread to include other biological and social sciences. Hybrid fields of research such as psychopharmacogenetics and neuro-behavioral-genetics are now recognized as part of its domain. Other disciplines involving biology and behavior such as ethology and sociobiology have some similar concerns, but their ties with behavior genetics are still weak. Ethologists deal mainly with species differences in behavior and their relationship to the environment. Sociobiologists mix ethology with evolutionary theory and use hypothetical genes as parts of evolutionary theory. Neither makes much use of genetical experimentation.

We shall not try to rank the importance of trends in behavior genetics over its brief existence as a recognized area of science. However, we shall list areas that seem to be increasing in popularity and those that are decreasing. On the decreasing list are the construction of Mendelian models to fit with differences between strains and their hybrids. Some experiments do provide simple fits, but turn out to be difficult to replicate. A single example is a summary of 10 studies of the mode of inheritance of audiogenic seizure susceptibility in mice over the period 1949–1973 (Fuller & Thompson, 1978, p. 102). Six of these favored polygenic control of susceptibility. Three favored single- or two-locus models. In one experiment replicate trials were run resulting in a tie: single locus on trial 1; polygenic on trial 2.

Adoption studies seem to be vying with twin studies, but both are still popular. Less common, but useful techniques are monozygotic twins reared apart (Bouchard, 1984), and the twin-family technique (Rose et al., 1979). The rarity of such subjects makes it difficult to obtain large samples, but progress is being made. Their advantage lies in the reduction of common environmental factors that are present in usual twin pairs. Developmental level is another dimension that is becoming important in human studies. One of the best examples is a study of fears in twins (Rose & Ditto, 1983).

On the animal side we expect to see more research with species other than *Mus, Rattus*, and *Drosophila*. As a start, this volume includes contributions on the blowfly (*Phormia regina*) and Japanese quail (*Coturnix coturnix*). We also expect rapid progress in psychoneurogenetics. For samples of the direction this field is taking see "Genetic Variability in Forebrain Structures Between Inbred Strains of Mice" (Wimer, Wimer, & Roderick, 1969), and "A Geneticist's Map of the Mouse Brain" (Wimer & Wimer, 1982). Another advance would be an increase in selection programs. Selection has been a classical technique, but it has not been used to best advantage. Recent programs have been run in duplicate in order to test for possible differences in results. Selectings for a given character in a variety of heterogeneous lines would also be of interest. Would the products of such selections be genetically and phenotypically similar? phenotypically similar and genetically different? or unlike in both phenotypes and genotypes?

We shall not try to predict the future of behavior genetics over the next quarter-century, but we have some ideas. In one direction it will become closer to the neurosciences. In another it will form bonds with the more biologically based social sciences such as anthropology. Either direction could lead to new techniques and their applications. We do have one suggestion for an added function in the Behavior Genetics Association. It is the establishment of a repository for the preservation and availability of data beyond that which is available in its journal. Authors would deposit copies of their records in a central place where photocopies would be available to qualified researchers for pooling, comparing, and analyzing large bodies of data.

ACKNOWLEDGMENT

The preparation of this chapter was aided by a grant from the Harry Frank Guggenheim Foundation to John L. Fuller.

REFERENCES

Atkinson, R. L., Atkinson, R. C., & Hilgard, E. A. (1983). *Introduction to psychology* (8th ed.). New York: Harcourt Brace Jovanovich.

Bliss, E. L. (1962). *Roots of behavior*. New York: Harper.

Broadhurst, P. L., Fulker, D. W., & Wilcock, J. (1974). Behavioral genetics. *Annual Review of Psychology, 25,* 389–415.

Bouchard, T. J., Jr. (1984). Twins reared together and apart: What they tell us about human diversity. In S. W. Fox (Ed.), *Individuality and determinism* (pp. 147–184). New York: Plenum Press.

Claridge, G., Canter, S., & Hume, W. I. (1973). *Personality differences and biological variations*. Oxford: Pergamon Press.

Darley, J. M., Glucksberg, S., Kamin, L. J., & Kinchla, A. A. (1984). *Psychology* (2nd ed.). Englewood Cliffs, NJ: Prentice-Hall.

DeFries, J. C., & Hegmann, J. P. (1970). Genetic analysis of open field behavior. In G. Lindzey & D. D. Thiessen (Eds.), *Contributions to behavior-genetic analysis* (pp. 23–56). New York: Appleton-Century-Crofts.

DeFries, J. C., & Plomin, R. (1978). Behavioral genetics. *Annual Review of Psychology, 29*, 473–525.

Dobzhansky, T. (1957). The biological concept of heredity applied to man. In *The nature and transmission of the genetics and cultural characteristics of human populations* (pp. 11–19). New York: Milbank Memorial Fund.

Ehrman, L., Omenn, G. S., & Caspari, E. (1972). *Genetics, environment, and behavior*. New York: Academic Press.

Ehrman, L., & Parsons, P. A. (1970). *The genetics of behavior*. Sunderland, MA: Sinauer.

Ehrman, L., & Parsons, P. A. (1981). *Behavior genetics and evolution*. New York: McGraw-Hill.

Fieve, R. R., Rosenthal, D., & Brill, H. (1975). *Genetic research in psychiatry*. Baltimore: Johns Hopkins University Press.

Fuller, J. L. (1957). The genetic base: Pathways between genes and behavioral characteristics. In *The nature and transmission of the genetics and cultural characteristics of human populations* (pp. 101–111). New York: Milbank Memorial Fund.

Fuller, J. L. (1960). Behavior genetics. *Annual Review of Psychology, 11,* 41–70.

Fuller, J. L. (1982). Psychology and genetics: A happy marriage? *Canadian Psychology: Psychologie Canadienne, 23*, 11–21.

Fuller, J. L. (1983). Sociobiology and behavior genetics. In J. L. Fuller & E. C. Simmel (Eds.), *Behavior genetics: Principles and applications* (pp. 436–477). Hillsdale, NJ: Lawrence Erlbaum Associates.

Fuller, J. L. (1984). Psychology and genetics. In M. H. Bernstein (Ed.), *Psychology and its allied disciplines* (Vol. 3, pp. 65–107). Hillsdale, NJ: Lawrence Erlbaum Associates.

Fuller, J. L., & Thompson, W. R. (1960). *Behavior genetics*. New York: Wiley.

Fuller, J. L., & Thompson, W. R. (1978). *Foundations of behavior genetics*. St. Louis: C. V. Mosby.

Glass, D. C. (1968). *Biology and behavior genetics*. New York: Rockefeller University Press.

Gleitman, H. (1983). *Basic psychology*. New York: Norton.

Goldsmith, H. H. (1983). Genetic influences on personality from infancy to adulthood. *Child Development, 54*, 331–355.

Hall, C. S. (1951). The genetics of behavior. In S. S. Stevens (Ed.), *Handbook of experimental psychology* (pp. 304–329). New York: Wiley.

Hay, D. (1985). *Essentials of behaviour genetics*. Melbourne, Australia: Blackwell.

Hebb, D. O. (1966). *A textbook of psychology* (2nd ed.). Philadelphia: Saunders.

Henderson, N. D. (1982). Human behavior genetics. *Annual Review of Psychology, 33*, 403–440.

Hilgard, E. R., & Atkinson, R. C. (1967). *Introduction to psychology* (4th ed.). New York: Harcourt, Brace & World.

Hirsch, J. (1967). *Behavior genetic analysis*. New York: McGraw-Hill.

Hirsch, J., & McGuire, T. R. (1982). *Behavior-genetic analysis*. Stroudsburg, PA: Dowden, Hutchinson & Ross.

Hutt, M. L., Isaacson, R. L., & Blum, M. L. (1966). *Psychology: The science of behavior*. New York: McGraw-Hill.

Kendler, H. H. (1968). *Basic psychology* (2nd ed.). New York: Appleton-Century-Crofts.

Kimble, G. A., & Garmezy, N. (1963). *Principles of general psychology* (2nd ed.). New York: Ronald.

Kimble, G. A., Garmezy, N., & Zigler, E. (1984). *Principles of psychology* (6th ed.). New York: Wiley.

Lewontin, R. C., Rose, S., & Kamin, L. J. (1984). *Not in our genes*. New York: Random House.

Lindzey, G., Loehlin, J., Manosvitz, M., & Thiessen, D. (1971). Behavioral genetics. *Annual Review of Psychology, 22*, 39–94.

Lindzey, G., & Thiessen, D. D. (1969). *Contributions to behaviorgenetic analysis*. New York: Appleton-Century-Crofts.

Loehlin, G., Lindzey, G., & Spuhler, J. N. (1975). *Race differences in intelligence*. San Francisco: W. H. Freeman.

Manosevitz, M., Lindzey, G., & Thiessen, D. D. (1969). *Behavioral genetics: Method and research*. New York: Appleton-Century-Crofts.

McClearn, G. E. (1963). The inheritance of behavior. In L. P. Postman (Ed.), *Psychology in the making* (pp. 144–252). New York: Random House.

McClearn, G. E., & DeFries, J. C. (1973). *Introduction to behavioral genetics*. San Francisco: W. H. Freeman.

McClearn, G. E., & Meredith, W. (1966). Behavioral genetics. *Annual Review of Psychology, 17*, 515–550.

Meade, J. E., & Parkes, A. S. (1965). *Biological aspects of social problems*. Edinburgh: Oliver & Boyd.

Meade, J. E., & Parkes, A. S. (1966). *Genetic and environmental factors in human ability*. Edinburgh: Oliver & Boyd.

Mednick, S. A., Schulsinger, F., Higgins, H., & Bell, B. (1974). *Genetics, environment, and psychopathology*. Amsterdam: North Holland.

Morgan, C. T. (1961). *Introduction to psychology* (2nd ed.). New York: McGraw-Hill.

Munn, N. L. (1966). *Psychology* (5th ed.). Boston: Houghton Mifflin.
Oliverio, A. (1977). *Genetics, environment, and intelligence*. Amsterdam: North Holland.
Plomin, R., DeFries, J. A., & McClearn, G. E. (1980). *Behavioral genetics: A primer*. San Francisco: W. H. Freeman.
Rose, R. J., & Ditto, W. B. (1983). A developmental-genetic analysis of common fears from early adolescence to early adulthood. *Child Development, 54*, 361–368.
Rose, R. J., Miller, J. Z., Dumont-Driscoll, M., & Evans, M. M. (1979). Twin family studies of perceptual speed ability. *Behavior Genetics, 9*, 71–86.
Royce, J.R., & Mos, L. P. (1979). *Theoretical advances in behavior genetics*. Alphen aan den Rijn, The Netherlands: Sijthoff & Noordhoff.
Sarbin, T. R., & Mancuso, J. L. (1980). *Schizophrenia: Medical diagnosis or moral verdict?* New York: Pergamon Press.
Schaie, K. W., Anderson, V. E., McClearn, G. E., & Money, J. (1975). *Developmental human behavior genetics*. Lexington, MA: Lexington Books.
Scott, J. P., & Fuller, J. L. (1965). *Genetics and the social behavior of the dog*. Chicago: University of Chicago Press.
Sperber, M. A., & Jarvik, L. F. (1976). *Psychiatry and genetics: psychosocial, ethical, and legal considerations*. New York: Basic Books.
Sprott, R. L., & Staats, J. (1975). Behavioral studies using genetically defined mice—a bibliography. *Behavior Genetics, 5*, 27–82.
Sprott, R. L., & Staats, J. (1978). Behavioral studies using genetically defined mice—a bibliography. *Behavior Genetics, 8*, 183–206.
Sprott, R. L., & Staats, J. (1979). Behavioral studies using genetically defined mice—a bibliography. *Behavior Genetics, 9*, 87–102.
Sprott, R. L., & Staats, J. (1980). Behavioral studies using genetically defined mice—a bibliography. *Behavior Genetics, 10*, 93–104.
Sprott, R. L., & Staats, J. (1981). Behavioral studies using genetically defined mice—a bibliography. *Behavior Genetics, 11*, 73–84.
Spuhler, J. N. (1967). *Genetic diversity and human behavior*. Chicago: Aldine.
Thoday, J. M., & Parkes, A. S. (1968). *Genetic and environmental influences on behavior*. Edinburgh: Oliver & Boyd.
Thompson, W. R. (1957). The significance of personality and intelligence tests in the evaluation of population characteristics. In *The nature and transmission of the genetic and cultural characteristics of human populations* (pp. 37–50). New York: Milbank Memorial Fund.
van Abeelen, J. H. F. (1974). *The genetics of behaviour*. Amsterdam: North Holland.
Vandenberg, S. G. (1965). *Methods and goals in human genetics*. New York: Academic Press.
Vandenberg, S. G. (1968). *Progress in human behavior genetics*. Baltimore: Johns Hopkins University Press.
Wilson, E. O. (1975). *Sociobiology*. Cambridge, MA: Belknap Press.
Wilson, J. R. (1973). *Behavioral genetics: Simple systems*. Boulder, CO: Colorado Associated University Press.
Wimer, R. E., & Wimer, C. C. (1982). A geneticist's map of the mouse brain. In I. Lieblich (Ed.). *Genetics of the brain*. (pp. 395–420). New York: Elsevier.
Wimer, R. E., & Wimer, C. C. (1985). Animal behavior genetics: A search for the biological foundations of behavior. *Annual Review of Psychology, 36*, 171–219.
Wimer, R. E., Wimer, C. C., & Roderick, T. H. (1969). Genetic variability in forebrain structures between inbred strains of mice. *Brain Research, 16*, 257–264.

2 Colorado Family Reading Study: An Overview

John C. DeFries
George P. Vogler
Michele C. LaBuda
Institute for Behavioral Genetics
University of Colorado, Boulder

Reading disability is an important public health problem because of its relatively high prevalence rate among school-age children and young adults. Although estimates of the prevalence rate vary widely across different studies, typical estimates are in the range of 5% to 10%. Almost all studies, however, indicate that boys are at higher risk for the disorder than girls, with sex ratios of three or four to one being commonly reported.

The problem was first described in the medical literature in 1896 by Morgan, who referred to the condition as "congenital word blindness." A number of other terms for the disorder have subsequently been used, including *dyslexia, specific developmental dyslexia,* and *specific reading disability.* We employ the simpler term *reading disability* for two reasons: First, the term *dyslexia* frequently evokes images of perceptual problems such as letter and word inversions (e.g., *was* versus *saw*). Most reading-disabled children have no such problems. Second, because reading ability is correlated with other verbal and nonverbal abilities, reading-disabled children also frequently manifest deficits in other cognitive abilities.

That reading disability may have a constitutional or genetic basis has long been recognized. For example, in 1905, Thomas described the familial nature of congenital word-blindness as follows: "It is to be noted that it frequently assumes a family type; there are a number of instances of more than one member of the family being affected, and the mother often volunteers the statement that she herself was unable to learn to read, although she had every opportunity" (p. 381). This biological perspective is also clearly exemplified by the World Federation of Neurology's definition of specific developmental dyslexia: "A disorder manifested by difficulty in learning to read despite conventional instruction, adequate intelligence, and socio-cultural opportunity. It is dependent upon fundamental cognitive disabilities which are frequently of constitutional origin" (Critchley, 1970, p. 11).

Case studies of identical and fraternal twins reviewed by Zerbin-Rüdin (1967) suggest that reading disability may be highly heritable. Among a sample of 17 identical twin pairs and 34 fraternal twin pairs, observed concordance rates were 100% and 35%, respectively. These cases had all been referred to clinics so that ascertainment bias is possible (Belmont & Birch, 1965). For example, referrals to clinics are almost certainly more severely affected than are subjects ascertained from school populations, and severely affected identical twins are more likely to be concordant than are those with less serious problems. This may account for the somewhat lower concordance rates observed in a non-clinic population by Bakwin (1973). He ascertained 338 pairs of like-sexed twins ranging in age from 8 to 18 years through mothers-of-twins clubs. Reading history was obtained through interviews with parents, telephone calls, and mail questionnaires. A positive history for reading disability was obtained in 97 of the 676 children, a prevalence rate of 14.3%. This rather high rate may be due to a higher risk for reading disability among twins than among singletons (Hay & O'Brien, 1982). On the other hand, it may merely indicate that parents of twins in this study were inclined to over-report problems among their children. In any case, observed prevalence rates were highly similar for identical and fraternal twins (14.0% and 14.9%, respectively), whereas pairwise concordance rates were considerably different (84% for identical twin pairs versus only 29% for fraternal twin pairs). Concordance rates were essentially equal for male and female identical twin pairs, but male fraternal twins had higher concordance rates than female fraternals (42% versus 8%). Although Bakwin's (1973) results suggest substantial genetic influence, his definition of reading disability "as a reading level below the expectation derived from the child's performance in other school subjects" (p. 184) is rather vague. In addition, the validity of his parental reports was not demonstrated. Thus, until objective test data are obtained from a large and representative sample of identical and fraternal twin pairs in which at least one twin is reading disabled, the twin literature should be regarded as only being suggestive of genetic influence.

Studies of reading disability in nuclear families have been much more intensive. Recent reviews of this extensive literature have been provided by Aman and Singh (1983), Benton (1975), Finucci (1978), Finucci and Childs (1983), Herschel (1978), Ludlow and Cooper (1983), Owen (1978), and Pennington and Smith (1983). (See Finucci, 1978, for an especially detailed critique.) The first large-scale family study of reading disability was reported by Hallgren (1950). In a sample of 112 families, Hallgren reported that 88% of the probands had one or more relatives who also were affected and concluded that dyslexia follows an autosomal dominant mode of inheritance. However, several problems with

Hallgren's study render this interpretation untenable. First, both parents were unaffected in 17% of the probands' families—a finding that is incompatible with simple autosomal dominance. Second, although some test data were available from certain family members, much of the extended family information was based upon self-reports. Third, a careful reading of Hallgren's case studies reveals an apparent preoccupation with familial transmission. Although it is difficult to document, the reader gains the distinct impression from reading these case studies that Hallgren was reluctant to diagnose a child as being dyslexic unless another member of the family also was affected. Such a bias toward familial transmission may have necessarily led Hallgren to accept the hypothesis of autosomal dominance. Other investigators have proposed alternative modes of inheritance for reading disability including autosomal dominance with partial sex limitation (Zahalkova, Vrzal, & Kloboukova, 1972), sex-linkage (Symmes & Rapoport, 1972), dominance with incomplete penetrance in males and recessive inheritance in females (Sladen, 1970), and autosomal dominance with polygenic modifiers (Lenz, 1970).

Finucci, Guthrie, Childs, Abbey, and Childs (1976) reported the first family study in which relatives as well as probands were administered an extensive test battery. The sample of probands was rather small (15 males and 5 females), but 75 first-degree relatives were classified as being either reading disabled or normal readers on the basis of objective test performance. The major conclusion from this study was that reading disability is not randomly distributed in the population, but clusters within families. Thirty-four of the 75 first-degree relatives of the probands were classified as being reading disabled. In 16 families in which both parents were evaluated, 3 had both parents affected and 10 had one parent affected, i.e., 81% of the probands had at least one affected parent, a value very similar to that reported by Hallgren (1950). In only 3 of the 20 families was the proband the only affected member of the family and in 2 of these 3 cases the reading status of only one parent was assessed. The apparent absence of a uniform transmission pattern among the different pedigrees was interpreted by the authors as being indicative of genetic heterogeneity.

Smith, Kimberling, Pennington, and Lubs (1983) recently reported a linkage analysis in families with apparent autosomal dominance for reading disability. Families were selected for testing if a history of reading disability occurred in three successive generations (proband, mother, and a maternal grandparent; or proband, father, and a paternal grandparent). Each family member was administered a series of standardized achievement tests. Children were diagnosed as being reading disabled if they had a full-scale IQ greater than 90 and a reading level at least 2 years below expected grade level. Adult reading status, however, was determined by self-reports of reading history if there was a discrepancy between test results and self-reported reading disability. Data from nine kindreds, including 84 tested individuals, were reported. Twenty-one genotyping markers, as well as chromosomal heteromorphisms, were used for linkage analysis. Results suggested linkage between reading disability and chromosome 15 heteromorphisms. A lod score of 3.241 was obtained, but about 70% of this was contributed by only one kindred. Although a lod score above 3.0 is usually considered sufficient to establish linkage, the authors indicate that confirmation by a second study will be required before linkage is accepted with confidence.

In general, results of twin and family studies strongly suggest that reading disability is heritable, and a number of different modes of inheritance have been proposed to account for familial transmission. Establishing the mode or modes of inheritance for reading disability would be of considerable theoretical significance and could suggest clues regarding mechanism and amelioration. However, results of carefully designed family studies can provide much additional important information regarding, for example, the risk of reading disability in children from families with an affected parent or sibling, long-term stability and prognosis for remediation, the etiology of covariation among transmissible influences, the possibility of heterogeneity of the disorder, and subtype validity. The primary objective of the present chapter is to provide an overview of the Colorado Family Reading Study (FRS), including results of recent risk, longitudinal, and bivariate path analyses.

OVERVIEW OF FRS

The primary objectives of the original FRS, a 3-year project funded by the Spencer Foundation, were as follows: to construct a battery of tests that differentiates children with diagnosed reading problems from controls; to assess possible cognitive and reading deficits in parents and siblings of children with reading problems; and, if such deficits are found, to study their transmission in families. Subjects were referred for testing by personnel of the Boulder Valley and St. Vrain Valley school districts in Colorado. The referral criteria employed for reading-disabled probands included an IQ score of 90 or above as measured by a standardized intelligence test; reading achievement level of one half of grade level expectancy or lower as measured by a standardized reading test (e.g., a child in the fourth grade who is reading at or below second-grade level); chronological age between 7.5 and 12 years; residence with both biological parents; no known emotional or neurological impairment; and no uncorrected visual or auditory acuity deficits. Control children were matched to reading-disabled children on the basis of age (within 6 months), sex, grade, school, and home neighborhood. Except for reading level, which was equal to or greater than current grade placement, each control child met all of the criteria for the selection of probands. In addition to the probands and matched control children, their parents and siblings (7.5–18 years of age) were also tested. Families were typically middle-class Caucasians and the primary language spoken in the home was always English.

The very simple design of the FRS, illustrated in Fig. 2.1, facilitates a number of informative comparisons. In addition to the obvious comparison of probands versus matched controls, it is possible to compare siblings of probands to siblings of controls, and parents of probands to parents of controls. To the extent that reading disability is heritable, relatives of reading-disabled children should manifest at least some deficits on reading-related tests.

During the initial phase of the study, a 3-hour battery of psychometric tests was individually administered by trained examiners to members of 58 matched pairs of families. The most discriminating and reliable tests were retained for a 2-hour battery that was employed during the remainder of the study. Tests in the reduced battery were individually administered in two 1-hr blocks, separated by a 15-min break for rest and refreshment.

During the original 3-year project, 125 probands, their parents and siblings, and members of 125 control families were tested. The total number of subjects tested in these 250 families was 1,044, making it the most extensive family study of reading disability conducted to date. Test descriptions and mean scores for individual tests were previously reported by Foch, DeFries, McClearn, and Singer (1977) and DeFries, Singer, Foch, and Lewitter (1978). In the present chapter, we present composite scores based on five tests (Peabody Individual Achievement Test [Dunn & Markwardt, 1970] Reading Recognition, Reading Comprehension, and Spelling; Wechsler Intelligence Scale for Children-Revised [Wechsler, 1974] Coding Subtest Form B; and the Colorado Perceptual Speed

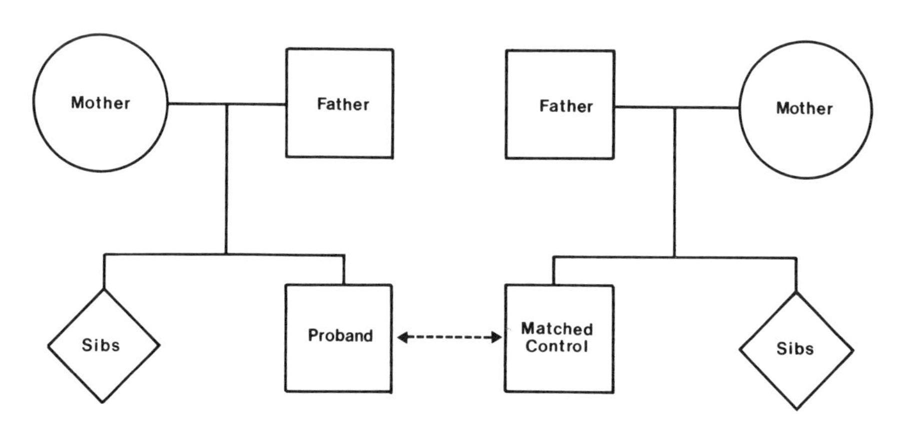

FIG. 2.1 Design of the Colorado Family Reading Study. From "Genetic Aspects of Reading Disability: A Family Study" (p. 258) by J. C. DeFries and S. N. Decker, 1982, *Reading Disorders: Varieties and Treatments*, edited by R. N. Malatesha and P. G. Aaron, New York: Academic Press. Copyright 1982 by Academic Press. Reprinted by permission.

Test: Rotatable Letters and Numbers) that were administered to subjects in both the original FRS and a follow-up study.

Individual test scores were age adjusted to facilitate comparisons among subjects of different ages. The sample was divided into the following age groups: children under 10 years of age, children 10 and older, and adults. Each subject's score was expressed as a deviation from expectation based upon a linear and quadratic regression equation estimated from data on the control sample. The resulting age-adjusted scores were transformed to *T*-scores with a mean of 50 and a standard deviation of 10 in each control age group. These scores were then intercorrelated and subjected to principal component analysis with Varimax rotation (Nie, Hull, Jenkins, Steinbrenner, & Bent, 1975). As described below, the unweighted mean of three tests (Reading Recognition, Reading Comprehension, and Spelling) was used as a general reading measure, whereas the unweighted mean of the other two tests (WISC Coding and Colorado Perceptual Speed) provided a measure of symbol-processing speed.

Approximately 5 years after being tested in the original FRS, 69 pairs of reading-disabled and control children participated in a follow-up study. The average ages of these children were 9.4 and 14.8 years at the time of their initial and follow-up tests, respectively. Of the 69 reading-disabled children, 51 had served as probands and 18 were siblings of probands in the original FRS. As discussed below, principal component analysis of correlation matrices estimated from test and retest data revealed highly similar factor structures.

Correlation Matrices

Separate correlation matrices were computed from data on probands, siblings, and parents within each family type tested in the original FRS. These six correlation matrices were then tested for homogeneity using the computer program LISREL V (Jöreskog & Sörbom, 1981). The null hypothesis of homogeneity is rejected with considerable confidence [χ^2 (50) = 144.0, $p < .0001$]. This is not due entirely to the inclusion of probands who have reduced variances for the reading measures (DeFries & Decker, 1982) because a test of the homogeneity of the five other correlation matrices (siblings and parents of reading-disabled children; and control children, their siblings, and parents) is also rejected [χ^2 (40) = 103.9, $p < .0001$]. The three correlation matrices for members of families of reading-disabled children (probands, siblings, and parents) are also heterogeneous [χ^2 (20) = 74.02, $p < .0001$], but those for members of control families are marginally homogeneous [χ^2 (20) = 29.7, p = .07]. However, for both family types, matrices of correlations for siblings and parents are homogeneous [parents and siblings of reading-disabled children: χ^2 (10) = 12.9, p = .24; those of controls; χ^2 (10) = 6.10, p = .81]. Thus, pooled LISREL estimates of the correlations for parents and siblings of probands are presented in Table 2.1, along with the observed correlations for the probands. A similar matrix for control

TABLE 2.1
Correlations Among Cognitive Measures in Families of Reading-Disabled Children (Probands Above Diagonal, N = 125; Pooled Estimates From Parents and Siblings Below Diagonal, N = 403)

	Reading Recognition (Rec)	*Reading Compre-hension (Comp)*	*Spelling (Spell)*	*WISC Coding (Coding)*	*Colorado Perceptual Speed (CPS)*
Rec	1.00	.53	.57	.00	.00
Comp	.73	1.00	.48	.10	.20
Spell	.73	.65	1.00	.06	.15
Coding	.40	.38	.38	1.00	.57
CPS	.47	.44	.56	.64	1.00

TABLE 2.2
Correlations Among Cognitive Measures in Families of Control Children (Matched Control Children Above Diagonal, N = 124; Pooled Estimates From Parents and Siblings Below Diagonal, N = 384)

	Reading Recognition (Rec)	*Reading Compre-hension (Comp)*	*Spelling (Spell)*	*WISC Coding (Coding)*	*Colorado Perceptual Speed (CPS)*
Rec	1.00	.50	.62	.11	.25
Comp	.60	1.00	.41	.17	.30
Spell	.60	.43	1.00	.19	.45
Coding	.27	.20	.29	1.00	.31
CPS	.40	.28	.51	.59	1.00

children and their family members is presented in Table 2.2. It may be noted from Table 2.1 that correlations involving probands are consistently smaller than those based upon data on their parents and siblings. Moreover, correlations obtained from data on relatives of reading-disabled children are consistently larger than those obtained from data on control children and their relatives. This may be due to the greater variance of reading-related measures observed among relatives of reading-disabled children (DeFries & Decker, 1982).

In spite of this apparent heterogeneity of correlational structure in the different groups, principal component analyses of the various correlation matrices yield highly similar results. Loadings of the five tests on two principal components for each of the four correlation matrices are presented in Table 2.3. In each case, the first component (reading) correlates highest with Reading Recognition, Reading Comprehension, and Spelling, whereas the second (symbol-processing speed)

TABLE 2.3
Varimax Rotated Principal Component Loadings for Reading (1) and Symbol-Processing Speed (2) Dimensions

	Families of reading-disabled children				*Families of control children*			
	Probands		*Parents and siblings*		*Probands*		*Parents and siblings*	
	1	2	*1*	2	*1*	2	*1*	2
Reading Recognition	.86	−.09	.89	.24	.88	.01	.86	.20
Reading Comprehension	.79	.16	.87	.20	.73	.15	.84	.04
Spelling	.83	.08	.83	.31	.80	.26	.72	.37
WISC Coding	.01	.88	.18	.90	−.01	.89	.07	.89
Colorado Perceptual Speed	.10	.88	.34	.83	.38	.68	.31	.83
Percent common variance	58	42	78	22	70	30	72	28

has highest loadings on WISC Coding and Colorado Perceptual Speed. Coefficients of congruence between corresponding loadings estimated from data on the four groups each exceed .94 and the median value is .98. As recently reported, this factor structure has also been found to be robust across various sex and test-retest subgroupings of the FRS data (DeFries & Baker, 1983).

In previous analyses (e.g., DeFries & Baker, 1983), principal component scores were computed for each subject from the sum of the cross-products of the standardized test scores and their corresponding factor score coefficients. For the present analyses, however, a general reading measure was obtained for each individual by calculating the unweighted mean of the age-adjusted Reading Recognition, Reading Comprehension, and Spelling *T* scores. A corresponding symbol-processing speed measure was computed from the mean of the age-adjusted Coding and Colorado Perceptual Speed *T* scores. This somewhat different method was employed in the present study to yield scores that characterize the reading and symbol-processing speed dimensions, but which retain covariation between them, as required for the bivariate analyses to be described in a later section.

Group Means

Average reading and symbol-processing speed composite scores of probands, matched controls, siblings, and parents are presented in Table 2.4. Results of Multivariate Analyses of Variance indicate significant differences between family types and sexes for each of the three comparison groups (probands versus controls; siblings of reading-disabled children versus siblings of controls; and parents of probands versus those of controls). As expected, the largest univariate difference is between probands and controls on the reading measure. Probands have average reading scores approximately two standard deviations below those of controls, which demonstrates that the FRS probands are indeed severely reading disabled. Probands also score about .8 of a standard deviation lower than controls on symbol-processing speed and girls obtain scores about .5 of a standard deviation higher than boys on this measure. It is of interest to note that there is no significant multivariate or univariate interaction between sex and group in the proband data, suggesting that proband girls as a group are no more or less impaired than proband boys. Therefore, there is little or no evidence of bias in diagnosis as a function of sex in the study, despite the marked difference in prevalence rate. As shown in Table 2.4, the sex ratio in the FRS is 3.3:1, a finding that is consistent with most previous studies.

A similar pattern of significant main effects is present in the sibling data. However, the difference in reading performance between brothers of probands and brothers of controls is larger (about one standard deviation) than that for sisters (about .3 standard deviation), resulting in a significant univariate interaction for the reading measure. This finding suggests that both a child's sex and

TABLE 2.4
Multivariate Analyses of Variance of Composite Reading and Symbol-Processing Speed Scores

	Mean scores				*F* values			
	Males		*Females*					
	Reading disabled	*Control*	*Reading disabled*	*Control*	Group	Sex	Group × sex	*df*
Probands								
Reading	29.79	50.35	28.78	52.51	465.99*	.24	1.84	1, 239
Symbol-processing speed	41.80	48.68	45.81	54.27	56.40*	17.60*	.48	1, 239
Multivariate					234.62*	8.92*	.98	2, 238
N	93	94	28	28				
Siblings								
Reading	39.17	49.99	44.90	47.93	23.40*	1.81	6.76*	1, 174
Symbol-processing speed	41.16	46.93	51.02	52.51	8.22*	35.09*	2.62	1, 174
Multivariate					11.94*	18.51*	3.49*	2, 173
N	50	46	43	39				
Parents								
Reading	41.55	49.66	45.14	50.86	53.11*	6.32*	1.59	1, 482
Symbol-processing speed	42.98	48.62	47.01	51.38	33.73*	15.48*	.54	1, 482
Multivariate					29.56*	7.92*	.80	2, 481
N	121	122	121	122				

family history should be considered for purposes of risk analysis. Brothers of reading-disabled children in the FRS are at higher risk for the disorder than sisters. A significant difference between siblings of reading-disabled children and those of controls is also evident for symbol-processing speed, as is the sex difference in favor of girls.

With regard to analyses of parental data, fathers and mothers of reading-disabled children obtain lower reading and symbol-processing speed scores than do parents of control children. Differences are slightly greater for fathers than for mothers, but not enough to result in significant interactions. In addition, mothers obtain significantly higher scores than fathers for both measures.

These sibling and parental data conclusively demonstrate the familial nature of reading disability. Siblings and parents of reading-disabled children both obtain lower average reading and symbol-processing speed scores than do those of controls. However, all relatives of reading-disabled children are not equally impaired. As indicated in the following section, parents of reading-disabled children who indicate that they encountered serious problems learning to read are more seriously impaired than those who report no positive history of reading problems.

Validity of Self-Reports

In addition to psychometric test data obtained in the FRS, each parent completed a questionnaire regarding reading habits and abilities. One question asked was whether he or she had encountered any serious difficulty learning to read. We have recently tested the validity of these self-reports by subjecting principal component reading scores of parents to an unweighted means analysis of variance (Decker, Vogler, & DeFries, in press). Main effects included family type (parents of reading-disabled children versus those of controls), sex of parent (mother versus father), and self-reported reading status (disabled or normal).

Twenty-three of 123 mothers of reading-disabled children for whom self-report data were available reported that they themselves had encountered serious difficulties learning to read, whereas only 6 of 124 mothers of control children reported reading problems. Thirty-six of 119 fathers of reading-disabled children reported a positive history for reading problems, versus 7 of 124 fathers of control children. In general, parents who reported problems learning to read obtained lower average reading scores than did those who encountered no such problem—about a .5 standard deviation difference on the average [$F(1,482) = 39.65, p < .001$]. For parents of reading-disabled children, the mean difference between those who self-reported reading problems and those who did not was about .8 standard deviation. However, for parents of controls this difference was only about .2 standard deviation, resulting in a significant interaction between group membership and self-reported reading status [$F(1,482) = 5.06, p < .05$]. It also was found that parents of reading-disabled children who did not report a history of reading problems, as well as those who did, obtained lower average scores than did parents of controls. Therefore, the significant mean difference between parents of reading-disabled children and those of controls [$F(1{,}482) = 24.70$, $p < .001$] is not due entirely to the lower performance of those parents with a positive history of reading problems.

Results of this analysis clearly demonstrate that parents who report that they encountered serious problems learning to read obtain significantly lower reading scores than do those who report no such problems. That this difference is greater for parents of reading-disabled children than for control parents suggests that parents of affected children are not simply reporting more reading problems because of a greater awareness of reading disability. Parents of reading-disabled children who self-report a positive history of reading difficulty have average reading performance scores over one standard deviation below those of parents of control children. Not only did these parents experience problems learning to read; they also continue to have reading problems well into middle age.

This comparison of the average reading scores of parents who report that they encountered serious problems learning to read versus those of parents who report no positive history clearly validates parental self-reports as an index of reading status. Because such data are easily obtained and are valuable for risk analysis, as discussed in the following section, family history of reading difficulty should be routinely collected by clinicians and special educators.

Family History as an Indicator of Risk

Familial resemblance for reading disability provides a powerful tool for the assessment of a child's risk for developing reading problems. Identification of young children at risk for reading disability could facilitate preventive intervention or early remediation prior to the onset of serious academic problems. Parental self-reports of difficulty learning to read, the validity of which was demonstrated in the previous section, are easily obtained and can be used as an index of the risk that a child will become reading disabled.

Using the principles of Bayesian inverse probability analysis, we recently estimated the probability that a child will become reading disabled when a parent is affected [$P(C/R)$] from the probability that a parent will have reported reading problems given that a child is affected [$P(R/C)$] (Vogler, DeFries, & Decker, 1985). The estimate of $P(C/R)$ is obtained from the following equation:

$$P(C/R) = \frac{P(C)P(R/C)}{P(C)P(R/C) + P(\overline{C})P(R/\overline{C})}, \tag{1}$$

where $P(C)$ is the prior probability that a child will become reading disabled (i.e., an estimate of the population incidence); $P(\overline{C})$ is the prior probability that a child will not be reading disabled; $P(R/C)$ is the likelihood that a parent will be disabled given that a child is disabled (i.e., the observed frequency of self-reported reading problems among parents of probands); $P(R/\overline{C})$ is the likelihood that a parent will be disabled given that a child is not disabled (i.e., the observed frequency of self-reported reading problems among parents of control children); and $P(C/R)$ is the posterior probability that a child will become reading disabled given that a parent reported difficulty learning to read.

Posterior probability estimates were obtained from an analysis of parental self-report data from the FRS sample and from a subsequent study in which only reading-disabled and control children were tested, but for which parental self-report data also were available. The total sample consisted of the parents of 130 male probands, 44 female probands, and 182 control children.

Separate population prevalences for males and females were calculated assuming a population sex ratio of 3.5 disabled males to 1 disabled female and an overall population rate of 5%. The self-reported frequencies of reading problems among the parents of probands were 29% and 17%, respectively, for fathers and mothers of male probands; 36% and 25%, respectively, for fathers and mothers of female probands. In control families, 4% of the fathers and 3% of the mothers reported difficulties. Posterior probabilities that a child will become reading disabled based on these data are presented in Table 2.5. The risk that a child will develop reading problems is clearly elevated if a parent reported difficulties in learning to read. For a male offspring, the risk is nearly 40% if his father reported problems (or nearly 7 times greater than if his father reported no difficulties) and 35% if his mother reported problems (or 5 times greater than if his mother reported no difficulties). For a female, the absolute risk of 17% to 18% is lower than that for a male, but this represents an increase of 10 to 12 times the risk if her parents reported an absence of reading problems.

Parental self-reported history of reading disability is clearly a powerful predictor of reading disability in the offspring of affected parents. Thus, this simple measure should be routinely employed by clinicians and educators for early identification of children at risk for reading disability.

TABLE 2.5
Posterior Probability That a Child Will Become Reading Disabled as a Function of Parental Self-Reported Reading Ability

Sex of child	*Father disabled*	*Father normal*	*Mother disabled*	*Mother normal*
Male	.391	.058	.342	.067
Female	.177	.015	.171	.017

Source: Adapted from "Family History as an Indicator of Risk for Reading Disability" by G. P. Vogler, J. C. DeFries, and S. N. Decker, 1985, *Journal of Learning Disabilities*, *18*, p. 421. Copyright 1985 by The Professional Press, Inc. Reprinted by special permission of The Professional Press, Inc.

Longitudinal Analyses

Family data can also be used to improve the accuracy of predicting the later reading performance of individual reading-disabled children. As previously indicated, 69 matched pairs of reading-disabled and control children who participated in the original FRS were administered a follow-up test approximately 5 yearss later. When principal component measures of reading performance and symbol-processing speed were subjected to a mixed-model multivariate analysis of variance (DeFries & Baker, 1983), significant effects due to group (reading disabled versus control), time (initial test session versus follow-up), and their interaction were found. In the case of symbol-processing speed, the rate of improvement across the 5-year test-retest interval was significantly lower for reading-disabled children than for controls. However, with regard to the reading measure, the rate of change was highly similar for the two groups. Reading-disabled and control children differed substantially on the average at both ages, clearly demonstrating the persistent nature of reading disability. Although the rate of improvement in reading performance across the 5-year interval is similar on the average for reading-disabled and control children, longitudinal stability (i.e., the correlation between initial and follow-up test performance) is lower for the reading-disabled group. Thus, the prediction of later reading performance based upon an earlier test score may be more tenuous for reading-disabled children than for controls.

Because of the familial nature of reading disability, we tested the hypothesis that the accuracy of predicting reading performance of reading-disabled children over a 5-year test-retest interval can be significantly improved by incorporating parental data in a prediction equation (DeFries & Baker, 1983). Age-adjusted scores of the 51 probands who participated in both phases of the study and scores of their parents were subjected to a hierarchical multiple regression analysis. The following model was assumed:

$$C_2 = B_1C_1 + B_2M + B_3F + A, \tag{2}$$

where C_2 is a child's expected score at retest, M is its mother's score, and F is the father's score. The regression of follow-up test score on initial test score, B_1, is a measure of longitudinal stability. B_2 is the partial regression of child's retest score on mother's score, B_3 is the partial regression of child's retest score on father's score, and A is the regression constant. The significance of the regression coefficients is tested sequentially, viz., B_1 is estimated from C_1 and C_2 during step 1; M and F are added to the equation during step 2. The change in the squared multiple correlation between steps provides a test of the gain in accuracy that is due to the inclusion of parental data in the prediction equation.

Results of this hierarchical multiple regression analysis indicate that the prediction of later reading performance of reading-disabled children may be significantly improved by incorporating parental data into a prediction equation. As hypothesized, there is a significant increase in the squared multiple correlation for reading performance of reading-disabled children when parental data are added to the regression equation [$F(2,46) = 8.78$, $p < .01$]. However, no significant increase occurs when parental data are added to the regression equation for predicting retest reading performance of control children or for predicting retest symbol-processing speed scores of either group. These results suggest that parental data may significantly improve the accuracy of long-term prognoses for reading disability and may justify the collection of parental reading data by clinicians, researchers, and educators of reading-disabled children.

Although parental data significantly improve the prediction of later reading performance of reading-disabled children, but not of controls, this does not necessarily imply that the regression coefficients are significantly different in the two groups. In order to test the hypothesis of a differential group effect of incorporating parental data into a prediction equation, the regression model was extended as shown in Table 2.6. A second main effect, viz., group (G), is included in step 1. In addition, the interactions between mother's score and group ($M \times G$) and between father's score and group ($F \times G$) are tested during step 3. A significant increase in the squared multiple correlation between steps 2 and 3 would indicate differential effects of incorporating parental data into the prediction equation for the two groups. Because the results did not differ for probands and their reading-disabled siblings, data from all 69 pairs of children included in the longitudinal sample were subjected to hierarchical multiple regression analysis. Combining data from the two groups in one regression analysis, as well as adding data from siblings, yields a more powerful test of the importance of parental data for predicting later performance than the analysis reported by DeFries and Baker (1983).

TABLE 2.6
Regression Model Applied to Reading Scores

Model

$$\hat{C}_2 = \underbrace{b_1G + b_2C_1}_{\text{(step 1)}} + \underbrace{b_3M + b_4F}_{\text{(step 2)}} + \underbrace{b_5M \times G + b_6F \times G + A}_{\text{(step 3)}}$$

Results

	SS	*df*	*MS*	*F*	R^2
Step 1	188.94	2	94.47	122.76*	.63
Step 2	10.63	2	5.31	6.90*	.67
Step 3	2.46	2	1.23	1.60	.67
Residual	97.73	127	.77		
Total	299.76	133			

*$p < .01$.

Results of the extended multiple regression analysis are also presented in Table 2.6. As may be seen, the addition of parental data (step 2) significantly improves the prediction of later reading performance for both groups on the average. However, the interactions are not significant, indicating that the improvement in predicting retest reading performance by incorporating parental data into a regression equation is not greater for reading-disabled children than for controls. Similar results were obtained for symbol-processing speed data. Thus, results of this analysis suggest that parental data may increase the accuracy of predicting later tests scores for both reading-disabled and control children.

Bivariate Familial Analysis

Reading disability is characterized by depressed reading scores, but general cognitive ability within the normal range. However, reading performance is correlated with both verbal and nonverbal cognitive abilities; thus, reading-disabled children are expected to manifest deficits in other cognitive domains (Burns, 1984). Simultaneous familial analysis of reading and related skills may provide some insight into the etiology of the covariation among the measures. In the FRS, family data were obtained for the reading and symbol-processing speed composites defined previously. As indicated in Table 2.4, both measures were significantly lower among probands, siblings, and parents in the reading-disabled group relative to the corresponding control groups. In each case, the mean group difference in the reading composite is greater than the mean difference in the symbol-processing speed composite. Thus, the symbol-processing speed deficit may arise from its correlation with the reading measure, for which the deficit is presumably primary.

To examine the etiology of the covariation between these two measures, a bivariate familial path analysis was undertaken (Vogler & DeFries, 1985) in which both the phenotypic variances and their covariation are partitioned into components due to transmissible familial (genetic and/or family environmental) influences and specific, nontransmissible environmental influences. The path model employed is a bivariate application of a multivariate generalization of the pseudopolygenic model of Rice, Cloninger, and Reich (1978, 1980). Assortative mating is assumed to be phenotypic, and it includes both univariate assortative mating for each of the two variables and heteromorphic assortative mating where the correlation between reading in the mother and symbol-processing speed in the father can differ from the correlation between symbol-processing speed in the mother and reading in the father. The path from the transmissible value for the parent to that for the child is constrained to be 1/2. In the absence of twin or adoption data, heritability unconfounded by the effects of cultural or environmental transmission cannot be estimated, so phenotypic variation and covariation within an individual is partitioned into components due to familial influences (genetic and/or transmissible environmental effects) and specific, nontransmissible environmental influences. Univariate and bivariate sibling correlations among the nontransmissible environmental influences are permitted, with the correlation between specific environmental influences on reading in the proband and on symbol-processsing speed in the sibling being permitted to differ from that between symbol-processing speed in the proband and reading in the sibling in reading-disabled families. In control families, the two sibling environmental correlations are assumed to be equal since the classification of offspring as child 1 and child 2 is arbitrary. Within individuals, the correlation between familial and nontransmissible environmental influences is assumed to be zero.

Figure 2.2 is the general multivariate path diagram for the model used in this analysis. The variables are defined as follows (where the subscript M refers to the mother; the subscript F refers to the father; O_1 and O_2 refer to the proband or matched control and sibling, respectively): **P** represents the observed phenotypic measures, **G** denotes transmissible familial influences on the phenotypes, and **E** represents specific nontransmissible influences. In the bivariate application reported here, each "variable" in the diagram denotes a (2×1) column vector where the first element represents the reading composite and the second element represents the symbol-processing speed measure. The "path coefficients" of Fig. 2.2 are (2×2) diagonal matrices: the path matrix **h** contains h for reading in element (1,1) and h for symbol-processing speed in element (2,2); the path matrix **e** consists of e for reading in position (1,1) and for symbol-processing speed in position (2,2); and the path matrix **1/2** contains *1/2*'s along the main diagonal. **M** is a full, nonsymmetric (2×2) matrix containing the assortative mating spouse correlations, with the isomorphic reading and symbol-processing speed spouse correlations on the diagonal, the correlation of maternal reading with paternal symbol-processing speed in element (1,2), and the correlation of maternal symbol-processing speed with paternal reading in element (2,1). The full (2×2) matrix **C** represents the correlations among the sibling nontransmissible environments. In reading-disabled families, **C** is nonsymmetric, with the isomorphic sibling environmental correlations on the diagonal and the two potentially different heteromorphic correlations off the diagonal. In control families, **C** is symmetric since the two heteromorphic correlations are equated. The parental phenotypic vectors are entered twice following the model of phenotypic assortative mating developed by Wright (1978) and adapted by Fulker and DeFries (1983) using reversed path analysis with double entry of the phenotype.

The observed covariance matrix is divided into submatrices as shown in Fig. 2.3. Expectations in matrix notation, derived for each of the unique submatrices using the conventions for multivariate path analysis developed by Vogler (1985), are presented in Table 2.7. The matrices $\mathbf{V}_{M}^{1/2}$ $\mathbf{V}_{F}^{1/2}$, $\mathbf{V}_{O1}^{1/2}$, and $\mathbf{V}_{O2}^{1/2}$ are diagonal matrices containing phenotypic standard deviations for the mother, father, proband or matched control, and sibling, respectively.

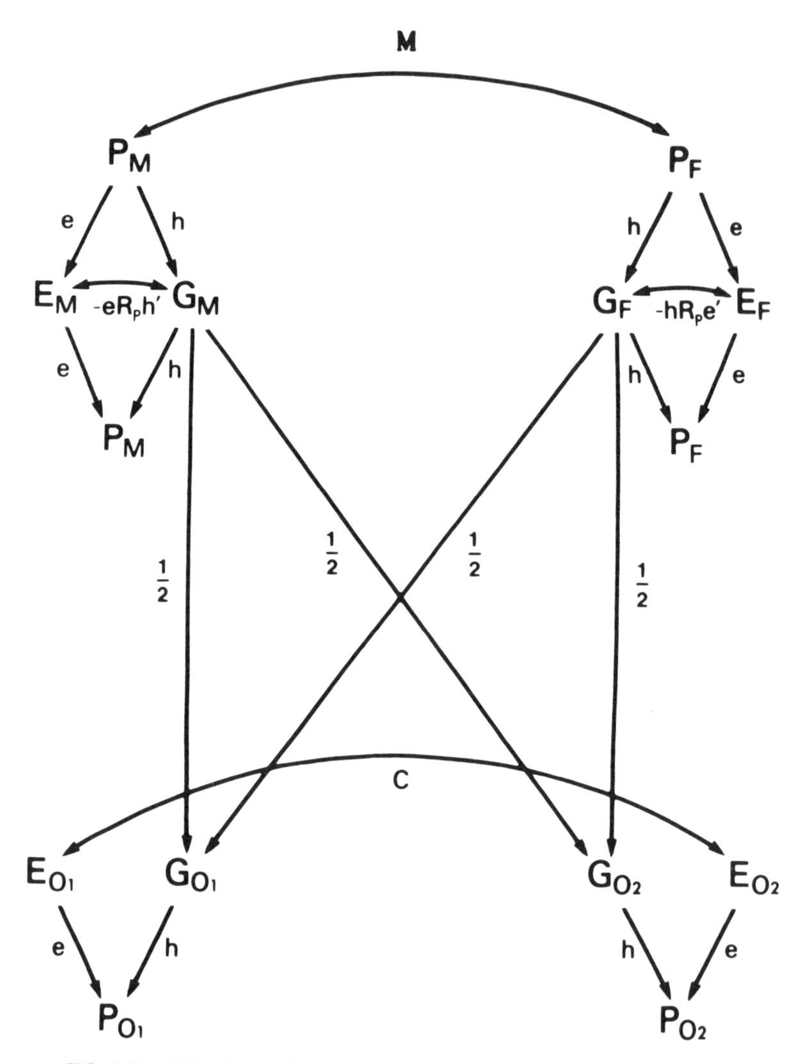

FIG. 2.2 Multivariate path model for the analysis of familial resemblance in nuclear families consisting of a mother, father, and two offspring.

Covariance matrices were computed separately for families of probands and families of controls, and separate matrices within family type were obtained for those families with complete data on the father, mother, and proband or matched control, and for those families for which data on an additional sibling were available. One sibling was randomly selected for inclusion in the analyses of data on families in which more than one sibling was tested. There were 121

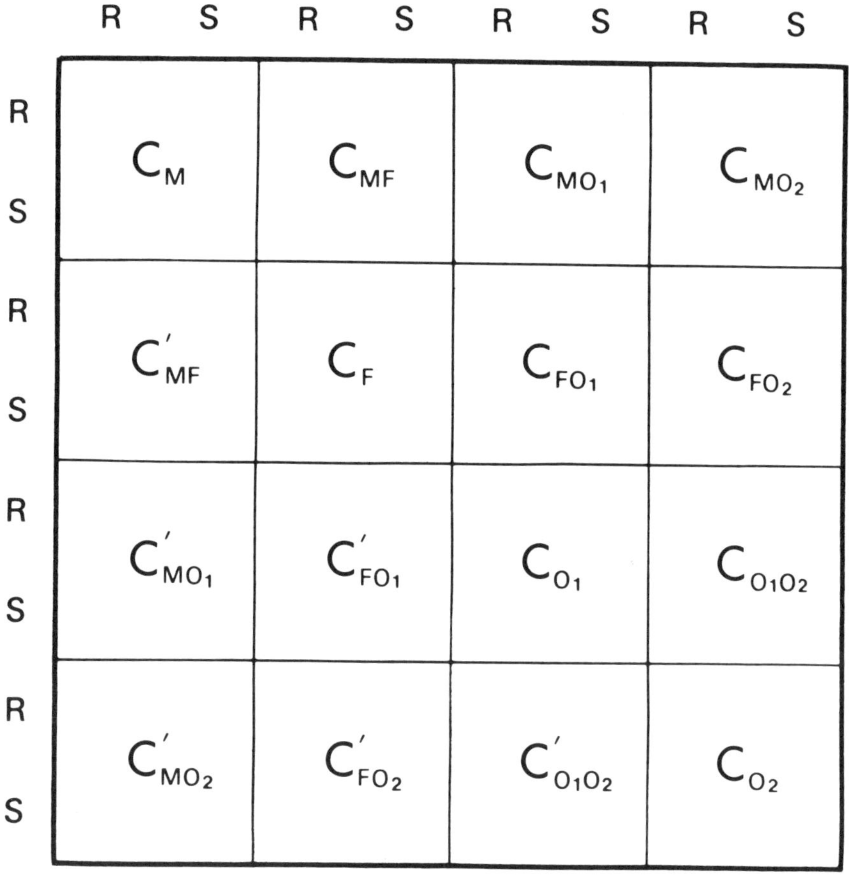

	R S	R S	R S	R S
R S	C_M	C_{MF}	C_{MO_1}	C_{MO_2}
R S	C'_{MF}	C_F	C_{FO_1}	C_{FO_2}
R S	C'_{MO_1}	C'_{FO_1}	C_{O_1}	$C_{O_1O_2}$
R S	C'_{MO_2}	C'_{FO_2}	$C'_{O_1O_2}$	C_{O_2}

FIG. 2.3 Submatrices of the observed covariance matrix of the reading composite (R) and the symbol-processing speed composite (S) measured on the mother (M), father (F), proband or matched control (O_1), and sibling (O_2).

families of probands (93 with a sibling and 28 without) consisting of 456 individuals, and 122 control families (85 with a sibling and 37 without) consisting of 451 individuals.

Separate analyses were conducted for the reading-disabled and control family types using a maximum-likelihood estimation procedure outlined by Jöreskog and Sörbom (1981) for the analysis of multiple matrices (families with a sibling and families without a sibling). The function, which was minimized using the generalized numerical optimization package MINUIT (CERN, 1977), yielded a log-likelihood ratio statistic which is distributed as chi-square.

Parameter estimates and their standard errors are presented in Table 2.8. There are moderate familial influences on both reading and symbol-processing speed

TABLE 2.7
Multivariate Expectations for the Submatrices of the Nuclear Family Model

Covariance matrix	*Expectation*
Phenotypic[a]	$\mathbf{C_P = V_P^{1/2} [hR_Gh' + eR_Ee']V_P^{1/2}}$
Mother-Father	$\mathbf{C_{MF} = V_M^{1/2}\, M\, V_F^{1/2}}$
Mother-Child 1	$\mathbf{C_{MO_1} = V_M^{1/2} [1/2\, (hR_G + Mh')h']V_{O_1}^{1/2}}$
Mother-Child 2	$\mathbf{C_{MO_2} = V_M^{1/2} [1/2\, (hR_G + Mh')h']\, V_{O_2}^{1/2}}$
Father-Child 1	$\mathbf{C_{FO_1} = V_F^{1/2}[1/2\, (hR_G + M'h')h']\, V_{O_1}^{1/2}}$
Father-Child 2	$\mathbf{C_{FO_2} = V_F^{1/2}[1/2\, (hR_G + M'h')h']\, V_{O_2}^{1/2}}$
Child 1-Child 2	$\mathbf{C_{O_1O_2} = V_{O_1}^{1/2}\{1/2\, h[R_G + h(1/2\, M + 1/2\, M')h']h' + eCe'\}\, V_{O_2}^{1/2}}$

[a]$\mathbf{V_P}^{1/2}$ is the appropriate matrix of phenotypic standard deviations for the father, mother, offspring 1, or offspring 2.

in both groups, with h^2 being about .3 for both phenotypes in control families and for symbol-processing speed in reading-disabled families, and .44 for reading in the reading-disabled families. Note that h^2 is not an estimate of heritability in these analyses; rather, it is an estimate of the contribution of *familial* influences on phenotypic variation. Familial influences on the two phenotypes are correlated substantially (.7 to .8) in both groups, whereas the correlation of the nontransmissible environmental influences is considerably lower (.31 in reading-disabled families and .25 in control families). Spouse correlations are highest for the reading measure, lower for symbol-processing speed, and lowest for the heteromorphic correlations. The sibling specific environmental correlations are generally nonsignificant.

Based on the results of the complete model, a reduced model was tested in which the two heteromorphic spouse correlations were equated and the sibling environmental correlations were fixed at zero. Parameter estimates for the reduced model are presented in Table 2.9. The unconstrained parameters are stable, and tests of the differences in log-likelihoods between the complete and reduced models are nonsignificant [$\chi^2(5) = 8.4$, $p = .14$ for reading-disabled families and $\chi^2 (4) = 2.8$, $p = .61$ for control families], indicating that the constraints are acceptable.

The expected phenotypic correlation matrix ($\mathbf{R_P}$) can be divided into components due to familial influences ($\mathbf{hR_Gh'}$) and specific environmental influences ($\mathbf{eR_Ee'}$) from the expectation $\mathbf{R_P = hR_Gh' + eR_Ee'}$, yielding expected values for the standardized components of both phenotypic variation (on the diagonal)

TABLE 2.8
Parameter Estimates ± S.E. for the Complete Model for the Reading-Disabled and Control Groups

		Reading disabled		*Control*	
Matrix		*Reading*	*Speed*	*Reading*	*Speed*
h	R	0.66 ± 0.06	0	0.52 ± 0.09	0
	S	0	0.54 ± 0.09	0	0.57 ± 0.08
e	R	0.75	0	0.86	0.
	S	0	0.84	0	0.82
R_G	R	1	0.70 ± 0.19	1	0.78 ± 0.22
	S	0.70 ± 0.19	1	0.78 ± 0.22	1
R_E	R	1	0.31 ± 0.11	1	0.25 ± 0.10
	S	0.31 ± 0.11	1	0.25 ± 0.10	1
M	R	0.30 ± 0.08	0.12 ± 0.03	0.36 ± 0.08	0.16 ± 0.08
	S	0.04 ± 0.09	0.21 ± 0.09	0.10 ± 0.09	0.15 ± 0.09
C	R	−0.16 ± 0.17	0.17 ± 0.15	0.17 ± 0.14	0.08 ± 0.11
	S	−0.13 ± 0.15	0.30 ± 0.12	0.08 ± 0.11	0.17 ± 0.14
$\chi^2(df)$		82.1 (37)		52.6 (40)	
p		< .001		.09	

Source: From "Bivariate Path Analysis of Familial Resemblance for Reading Ability and Symbol Processing Speed" by G. P. Vogler and J. C. DeFries, 1985, *Behavior Genetics, 15*, p. 118. Copyright 1985 by Plenum Publishing Corporation. Reprinted by permission.

and covariation (off the diagonal). These expectations from the reduced model are as follows for reading-disabled families:

$$\begin{bmatrix} 1.00 & 0.45 \\ 0.45 & 1.00 \end{bmatrix} = \begin{bmatrix} 0.45 & 0.25 \\ 0.25 & 0.34 \end{bmatrix} = \begin{bmatrix} 0.55 & 0.20 \\ 0.20 & 0.66 \end{bmatrix}.$$

The analogous partitioning of $\mathbf{R}_P$ for control families is:

$$\begin{bmatrix} 1.00 & 0.40 \\ 0.40 & 1.00 \end{bmatrix} = \begin{bmatrix} 0.29 & 0.24 \\ 0.24 & 0.36 \end{bmatrix} + \begin{bmatrix} 0.71 & 0.16 \\ 0.16 & 0.64 \end{bmatrix}.$$

In both family types, the contribution of specific environmental influences to phenotypic variation is substantially greater than the contribution of familial influences. However, the phenotypic covariation results to a greater extent from familial influences than from specific environmental influences.

Although the fit of the model to the control data is acceptable, the model clearly fails for reading-disabled families even though parameter estimates for

TABLE 2.9
Parameter Estimates ± S.E. for the Reduced Model for the Reading-Disabled and Control Groups

Matrix		Reading-disabled		Control	
		Reading	Speed	Reading	Speed
h	R	0.67 ± 0.06	0	0.54 ± 0.08	0
	S	0	0.58 ± 0.07	0	0.60 ± 0.07
e	R	0.74	0	0.84	0
	S	0	0.82	0	0.80
R_G	R	1	0.64 ± 0.16	1	0.74 ± 0.18
	S	0.64 ± 0.10	1	0.74 ± 0.18	1
R_E	R	1	0.33 ± 0.10	1	0.24 ± 0.10
	S	0.33 ± 0.10	1	0.24 ± 0.10	1
M	R	0.31 ± 0.08	0.09 ± 0.06	0.37 ± 0.08	0.14 ± 0.07
	S	0.09 ± 0.06	0.20 ± 0.09	0.14 ± 0.07	0.14 ± 0.09
C	R	0	0	0	0
	S	0	0	0	0
$\chi^2(df)$		90.5 (42)		55.4 (44)	
p		< .001		.12	

Source: Adapted from "Bivariate Path Analysis of Familial Resemblance for Reading Ability and Symbol Processing Speed" by G. P. Volgler and J. C. DeFries, 1985, *Behavior Genetics, 15*, p. 118. Copyright 1985 by Plenum Publishing Corporation. Reprinted by permission.

the two groups are similar. The failure of the model in the reading-disabled group may result from deviations from multivariate normality and aberrations in the covariance structure of the data due to the inclusion of a number of etiologically heterogeneous subtypes of reading disability in our sample or due to major gene influences. Greater phenotypic variance among parents and siblings in the reading-disabled sample relative to the variance in the control sample suggests that etiological heterogeneity, possibly including major gene influence, is present in our sample of families containing a reading-disabled child. These issues are discussed in the context of the FRS sample in the following sections.

Genetic Models

Results of FRS analyses conclusively demonstrate the familial nature of reading disability. To test for genetic influence, familial resemblance is necessary, but not sufficient. FRS data have been used to test the adequacy of various genetic models to account for familial transmission, and these analyses were recently summarized by DeFries and Decker (1982). One model for which the FRS data provide some support is the polygenic threshold model (Carter, 1973). This model assumes an underlying continuous liability or predisposition toward a condition that is a function of both genetic and environmental influences. Individuals beyond a "risk threshold" are assumed to express the condition. To the extent that the condition is heritable, relatives of affected individuals should have a higher average liability. For conditions like reading disability in which there are different prevalence rates for males and females, different thresholds are assumed. If the prevalence rate is higher in males, then females must have a higher threshold; i.e., for females to be affected, they must have a higher liability than males. Therefore, the polygenic threshold model predicts that a greater proportion of the relatives of female probands should be affected than those of male probands. In order to test this model, principal component reading scores were dichotomized. A comparison of the distributions of reading scores of probands and controls indicated a natural break at about −.50 standard deviations; thus, scores below this point were assumed to be indicative of a reading disability, whereas higher scores were considered to be within the normal range. Using this classification system, the proportions of affected relatives of male and female probands were compared. For each possible comparison (fathers, mothers, brothers, and sisters), a higher proportion of the relatives of reading-disabled girls were found to be affected. Therefore, these results are highly consistent with the polygenic threshold model.

Although the polygenic threshold model assumes multiple genetic and environmental factors, single-locus and environmental threshold models actually could account for the pattern of observed results. Classical single-gene models, on the other hand, are more parsimonious and are more likely to facilitate mechanism-related research. In an attempt to fit such models, the FRS data were subjected to segregation analysis using Elston and Yelverton's (1975) computer program GENSEG for analysis of a continuously distributed character (Lewitter, DeFries, & Elston, 1980). Five hypotheses were tested: a single autosomal dominant locus with two alleles; a single autosomal recessive locus with two alleles; two rather than three phenotypic distributions of the character; a single autosomal locus with two alleles; and no specific familial transmission, i.e., a within-family environmental hypothesis. When data from all proband families were analyzed, chi-square tests of goodness of fit indicated that each of the five hypotheses must be rejected: Neither the single-locus models nor the within-family environmental model adequately accounts for the familial transmission of reading disability. Similar results were obtained when data from only male proband families were analyzed. However, when data from families of female probands were subjected to segregation analysis, the hypothesis of recessive inheritance could not be rejected and the reduction in chi-square was greater than could be accounted for on the basis of sample size alone. Thus, results of these segregation analyses provide at least some evidence for autosomal recessive inheritance in families of reading-disabled girls.

It has also been suggested that the higher prevalence rate for reading disability in boys than girls may be indicative of sex-linked recessive inheritance (Symmes & Rapoport, 1972). However, in order to account for the observed sex ratio of about 3.5:1, the frequency of the hypothesized sex-linked recessive allele would have to be about .3 and this frequency is clearly inconsistent with observed prevalence rates (DeFries & Decker, 1982). A more rigorous test employing hierarchical multiple regression analysis (DeFries et al., 1979) also provided little or no evidence for the hypothesis that reading disability is due to a sex-linked recessive gene (DeFries & Decker, 1982). Thus, although segregation analyses of FRS data provide some evidence to support a hypothesis of autosomal recessive inheritance in families of female probands, no single-locus model (autosomal or sex-linked) appears to be adequate to account for the observed familial transmission of reading disability in the full sample.

Subtypes of Reading Disability

That no single-gene model adequately accounts for the transmission of reading disability in the total FRS data set may be due to the heterogeneity of the disorder. That is, there may be several etiologically distinct forms of reading disability, some heritable and some not. In order to explore this possibility, Decker and DeFries (1981) classified probands into four subtypes on the basis of principal component score profiles (reading, spatial/reasoning, and symbol-processing speed). The validity of this typology was then assessed by classifying affected siblings and parents using the same procedure and then cross-tabulating them as a function of proband's subtype. It was predicted that relatives of reading-disabled subjects of a given subtype should be more likely to be of the same subtype than expected on the basis of chance alone. Although some evidence was obtained for profile similarity between probands and their affected siblings, this did not occur for the parental data. Thus, unless reading disability is manifested by different profiles in the different age groups, it would appear that this particular typology does not meet the validity criterion.

An alternative approach to subtype identification has recently been reported by Pennington, Smith, McCabe, Kimberling, and Lubs (1984). Subjects were from the families with apparent autosomal dominant transmission ascertained by Smith et al. (1983) and briefly described in the first section of this chapter. When the average test scores of the familial dyslexics (N = 63) were compared to those of unaffected relatives (N = 41), it was found that the affecteds exhibited markedly depressed scores on reading recognition and spelling, but mathematics and general comprehension scores were within the normal range. Deficits with regard to reading comprehension were found to be intermediate in level of severity. Pennington et al. (1984) report that this profile of achievement test scores is almost diagnostic for specific dyslexia in this population and in a large clinic sample of learning-disabled individuals. The investigators then incorporated this profile into a diagnostic algorithm and applied it to all subjects in their sample. In brief, it was found that the algorithm identified 88% of those subjects previously classified as affected on the basis of history and rejected 93% of those previously classified as unaffected.

Because the test battery employed in the FRS includes many of the same tests utilized by Pennington et al. (1984), it is possible to apply a very similar diagnostic algorithm to subjects in the FRS sample. This algorithm employs Z-transformed PIAT standard scores and is as follows:

1. PIAT Mathematics Z score $\geq -.5$ and $\geq$ both PIAT Reading Recognition and PIAT Spelling.
2. The Z score of PIAT Mathematics exceeds that of either PIAT Reading Recognition or PIAT Spelling by $\geq +1.0$.
3. The Z score for PIAT Reading Comprehension is $\geq$ the Z score for either PIAT Reading Recognition or Spelling.
4. PIAT Reading Recognition or Spelling Z score must be $\leq +.5$, unless the discrepancy found for Criterion 2 is ≥ 2.0.

Somewhat surprisingly, only 35 (28%) of the 125 FRS probands fit these diagnostic criteria. This suggests that the different ascertainment criteria employed in the two studies may have resulted in markedly different subject populations. Moreover, when the relatives of these 35 probands were classified, it was found that at least one parent is similarly affected in only 20 cases and only 9 (20%) of 45 siblings meet the diagnostic criteria. Of course, if these criteria were diagnostic for a subtype of reading disability that is inherited as an autosomal dominant with full penetrance, at least one parent should be similarly affected in all cases and about half of the siblings should meet the diagnostic criteria. Thus, these results provide little evidence to support the hypothesis that a subtype of reading disability characterized by this particular profile of achievement test scores has an autosomal dominant mode of inheritance. Nevertheless, this study illustrates how family data may facilitate the discovery and characterization of possible reading-disability subtypes.

CONCLUSIONS

Reading disability is clearly familial, and results of various twin and family studies strongly suggest at least some genetic influence. Analyses of FRS data yield results that are consistent with expectations based upon the polygenic threshold model; however, no single-locus model has been found to account adequately for observed patterns of familial transmission. This lack of an adequate fit of any individual single-gene model to the total FRS data set may be due to the heterogeneity of the disorder. However, although the possibility of etiologically distinct subtypes of reading disability seems eminently reasonable, evidence for heterogeneity is somewhat equivocal. For example, results of recent analyses by Rodgers (1983) suggest that reading disability may merely represent the lower extreme of a normal continuum of achievement. Regardless of mode or modes of inheritance, results of recent FRS analyses demonstrate the importance of family data for risk analysis, for predicting the long-term consequences of reading disability, for studies of subtype validity, and for analyses of the etiology of covariation among reading-related measures.

ACKNOWLEDGMENTS

This work was supported in part by grants from the Spencer Foundation and NICHD (HD-11681) to J. C. DeFries. M. C. LaBuda is a predoctoral trainee supported in part by NIMH grant MH-16880.

We wish to acknowledge the invaluable contributions of staff members of the Boulder Valley and St. Vrain Valley school districts and of the families who participated in the study. We also thank Samuel A. Kirk, Winifred D. Kirk, and William Meredith for their help in planning the study and Rebecca G. Miles for expert editorial assistance.

REFERENCES

Aman, M. G., & Singh, N. N. (1983). Specific reading disorders: Concepts of etiology reconsidered. In K. D. Gadow & I. Bailer (Eds.), *Advances in learning and behavioral disabilities* (Vol. 2, pp. 1–47). Greenwich, CT: Jai Press.

Bakwin, H. (1973). Reading disability in twins. *Developmental Medicine and Child Neurology, 15,* 184–187.

Belmont, L., & Birch, H. G. (1965). Lateral dominance, lateral awareness and reading disability. *Child Development, 36,* 57–72.

Benton, A. L. (1975). Developmental dyslexia: Neurological aspects. In W. J. Friedlander (Ed.), *Advances in neurology* (Vol. 7, pp. 1–43). New York: Raven Press.

Burns, E. (1984). The bivariate normal distribution and the IQ of learning disability samples. *Journal of Learning Disabilities, 17,* 294–295.

Carter, C. O. (1973). Multifactorial genetic disease. In V. A. McKusick & R. Clairborne (Eds.), *Medical genetics* (pp. 199–208). New York: HP Publishing.

CERN (1977). *MINUIT: A system for function minimization and analysis of the parameter errors and correlations.* Geneva: CERN.

Critchley, M. (1970). *The dyslexic child.* New York: Heinemann Medical Books.

Decker, S. N., & DeFries, J. C. (1981). Cognitive ability profiles in families of reading-disabled children. *Developmental Medicine and Child Neurology, 23,* 217–227.

Decker, S. N., Vogler, G. P., & DeFries, J. C. (in press). Validity of self-reported reading disability by parents of reading-disabled children. *Acta Paedologica.*

DeFries, J. C., & Baker, L. A. (1983). Colorado Family Reading Study: Longitudinal analyses. *Annals of Dyslexia, 33,* 153–162.

DeFries, J. C., & Decker, S. N. (1982). Genetic aspects of reading disability: A family study. In R. N. Malatesha & P. G. Aaron (Eds.), *Reading disorders: Varieties and treatments* (pp. 255–279). New York: Academic Press.

DeFries, J. C., Johnson, R. C., Kuse, A. R., McClearn, G. E., Polovina, J., Vandenberg, S. G., & Wilson, J. R. (1979). Familial resemblance for specific cognitive abilities. *Behavior Genetics, 9,* 23–43.

DeFries, J. C., Singer, S. M., Foch, T. T., & Lewitter, F. I. (1978). Familial nature of reading disability. *British Journal of Psychiatry, 132,* 361–367.

Dunn, L. M., & Markwardt, F. C. (1970). *Examiner's manual: Peabody Individual Achievement Test.* Circle Pines, MN: American Guidance Service.

Elston, R. C., & Yelverton, K. C. (1975). General models for segregation analysis. *American Journal of Human Genetics, 27,* 31–45.

Finucci, J. M. (1978). Genetic considerations in dyslexia. In H. R. Myklebust (Ed.), *Progress in learning disabilities* (Vol. 4, pp. 41–63). New York: Grune & Stratton.

Finucci, J. M., & Childs, B. (1983). Dyslexia: Family studies. In C. L. Ludlow & J. A. Cooper (Eds.), *Genetic aspects of speech and language disorders* (pp. 157–167). New York: Academic Press.

Finucci, J. M., Guthrie, J. T., Childs, A. L., Abbey, H., & Childs, B. (1976). The genetics of specific reading disability. *Annals of Human Genetics, 40,* 1–23.

Foch, T. T., DeFries, J. C., McClearn, G. E., & Singer, S. M. (1977). Familial patterns of impairment in reading disability. *Journal of Educational Psychology, 69,* 316–329.

Fulker, D. W., & DeFries, J. C. (1983). Genetic and environmental transmission in the Colorado Adoption Project: Path analysis. *British Journal of Mathematical and Statistical Psychology, 36,* 175–188.

Hallgren, B. (1950). Specific dyslexia ("congenital word-blindness"): A clinical and genetic study. *Acta Psychiatrica et Neurologica Scandinavica,* Suppl. 65.

Hay, D. A., & O'Brien, P. J. (1982). Problems of twins in developmental behavioral genetics. *Behavior Genetics, 12,* 587. (Abstract)

Herschel, M. (1978). Dyslexia revisited: A review. *Human Genetics, 40,* 115–134.

Jöreskog, K. G., & Sörbom, D. (1981). *LISREL: Analysis of linear structural relationships by the method of maximum likelihood (Version V).* Chicago: International Educational Services.

Lenz, W. (1970). *Medizinische Genetik-Grundlagen, Ergebnisse und Probleme* (2nd ed.). Stuttgart: Thieme.

Lewitter, F. I., DeFries, J. C., & Elston, R. C. (1980). Genetic models of reading disability. *Behavior Genetics, 10,* 9–30.

Ludlow, C. L., & Cooper, J. A. (1983). Genetic aspects of speech and language disorders: Current status and future directions. In C. L. Ludlow & J. A. Cooper (Eds.), *Genetic aspects of speech and language disorders* (pp. 3–20). New York: Academic Press.

Morgan, W. P. (1896). A case of congenital word-blindness. *British Medical Journal, 11,* 1378.

Nie, N. H., Hull, C. H., Jenkins, J. G., Steinbrenner, K., & Bent, D. H. (1975). *Statistical package for the social sciences* (2nd ed.). New York: McGraw-Hill.

Owen, F. W. (1978). Dyslexia—genetic aspects. In A. L. Benton & D. Pearl (Eds.), *Dyslexia: An appraisal of current knowledge* (pp. 265–284). New York: Oxford University Press.

Pennington, B. F., & Smith, S. D. (1983). Genetic influences on learning disabilities and speech and language disorders. *Child Development, 54,* 369–387.

Pennington, B. F., Smith, S. D., McCabe, L. L., Kimberling, W. J., & Lubs, H. A. (1984). Developmental continuities and discontinuities in a form of familial dyslexia. In R. N. Emde & R. J. Harmon (Eds.), *Continuities and discontinuities in development* (pp. 123–151). New York: Plenum Press.

Rice, J., Cloninger, C. R., & Reich, T. (1978). Multifactorial inheritance with cultural transmission and assortative mating. I. Description and basic properties of the unitary models. *American Journal of Human Genetics, 30,* 618–643.

Rice, J., Cloninger, C. R., & Reich, T. (1980). Analysis of behavioral traits in the presence of cultural transmission and assortative mating: Applications to IQ and SES. *Behavior Genetics, 10,* 73–92.

Rodgers, B. (1983). The identification and prevalence of specific reading retardation. *British Journal of Educational Psychology, 53,* 369–373.

Sladen, B. K. (1970). Inheritance of dyslexia. *Bulletin of the Orton Society, 20,* 30–39.

Smith, S. D., Kimberling, W. J., Pennington, B. F., & Lubs, H. A. (1983). Specific reading disability: Identification of an inherited form through linkage analysis. *Science, 219,* 1345–1347.

Symmes, J. S., & Rapoport, J. L. (1972). Unexpected reading failure. *American Journal of Orthopsychiatry, 42,* 82–91.

Thomas, C. J. (1905). Congenital "word-blindness" and its treatment. *Ophthalmoscope, 3,* 380–385.

Vogler, G. P. (1985). Multivariate path analysis of familial resemblance. *Genetic Epidemiology, 2,* 35–53.

Vogler, G. P., & DeFries, J. C. (1985). Bivariate path analysis of familial resemblance for reading ability and symbol processing speed. *Behavior Genetics, 15,* 111–121.

Vogler, G. P., DeFries, J. C., & Decker, S. N. (1985). Family history as an indicator of risk for reading disability. *Journal of Learning Disabilities, 18,* 419–421.

Wechsler, D. (1974). *Examiner's manual: Wechsler Intelligence Scale for Children (Revised).* New York: Psychological Corporation.

Wright, S. (1978). *Evolution and the genetics of populations: Vol. 4. Variability within and among natural populations.* Chicago: University of Chicago Press.

Zahalkova, K., Vrzal, V., & Kloboukova, E. (1972). Genetical investigations in dyslexia. *Journal of Medical Genetics, 9,* 48–52.

Zerbin-Rüdin, E. (1967). Kongenitale Wortblindheit oder spezifische Dyslexie (congenital word-blindness). *Bulletin of the Orton Society, 17,* 47–56.

3 Audiogenic Seizures in Relation to Genetically and Experimentally Produced Cochlear Pathology

Kenneth R. Henry
University of California
Davis, CA

Audiogenic seizures have captured the imagination of scores of investigators for the past 60 years. Donaldson's 1924 description of sound-produced convulsions in rats has led to over a thousand articles on this topic, describing experiments conducted throughout the globe. Much of the fascination involves the behavior itself: a massive convulsion, precipitated by something as common (and presumably innocuous) as sound. But the audiogenic seizure is also considered a potentially useful model for investigating the ways in which genes, early experience, and auditory and neural events interact to produce an apparently simple behavior.

The present chapter reviews major factors that influence audiogenic seizures, but it is not intended to be comprehensive. The reader who wishes to be overwhelmed by the massive literature should start with the general reviews of Finger, 1947; Bevan, 1955; and Lehmann and Busnell, 1963. The subsequent reviews tend, of necessity, to be more specialized. They emphasize the relationship of audiogenic seizures to topics such as genetics, biochemistry, epilepsy, or pharmacology. The present review stresses the relationship of genetic and environmental factors to the development and function of the auditory system.

In spite of the numerous attempts, we know next to nothing about the genetics of audiogenic seizures. This failure stems, I believe, from a basic flaw in these experiments: concentrating on the easily observable seizure, while ignoring the auditory component of the behavior. Because most "nonsusceptible" mice can be made susceptible to audiogenic seizures by early disruptions of cochlear function, it is argued that most mice carry the "audiogenic seizure genes." All that is necessary to activate these genes is environmentally or genetically determined actions upon the cochlea.

Description of the Behavior

Audiogenic seizures can be produced by merely exposing a susceptible animal to 10–120 s of the acoustic stimulus. The subjects are individually tested in an enclosure that allows the experimenter good visibility (a 12-in-diameter × 18-in-high glass chromatography jar is ideal). The acoustic source (a loudspeaker driven by a white or filtered noise source, an electric bell, or even mechanically shaken keys) is then mounted atop the test chamber. After a brief period of adjustment to its new surroundings, the animal is exposed to the acoustic stimulus.

Immediately following the onset of a loud sound, the susceptible animal typically ceases ongoing behaviors. It may freeze, display emotional reactions, or begin to walk or jump. After a latency period, it will begin a wild-running response, which often leads to the myoclonic convulsive phase, in which the limbs conspicuously jerk back and forth. This may progress to the myotonic stage, beginning when the hind limbs are drawn forward, nearly touching the face, and continuing as they are extended away from the body. Double motor components (seizure, recovery, seizure) may occur. The animal may then recover, or death may ensue as a result of respiratory paralysis.

The behavior may be easily quantified by merely counting the number of animals who enter each of the consecutive phases, or by ascribing a severity score to each phase; for example, wild running = 1, clonic seizure = 2, and tonic seizure = 3 points. Some persons consider death to be part of the syndrome and quantify it as well. But this appears to be a genotype-specific characteristic, since not all strains and species die following the myotonic convulsion. The latency of each phase of the syndrome can also be recorded, providing a parametric measure of seizure severity.

In the most recent attempt to standardize the nomenclature, Schreiber, Lehmann, Ginsburg, and Fuller (1980) compared seizures in DBA/2, RB1 and RB2 mice. They agreed upon seven categories: No response, Wild run, Spasm, Clonic, Clonic-tonic, Tonic, and Death.

Sound-produced seizures are most often studied in mice and rats. Various inbred and outbred strains have been characterized in terms of their innate susceptibility. A few lines of susceptible rabbits exist. Idiosyncratic or experimental conditions may also produce susceptibility in hamsters and dogs. Similar behaviors have been reported in goats, cats, chickens, and humans, although people do not express a wild-running phase (Iturrian & Johnson, 1971; Jobe, 1981; Minami, 1981).

Age Factors in Audiogenic Seizures

Audiogenic seizures are typically described as being age-dependent in mice. The classic studies of Vicari (1951) categorized the DBA/2 as being nonsusceptible prior to 19 days, with approximately 60% convulsing at 20–24 days, and 90%

seizing at 30–34 days of age. By 80 days, mice of this strain were no longer susceptible. Schlesinger, Boggan, and Freedman (1965) described these stages as occurring somewhat earlier in their DBA/2 and B6D2F1 (cross of the DBA/2 and C57BL/6 mice). Plotnikoff (1960) described Swiss-O'Grady mice as being 95% susceptible between 21 and 42 days. Factors that influence developmental rate (diet, housing, etc.) can apparently modify these ages.

But some mice maintain susceptibility for a very long time. The A strain of mice are susceptible from 45 to 325 days (Vicari, 1957), and the Frings strain is still slightly susceptible (17% myoclonic seizures) at 365 days (Castellion, Swinyard, & Goodman, 1965).

The KM and UAZ lines of rats develop susceptibility at approximately the same ages as the mice described above. These strains do not typically die during testing, and maintain susceptibility throughout their life spans (Consroe, Picchioni, & Chin, 1979; Jobe, 1981; Kruschinsky et al., 1970; Sterc, 1963).

Sensitivity of the rabbit appears to be age-dependent. Ross, Sawin, Denenberg, and Volow (1963) described the decline of susceptibility in the ACAEP rabbit. At 22–31 days, 73% of the rabbits convulsed. This declined to 9% by 100 days.

A later section describes how these age-related changes of susceptibility are associated with development and decline of cochlear function.

Relationship to Other Types of Seizures

Susceptibility to both audiogenic and electroconvulsive seizures appears to develop at the same rate in the mouse. Castellion et al. (1965) traced the maturation of responsiveness to acoustic and electroconvulsive stimulation in Frings, O'Grady, and CF#1 mice. At approximately 20 days postpartum, maximal responsiveness developed to both forms of convulsions. However, susceptibility to sound-produced convulsions declined in mice older than 3 weeks, whereas no change was noted for electrically induced seizures. Deckard, Lieff, Schlesinger, and DeFries (1976) noted a similar developmental pattern in their six inbred strains. But they noted an even larger correlation (.91) between strains, indicating that genotypes that are highly susceptible to one type of seizure are also highly susceptible to the other.

In the rat, direct electrical stimulation of subcortical auditory structures, especially the inferior colliculus, will produce a seizure. This seizure pattern appears to be similar, if not identical, to that seen in response to sound (Duplisse, 1976; Huxtable & Laird, 1978). The single line of rats that was susceptible to audiogenic seizures was considerably more sensitive than the single nonsusceptible line to seizures induced by electrical stimulation of the auditory midbrain nuclei (Laird & Huxtable, 1978).

Nonetheless, these two types of seizures can be differentially affected by hypothermia. Reducing the core (rectal) temperatures of otherwise susceptible rats and mice to below 27°C will reversibly protect them from audiogenic seizures produced either by sound or by direct electrical stimulation of the auditory cortex. This treatment has no effect on electroconvulsive seizures (Bures, 1963). Perhaps this is related to the observation that hypothermia increases transmission time in the auditory brainstem, with the inferior colliculi being especially affected (Henry, 1980). It also elevates cochlear thresholds, with this effect being greatest at the high frequency end (Henry & Chole, 1984).

Acoustic Stimulus Parameters

Some of the earlier studies doubted the essential nature of sound in precipitating what is now termed the audiogenic seizure; Morgan and Morgan (1939) challenged the view that this behavior was a "neurotic response to conflict." They suggested that the high-pitched sounds of the air blast stimulus was essential to the convulsive reaction in rats.

Dice and Barto (1952) tested various "races" of deermice (genus *Peromyscus*) over a wide range of frequencies (500 to 95,00 Hz), finding sounds of 5–16 kHz most effective in producing convulsions. Frings and Frings (1952) tested their line of susceptible albino mice at frequencies from 6 to 25 kHz, reporting 10–12 kHz as most effective. Schreiber (1978) found frequencies from 15 to 30 kHz to be very effective for the DBA/2 mouse, with severities greatest at 20 kHz. Ralls (1967) used auditory-evoked responses, recorded from the inferior colliculi of several strains of mice (both *Mus musculus* and several lines of *Peromyscus*), to determine their auditory sensitivity. She found the *Mus* were typically most sensitive at approximately 15 kHz, whereas *Peromyscus* were most responsive over a wider range of frequencies (10–40 kHz). Therefore, susceptibility to audiogenic seizures appears to be most severe at or near those frequencies that are most readily detected by the subject.

Increasing the stimulus intensity typically increases seizure severity (Frings & Frings, 1952). Alexander and Gray (1972) found that increasing the intensity of the stimulus would allow seizures to be produced by a wider range of pure tones.

Mixed sounds (an electric bell, jingling keys) are more effective than pure tones (Bevan, 1955). But they are difficult to replicate from lab to lab; their complex spectral characteristics are often so idiosyncratic that a louder bell may be less effective than a smaller, more quiet one (Schreiber, 1978). One good compromise is the use of a well-defined electronically produced noise, such as the octave band used by Bock and Saunders (1977).

Nonauditory Neural Factors

Several studies have suggested that certain forebrain structures influence audiogenic seizures. Aluminum hydroxide gel applied to the cerebral cortex can produce susceptibility to audiogenic seizures in rats (Servit & Sterc, 1958).

Chocholova (1962) applied a KCl solution to the cerebral cortex of the rat, thereby producing a temporary neural insensitivity. The susceptibility to audiogenic seizures was reduced in these rats with temporarily lesioned cerebral cortices. Kesner, O'Kelly, and Thomas (1965) obtained similar results. But they were more cautious in their interpretations, since Buresova, Fures, and Fifkova (1962) had shown that spreading cortical depression can spread to subcortical regions. Kim (1961) bilaterally aspirated either the hippocampus and overlying neocortex, or the overlying neocortex in susceptible albino rats, obtaining results somewhat at variance with these other reports: The neocortical lesions resulted in increased susceptibility to audiogenic seizures. Perhaps these differences can be accounted for by the heterogeneity of the cerebral cortex, with some areas having facilitating and others having inhibitory influences. Hippocampal ablations were even more effective in increasing susceptibility to audiogenic seizures in rats (Kim, 1961). Reid, Bowler, and Weiss (1983) reported an increase in the severity of audiogenic seizures in hippocampally lesioned susceptible mice. Reid, Mamott, and Bowler (1983) also reported audiogenic seizures in about half of their lesioned SJL mice, whereas their sham-lesioned mice were almost totally resistant to audiogenic seizures.

Kesner (1966) reported the effects of lesions at several subcortical loci. In his rats, hippocampal and amygdalar lesions had little influence, but caudate nucleus lesions increased the severity and incidence of audiogenic seizures. Perhaps most significantly, lesions of the reticular formation eliminated clonic-tonic convulsions but not the wild-running phase of the syndrome.

The cerebellum can also influence audiogenic seizures. Infantellina, Sanseverino, and Urbano (1964) applied strychnine to the cerebellar auditory projection area of the cat, thereby inducing susceptibility to audiogenic seizures. Willott and Urban (1978) made radio frequency paleocerebellar lesions in the DBA/2 mouse, noting an increase of both the incidence and severity of audiogenic seizures, when compared to their sham-operated controls.

Willott (1976) also performed unilateral lesions of the spinal cord of the DBA/2 mouse, in order to determine the influence of bodily movements on audiogenic seizures. These unilateral spinal cordotomies were combined with unilaterally occluded outer ears. Hemisection ipsilateral to the open ear reduced the severity of seizures, but contralateral hemisections had no effect. He suggested that this effect was due to the interruption of somatic afferent information to the brain.

These experiments do not necessarily mean that the described structures specifically affect audiogenic seizures. A lesion may disrupt not just the neurons at the site of the intended dysfunction, but also fibers of passage. They may also temporarily traumatize adjacent structures. In addition, the effects observed may not be specific to audiogenic seizures, but may influence convulsions induced by other means. Nonetheless, they do provide suggestions about the interrelationship of several brain areas which influence the audiogenic seizure.

A few studies have measured ongoing electroencephalographic (EEG) activity in mice undergoing audiogenic seizures. Niaussat and Laget (1963) noted an absence of paroxysmal activity from the cortex of RB mice undergoing audiogenic seizures, although chemoconvulsions did produce cortical spike waves. Maxson and Cowen (1976) performed similar measures on the susceptible C57BL/6Bg-Gad, DBA/1Bg, DBA/2Bg, Rb/1/Bg, and HAS/Bg mice. They, too, found no evidence of seizure activity in the cerebral cortex of mice undergoing audiogenic seizures, although it was present during seizures produced by picrotoxin or semicarbizide. They suggested that the audiogenic seizure was a type of brainstem epilepsy. Maxson and Sze (1976) also induced susceptibility to audiogenic seizures in mice by the method of alcohol withdrawal. Although alcohol withdrawal sometimes resulted in cortical seizure activity, this was not found in the C57BL/6J mice who were being tested for audiogenic seizures during alcohol withdrawal.

Auditory Lemniscal Pathways

The integrity of the lower portion of the auditory system is necessary in order that audiogenic seizures be expressed. Chemical destruction of the organ of Corti of the cochlea abolishes susceptibility to this behavior in the otherwise susceptible O'Grady mouse (Kornfield, Geller, Cowen, Wolf, & Altman, 1970).

A few studies of higher auditory neural structures provide evidence that the necessary pathways involve those that ascend to the inferior colliculus. Although Bures (1963) did not observe a suppression of audiogenic seizures following inferior collicular lesions in rats and mice, Servit (1960) and Kesner (1966) noted a decrease or cessation in seizure susceptibility following such a lesion. Wada, Terao, White, and Jung (1970) found that bilateral inferior collicular lesions could abolish audiogenic seizures in the rat, and that lesions confined to the ventral portion were most effective.

Willott and Lu (1980) made the most detailed histological study of the influence of the auditory midbrain on audiogenic seizures. Lesions of the central nucleus of the inferior colliculus typically abolished the seizure syndrome in the DBA mouse, whereas lesions confined to the external IC nucleus had little effect. Lesions of the deep superior colliculus and tegmentum, including the central gray, often terminated the seizures at the wild-running or clonic stage. Lesions of the dorsal superior colliculus were ineffective.

Auditory pathways above the level of the midbrain may not be necessary to support audiogenic seizures. Koenig (1957) found little effect of medial geniculate lesions in the rat, nor did Willott and Lu (1980), in the mouse.

Auditory Functions in Susceptible and Nonsusceptible Strains

Physiological and histological abnormalities have been noted in the inner ears of susceptible Swiss/RB mice. Niaussat and Legouix (1967) reported a low amplitude CM (cochlear microphonic, the presumed receptor potential that is produced by the cochlear hair cells) in this line, even though the cochlear action potential (AP, or auditory nerve evoked potential) appeared normal. The hair cells at the base (high-frequency portion) of the organ of Corti were degenerated in the susceptible RB line, and normal in the resistant RB line. Hair cells at the low-frequency (apical) region were normal in the susceptible, and damaged in the resistant line (Darrouzet & Guilhaume, 1967; Darrouzet, Niaussat, & Legouix, 1968). By 6 weeks of age, both lines showed degenerating outer hair cells, although progressing at different rates, with no apparent abnormality of inner hair cells (Portmann, Darrouzet, & Niaussat, 1971). This ultrastructural study also revealed considerable loss of both afferent and efferent fibers. Portmann et al. (1971) concluded that these changes all resulted from an enzymatic disturbance of protein synthesis in the hair cells and neurons.

Ebel, Stefannovic, Simler, Randrainarisoa, and Mandel (1974) compared choline acetyltransferase (ChAc) and acetylcholinesterase in the cochleas of susceptible and resistant RB Swiss mice. They reported that ChAc activity was higher in mice that convulsed after the acoustic exposure. They suggested that this might be involved with stronger excitatory impulses in the spiral ganglia of susceptible mice.

In the susceptible DBA/2 mouse, cochlear AP thresholds have been related, over time, with changing patterns of susceptivity. Ralls (1967) observed elevated thresholds for high-frequency inferior collicular evoked potentials in the young DBA/2 mouse. Even though she did not relate this to susceptibility for audiogenic seizures, she did hypothesize that the rapid age-related auditory losses in these mice were responsible for their eventual loss of susceptibility. These early losses of high-frequency sensitivity were later found to be of a cochlear origin, and were hypothesized to be causally involved with its susceptibility at 21 days (Henry & Haythorn, 1975; Henry, McGinn, Berard, & Chole, 1981).

Less is known about the cochlea of the susceptible rat. Glenn, Brown, Jobe, Penny, and Bourn (1980) compared CM and AP measures in two lines of Spague-Dawley rats. They found both measures to be elevated by 25–35 dB in the susceptible line. Penny et al. (1983) examined the cochleas of susceptible rats. Their strain showed a high amount of variability (one animal had 1,000 more hair cells than the controls, even though the typical pattern was for them to have fewer inner and/or outer hair cells). Abnormally long and misshapen stereocilla were also seen.

Niaussat (1969) made an interesting comparison of audiogenic seizure susceptibility and threshold to the Preyer acoustic reflex. Even though the susceptible RB line had evoked potential thresholds that were approximately 40 dB higher than the resistant line, the Preyer thresholds of their two lines were nearly the same. She suggested that a "recruitment" phenomenon existed in the susceptible mouse. This will be elaborated upon in the section relating to acoustic priming.

Willott (1981) obtained evidence of decreased neural inhibition in neurons of the inferior colliculus of the DBA/2 mouse. After-discharges (prolonged neural firings) were seen, only in response to higher intensity sounds, and only at ages when this mouse was susceptible to audiogenic seizure. This altered pattern of neural firing was only found in the ventral portion of the central nucleus of the inferior colliculus. As mentioned earlier, lesions of this area abolished susceptibility to audiogenic seizures (Willott & Lu, 1980).

STRAIN CORRELATES OF NEUROCHEMISTRY AND AUDIOGENIC SEIZURES

This type of comparison has a large literature, spanning many years. Most of it has been exhaustively reviewed elsewhere, and is not mentioned here. Only some of the more recent studies are discussed. Many of these correlational studies continue for years, interrupted by a failure to replicate in different strains.

Seyfried, Glaser, and Yu (1978) compared gangliosides between susceptible DBA/2 and nonsusceptible C57BL/6 strains. They hypothesized that the more heavily myelinated brain of the DBA/2 contributes to its susceptibility. This comparison was later extended to other nonsusceptible strains (LG, C3H/He, and BALB/c), reinforcing their earlier hypothesis (Seyfried, Glaser, & Yu, 1979a). However, subsequent experiments showed the B6D2F1 hybrid to have greater levels of myelin glycolipids than either parental strain, causing them to reject this hypothesis (Seyfried & Yu, 1980).

Seyfried, Glaser, and Yu (1979b, 1981) later found levels of thyroxine higher in DBA/2 than in C57BL/6 mice, and suggested an association of this thyroid hormone with abnormal CNS development in the DBA/2, making it susceptible to audiogenic seizures. In support of this hypothesis, antithyroid treatment protected the DBA/2, and thyroxine supplementation to the infant C57BL/6 rendered it susceptible. In opposition to their hypothesis is the long history of the influence of these treatments on the ear and audiogenic seizures. Both excesses and deficiencies of this hormone damage the ear; in a susceptible rodent, increasing the damage to its aberrant ear can increase the hearing loss, thereby protecting it from audiogenic seizures; in the nonsusceptible rodent, thyroxine manipulations can cause cochlear damage that has a similar effect as that caused genetically in the DBA/2 (Gusic, Femenic, Konic-Carnellutti, 1970; Hamburgh & Vicari, 1960; Henry et al., 1981; Van Middlesworth, 1977; Van Middlesworth & Norris, 1980; Vicari, 1953). This is expanded upon later.

Kristt, Shirley, and Kasper (1980) used 5-hydroxydopamine to localize monoamine synaptic boutons in the somatosensory and temporal cortex of 6-day-old DBA/2 and Swiss mice. They found less labeling in the temporal cortex of the DBA/2 mice and cautiously suggested that this might be related to their seizure susceptibility. Their caution appears justified, in view of recent studies relating monoamines (serotonin, norepinepherine, dopamine) to genetically determined susceptibility to audiogenic seizures in the mouse.

Several earlier investigators consistently found associations between monoamines and susceptibility to audiogenic seizures; strains that were deficient in these presumably inhibitory neurotransmitters were typically more susceptible than those that had them in greater concentrations (see Jobe, 1981, for a recent review). Several more recent studies, however, have cast doubt on this interpretation and even, in some cases, on the original observations (Alexander & Kopeloff, 1976; Bakhit, Shenoy, Swinyard, & Gibb, 1982; Lehmann, 1977; Lints, Willott, Sze, & Nenja, 1980).

Jobe (1981) feels that the evidence is better for rats, and that a deficiency in NE and/or 5HT may be an important contributory factor in susceptibility to audiogenic seizures in this species. This may be correct, but one should remember that fewer studies have been conducted on the rat, and fewer strains are available for comparison.

Genetic Analysis of Audiogenic Seizures

Most searches for the audiogenic seizure gene(s) have used the DBA/2 and C57BL/6 as the susceptible and resistant parental lines. The choice of these lines stems from studies conducted over 30 years ago, in which they were characterized as being very different in terms of this behavior (Fuller, 1949; Hall, 1947; Vicari, 1947). With the advantage of 20–20 hindsight and modern technology that was denied to these pioneers, I feel that this was a poor choice. Both these genotypes are aberrant in terms of cochlear function, a factor that was unknown until more recently (Henry & Chole, 1980; Ralls, 1967), and that is likely to influence the interpretation of studies using these strains. A better choice might be between some strain that maintains susceptibility for a long time (such as A strain mice) and a nonsusceptible one that maintains good cochlear function, such as the CBA (Henry & Chole, 1980).

Witt and Hall (1949) hypothesized a single autosomal dominant gene as being primarily responsible for audiogenic seizures with other genes modifying its effect. Fuller and colleagues (Fuller, Easler, & Smith, 1950; Fuller & Thompson, 1960; Fuller & Williams, 1951) stressed environmental factors in their threshold model, in which the polygenic trait is influenced by a variety of nongenetic factors that modify its penetrance. The years have been better to this model than to most.

Several earlier studies implicated the dilute or closely linked genes of the DBA/2. An initial biochemical association was phenylalanine hydroxylase activity, which could affect brain development and production of certain neural transmitter substances. Interaction with pyridoxine utilization was later postulated (Coleman & Schlesinger, 1965; Huff & Huff, 1962). Huff and Fuller (1964) found no support for this concept, and others concur that it has no direct influence on audiogenic seizures in the DBA/2 mouse (Guttman & Lieblich, 1964; Schlesinger & Griek, 1970).

Another study using crosses of the DBA/2, C57BL/6, and their progency concluded that innate susceptibility is greatly influenced by a single pair of alleles located on Linkage Group VIII (Collins, 1970; Collins & Fuller, 1968). This recessive *asp* (auidogenic seizure prone) gene was described as residing near the centromere, close to the *b* (brown) gene.

Chen and Fuller (1976) used a broader genetic baseline for their selection experiment: A heterogeneous stock, originally derived from an eight-way cross of eight inbred lines (which included the DBA/2 and C57BL/6 strains). They concluded that their data were not consistent with the hypothesis that all susceptible mice are homozygous for the recessive *asp* gene.

Seyfried, Yu, and Glaser (1980) used the DBA/2 and C57BL/6 strains as the foundation for 21 recombinant inbred (RI) strains, in a study of susceptibility to audiogenic seizures in the 21-day-old mouse. They were unable to find an association of susceptibility with genes of Linkage Group VIII. The large phenotypic variability of the lines led to their concluding that audiogenic seizures are a polygenically controlled threshold character. This threshold model is very similar to that which has been developed and expanded by Fuller and associates (Fuller et al., 1950; Fuller & Thompson, 1978).

Seyfried and Glaser (1981) also found an association between susceptibility to audiogenic seizures and the Ah locus. They concluded that the dominant AH^b gene of the C57BL/6 and five other strains conveyed protection, whereas the recessive AH^d gene of the DBA/2 and six other strains did not. The protective gene of the C57BL/6 was tentatively labeled *Ias* (inhibition of audiogenic seizures; Seyfried, 1981; Seyfried & Glaser, 1981). When Seyfried (1983) extended this analysis beyond 21 days of age in his RI lines, he found no influence of the *Ias* gene on susceptibility to audiogenic seizures. In adult C57BL/6 × DBA/2 RI and congenic lines, there was no correlation between the Ah locus and audiogenic seizures. One of his RI lines even showed biphasic susceptibility, with incidence declining and then increasing as a function of age.

They were also compelled to reject their earlier hypothesis relating susceptibility to genetically determined differences of thyroxine (Seyfried et al., 1979b). They had noted high levels of this hormone in the preweanling DBA/2, suggesting this altered the CNS, thereby predisposing this strain to audiogenic seizures. They tested this hypothesis with seven of their RI and their D2-*Ias* congenic lines, finding no association of susceptibility and serum thyroxine in these mice (Seyfried, Glaser, & Yu, 1984).

Dozens of experimenter years must have been spent on the search for the genetic determinants of audiogenic seizures, and I'm not sure that much progress has been made. But I feel that a few generalizations can be made from this review so far:

1. Age is a factor that must be considered in any genetic analysis. Too many studies have considered an examination of the mouse at 21 days of age as being

sufficient to determine whether or not it has the "susceptible genotype." Some strains are susceptible only at or around this age, but others maintain susceptibility for many months; some have a bimodal span of susceptibility, whereas others may only become susceptible later in life.

2. Susceptibility and resistance are not functions of just the DBA/2 and C57BL/6 lines, and the inclusion of parental lines from other strains would certainly reduce the chance of spurious correlations.

3. One of the few certainties in this research is that the auditory system is necessary for the audiogenic seizure to be expressed, and pathologies found in auditory structures of susceptible mice suggest that there may be a causal relationship between certain types of auditory dysfunction and audiogenic seizures.

The next section deals with audiogenic seizures of a "nongenetic" nature. It is hoped that it will allow an examination of these three factors from a slightly different context. And perhaps it will even lead to a somewhat more productive means of examining genetic contributions to audiogenic seizure susceptibility.

ACOUSTIC PRIMING FOR AUDIOGENIC SEIZURES

The hypothesis proposed in this section is that acoustic priming produces a behavioral phenocopy for audiogenic seizures, creating a cochlear dysfunction similar to that found in the DBA/2. Its advantage is that it allows studies to be performed with susceptible and nonsusceptible mice of the same genotype, unconfounded by other genetic differences. It is even possible to compare the primed and nonprimed ears of the same mouse, providing a very tight control over environmental factors.

If the genetically "nonsusceptible" C57BL/6 mouse is exposed to a loud noise at approximately 14–20 days postpartum, it is very unlikely to express an audiogenic seizure. But if this animal is reexposed to the same sound about 5 days later, it will rapidly display a very severe audiogenic seizure (Henry, 1966, 1967). This initial exposure, termed *acoustic priming,* is capable of inducing susceptibility in many strains of mice (Fuller & Collins, 1968a; Henry, 1984b; Iturrian & Fink, 1967, 1968). It can enhance the susceptibility of innately susceptible strains (Henry & Bowman, 1970a). Priming is also effective in the hamster (Iturrian & Johnson, 1971) and the rat (Brown, Jobe, Bairnsfather, Mims, & Woods, 1980). Acoustic priming appears to be specific to audiogenic seizures, having no influence on other types of convulsions (Deckard, 1977; Henry & Bowman, 1969). Nonacoustic methods that reduce transduction of the ear can also prime otherwise nonsusceptible mice (Gates, Chen, & Bock, 1973). By contrast, general stressing agents (electroconvulsive shock, hypothermia) administered after priming do not abolish the effects of acoustic priming (Henry, 1967).

Acoustic priming only appears to be effective if it occurs after the onset of cochlear function (Henry, 1967). This association has been supported in several genotypes of mice (Boggan, Freedman, Lovell, & Schlesinger, 1971; Henry, 1984a,b; Henry & Bowman, 1970a; Iturrian & Fink, 1968; Schreiber & Graham, 1976).

A certain amount of time must transpire after acoustic priming before the mouse expresses susceptibility to audiogenic seizures. A few mice may convulse when tested 1 hr later, although subjects of another strain or experimental condition may require days before they express susceptibility (Boggan et al., 1971; Chen, 1980; Henry, 1967).

The amount of time following priming before maximal susceptibility is expressed also depends upon the age of the mouse during priming (Henry, 1967; Henry & Bowman, 1970b). In the C57BL/6 mouse primed at about 20 days, maximal seizures are seen when testing occurs approximately 11 days later (Boggan et al., 1971; Schreiber & Graham, 1976). In the SJL mouse, this time is 36 to 42 hr (Fuller & Collins, 1968a).

The period of maximal effectiveness also occurs early in life, but this age is much more variable among genotypes. In the C57BL/6, this is a sharply defined span, from 16 to 20 days (Boggan et al., 1971). It is somewhat later in the DBA/2 and B6D2F1 (Henry & Bowman, 1970a). It is approximately 3 weeks in the CF#1 and BALB/c mice (Chen, 1973; Iturrian & Fink, 1968). In the CBA mouse, priming is most effective at 30 to 36 days (Henry, 1984b). In the hamster, this range is 22 to 42 days postpartum (Iturrian & Johnson, 1971; Stanek, Bock, Goran, & Saunders, 1977). In the BALB/c mouse, nonacoustic priming (tympanic membrane destruction) is effective at an earlier age than is acoustic priming (Chen et al., 1973).

The acoustic stimuli that are most effective in acoustically priming the mouse appear to be similar to those that are most effective in producing audiogenic seizures. The C57BL/6 is rendered more susceptible by 15 or 17.5 kHz priming than they are by lower frequencies, and broadband sounds are more effective than pure tones (Bock & Chen, 1972; Henry, Thompson, & Bowman, 1971). Schreiber (1975) varied the intensity of a broadband noise in four steps, from 108 to 127 dB. The severity of audiogenic seizures increased as a function of intensity, and those primed with the 127 dB bell remained susceptible longer than the other groups.

AUDITORY ANATOMICAL SIMILARITIES OF PRIMED AND GENETICALLY SUSCEPTIBLE MICE

Fuller and Collins (1968b) performed the first crucial experiment that localized the effects of acoustic priming to certain parts of the nervous system. They primed SJL mice in either the right or the left ear, by the simple method of having the nonprimed ear full of glycerine during the initial acoustic exposure. Reexposure to the same acoustic stimulus, 48 hours later, resulted in audiogenic seizures only if the mouse was exposed in the ear in which it had been primed. This lateralization has also been seen in other strains of mice (Henry, Bowman, English, Thompson, & LeFever, 1971). Fuller and Collins (1968b) interpreted this as indicating that the site of sensitization most likely resided either in the ear itself or in portions of the auditory system that receive inputs from only one side. Ward (1971) traced this effect up to the level of the midbrain. By priming a single ear of the SJL mouse, he observed that audiogenic seizures could be blocked if the contralateral inferior colliculus were lesioned, whereas lesioning the ipsilateral inferior colliculus had no effect. Henry, Wallick, and Davis (1972) replicated this in the C57BL/6 strain, but found no evidence for lateralization of the midbrain in the innately susceptible DBA/2 strain. This suggested that both sides of the auditory brainstem and/or cochlea were involved in audiogenic seizures of the DBA/2. Ward and Sinnett (1971) also found no evidence for involvement at the level of the cerebral cortex in either the DBA/2 or the primed SJL, as determined by spreading depression.

These studies suggest that similar or identical auditory structures are involved in both genetically determined and priming induced audiogenic seizures. These include the ear and inferior colliculus. These two structures are next compared in primed, nonprimed, and innately susceptible mice.

COCHLEAR FUNCTIONAL CHANGES IN PRIMED MICE

Saunders and his colleages provided the first evidence of cochlear dysfunction in the primed mouse. Saunders, Bock, Chen, and Gates (1972) reported that priming elevated both cochlear microphonic (CM) and action potential (AP) thresholds in the BALB/c mouse. Saunders and Hirsch (1976) found a similar effect in the C57BL/6 strain. This cochlear dysfunction was reflected at the level of the brainstem, where elevated thresholds of the auditory-evoked potential (EP) were found in the cochlear nucleus and inferior colliculus (Saunders, Bock, James, & Chen, 1972). Threshold elevations of the inferior collicular auditory EP were subsequently found in the primed C57BL/6 mouse by Henry and Saleh (1973). These findings were reminiscent of Niaussat's earlier (1969) report of elevated inferior collicular EP thresholds in the genetically susceptible RB mouse.

The waveform and latencies of the AP have also been described as aberrant in mice susceptible to audiogenic seizures. Haythorn (1979) described the click-evoked AP waveform in the young C57BL/6 as having two separate components. Priming resulted in the first of these disappearing, and the latency of the second one decreasing. The AP of the DBA/2 looked very much like that of the primed C57BL/6. Furlow (1981) also noted a difference between the AP of the C57BL/6 and the DBA/2, very similar to that described earlier by Haythorn. Subsequent studies of the click-evoked AP suggested that the first component originated from the basal (high-frequency) portion of the cochlea, with the later one being a response of the low-frequency apical end (Henry, 1980).

These early descriptions of cochlear dysfunctions in primed mice were limited by experimental procedures that were considered appropriate at that time. For example, some of the earlier reports used click stimuli, which prevented a direct analysis of frequency-specific dysfunctions. In some reports, the cochlear temperature was not carefully controlled, resulting in a loss of high-frequency sensitivity, as has been described elsewhere (Henry & Chole, 1984).

With the benefit of more recent techniques, we have noted the similarity of cochlear dysfunctions of the innately susceptible DBA/2, the acoustically primed C57BL/6, and the nonacoustically (neonatal thyroxine) primed C57BL/6 mice (Henry et al., 1981). The susceptibility of the thyroxine-treated C57BL/6 mouse appears due to its conductive losses, with thresholds being elevated at all frequencies (Fig. 3.1). But the major influence of thyroxine treatment is probably

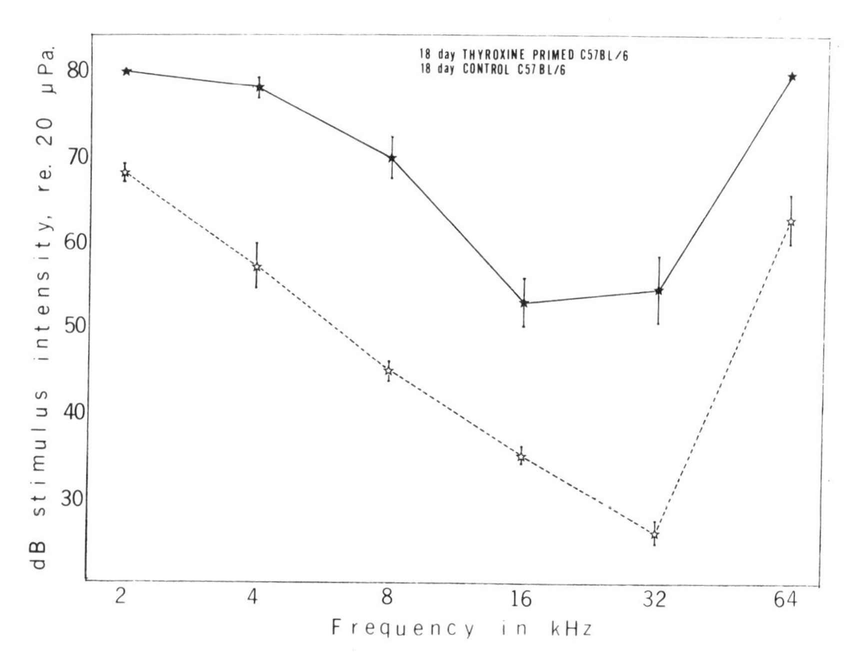

FIG. 3.1 Effects of neonatal (5 to 8 days postpartum) injections of thyroxine on auditory nerve thresholds of the C57BL/6 mouse. The resultant conductive hearing loss produced susceptibility to audiogenic seizures in these mice (solid stars). By contrast, the control mice (open stars) had normal auditory thresholds and were not susceptible to audiogenic seizures. (Henry, McGinn, Berard, & Chole, 1981, copyright 1981 by the American Psychological Association. Reprinted by permission.)

most closely related to the high-frequency dysfunctions, as suggested by a comparison with DBA/2 and acoustically primed C57BL/6 mice (Fig. 3.2).

Both the DBA/2 and the primed C57BL/6 mice have large (30–40 dB) elevations of the AP at 32 and 64 kHz (Fig. 3.2). The latency-amplitude function of the AP is normal in response to an 8 kHz tone in both of these susceptible mice, indicating normal functioning of this apical region of the organ of Corti (Fig. 3.3). But this function is clearly abnormal in response to a 32 kHz tone in these mice, indicating a basal cochlear dysfunction that is similar, but not identical, in the DBA/2 and the primed C57BL/6 mouse (Fig. 3.4).

BRAINSTEM AND BEHAVIORAL SIMILARITIES OF ACOUSTICALLY PRIMED AND GENETICALLY SUSCEPTIBLE MICE

Saunders, Bock, James, and Chen (1972) were the first to describe a significant change of the intensity dynamics of the amplitude of the auditory brainstem EPs of the primed mouse; although these thresholds of the cochlear nucleus and inferior colliculus were elevated, at higher sound levels, the EP amplitudes were

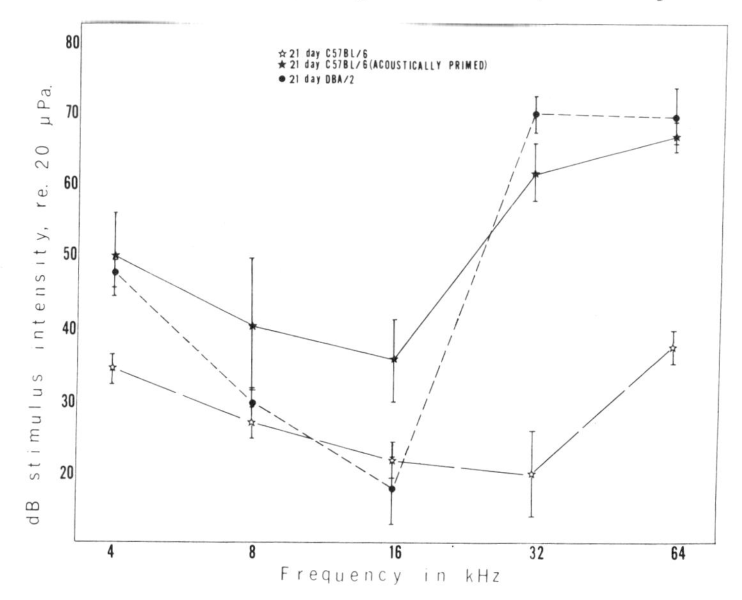

FIG. 3.2 Comparison of auditory nerve thresholds in genetically susceptible DBA/2 and experimentally susceptible (acoustically primed C57BL/6) mice. (Henry et al., 1981, copyright 1981 by the American Psychological Association. Reprinted by permission.)

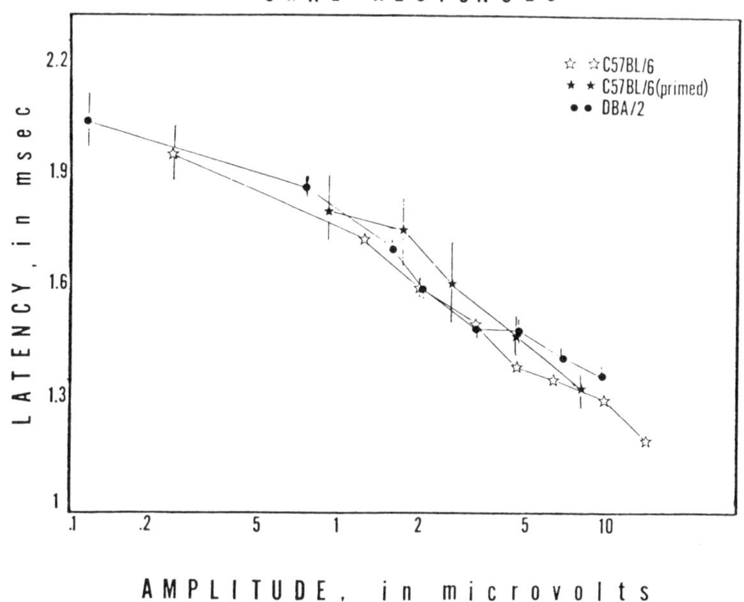

FIG. 3.3 Amplitude-latency auditory nerve evoked response functions in susceptible (DBA/2, primed C57BL/6) and nonsusceptible (C57BL/6) mice in response to an 8 kHz tone. All genotypes have similar, normal responses. (Henry et al., 1981, copyright 1981 by the American Psychological Association. Reprinted by permission.)

even larger than those of the nonprimed mouse. They noted that the EP amplitude-intensity functions of the primed mice resembled the steep slope of the loudness function that occurs in persons with recruitment of loudness. Saunders, Bock, James and Chen (1972) hypothesized that priming reduces cochlear responsiveness so that neural structures that are deprived of their normal auditory input develop a supersensitivity to subsequent acoustic input. Henry and Saleh (1973) also noted this recruitment-like pattern in the inferior collicular EP of the primed C57BL/6 mouse.

Niaussat (1969) had earlier suggested that recruitment of loudness occurs in the innately susceptible RB mouse. She noted the contrast of high inferior colliculus EP thresholds with low thresholds to the Preyer acoustic startle response. Henry (1972b) reported that acoustic priming lowers the threshold of this acoustic startle response by 15 dB, and that this effect is unilateral in the C57BL/6 mouse, as determined by the Fuller and Collins (1968a) monaural priming technique. Therefore, increased sensitivity to both audiogenic seizures and to the acoustic startle reflex could be considered as behavioral evidence for recruitment of loudness, and the increased amplitude of the auditory EP at high-stimulus intensities could be considered as neural evidence of recruitment (or, more precisely,

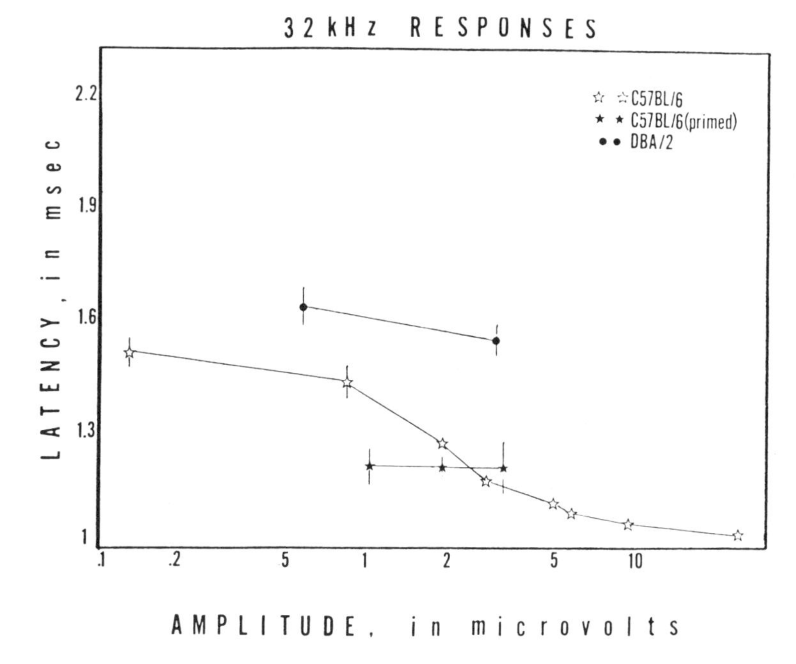

FIG. 3.4 Amplitude-latency auditory nerve functions in response to a high frequency (32 kHz) tone. Only the nonsusceptible (C57BL/6) group has a normal response, indicating high frequency cochlear dysfunction in the genetically and experimentally induced susceptible mice. (Henry et al., 1981, copyright 1981 by the American Psychological Association. Reprinted by permission.)

overrecruitment). But developmental studies of these changes show that they appear at different times. Priming the 16-day-old C57BL/6 mouse lowers the Preyer reflex threshold within 24 hr, but a longer time is required before it affects susceptibility to audiogenic seizures (Henry, 1972a).

Willott and Henry (1974) observed that the EP "recruiting" response was evident within minutes of priming in the C57BL/6, which is sooner than this strain shows susceptibility to seizures. In addition, C57BL/6J mice that have been nonacoustically primed by earplugging at 17 days develop susceptibility to audiogenic seizures before development of this "recruiting" response of the inferior collicular EP (McGinn, Willott, & Henry, 1973). Earplugging at 42 days of age can also increase the EP amplitude, although the C57BL/6 mice are too old to show audiogenic seizures (McGinn & Henry, 1975). Therefore, the "recruitment" pattern of the EP may develop before, after, or in the absence of susceptibility to audiogenic seizures.

Willott and colleagues have subsequently examined in more detail these inferior collicular changes of the primed mice. Willott, Henry, and George (1975) suggested that patterns of neuronal firing might be disrupted during priming. Urban and Willott (1979) recorded extracellularly from single neurons of the inferior colliculi of primed and control C57BL/6 mice. They found evidence that priming disrupted inhibitory mechanisms, including a prolongation of after discharges. A subsequent study (Willott & Lu, 1982) showed that some of these changes were evident within seconds following the noise exposure, and that discharge pattern changes were most pronounced in the high-frequency portion of the neuron's response areas. This was similar to changes that Willott and Lu (1980) had earlier found in the inferior colliculi of innately susceptible DBA/2 mice.

Genetic Studies of Acoustic Priming in Mice

Strains of mice differ in their ability to be acoustically primed. But ranking them in terms of their "primability" is fraught with the same types of dangers as were described earlier for ranking strains in terms of innate susceptibility to audiogenic seizures. For example, strains may differ in the age at which they are maximally susceptible to priming, or in the optimal stimulus properties required (frequency, intensity, duration). Therefore, any study that compares strains, or performs other genetic studies on priming, can only make statements that are limited to the particular choice of age or stimulus parameters that it has investigated. An example of some of these complexities may be found in Fig. 3.5.

In this study, C57BL/6, DBA/2, and their B6D2F1 hybrids were primed at 1 of 15 discrete ages, from birth to 28 days, and tested at 28 days of age. Either parental strain could be described as "dominant" for audiogenic seizures, depending upon the age chosen, and the F1 mice had a bimodal period of susceptibility. (This figure combines innate and priming-induced susceptibility to audiogenic seizures. An estimate of sensitivity to priming can be obtained by substracting the baseline scores of the mice primed at 0–6 days of age from the scores of all mice of that strain.) Although this study made an attempt to examine a simple interrelationship between the age when acoustically primed and genotype, it ignored the possibility that different priming stimuli would have a different effect on the three genotypes.

Strain differences also occur for nonacoustic priming. Maxson (1978) noted that C57BL/6Bg and DBA/1Bg-ras mice develop susceptibility at different rates following priming and earplugging, although only a single age was used for priming these mice.

Collins and Fuller (1968) provided the first evidence of the independence of genes involved in spontaneous and priming-induced audiogenic seizures. They utilized the standard C57BL/6 and DBA/2 parental strains, and their hybrids and backcrosses. Fuller (1975) found familial independence of these two traits in a heterogenous stock with a much broader genetic background. These studies only examined priming at a single age. Others have selectively bred lines of mice for the ability to be acoustically primed at a single age (Chen & Fuller, 1976; Deckard, Tepper, & Schlesinger, 1976).

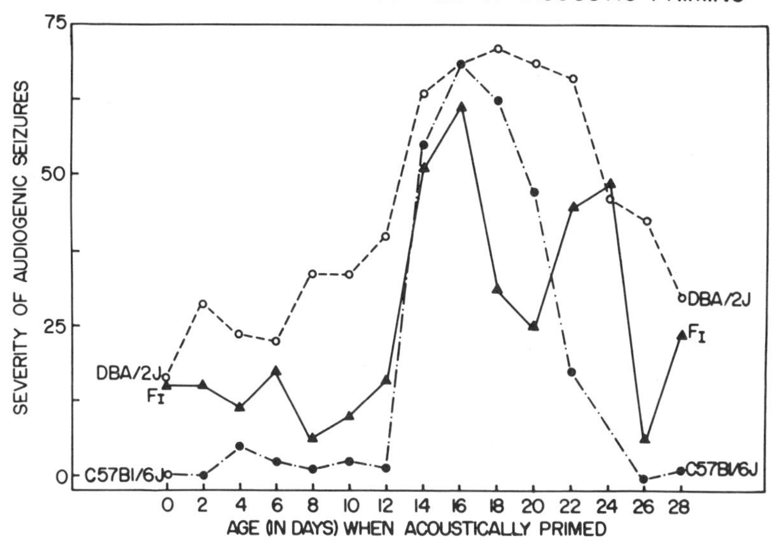

FIG. 3.5 Audiogenic seizure severity in acoustically primed DBA/2, C57BL/6 and their B6D2F1 hybrids. The mice were all acoustically primed at 1 of the 14 ages from 0 to 26 days of age, and tested at 28 days of age. The F1 could be described as being either more, less, intermediate to or equally severe to the parental strains, depending upon the age of priming. This illustrates the difficulty inherent in describing the "genetics of priming" or "genetics of audiogenic seizures" if developmental factors are not considered. (Henry and Bowman, 1970a, copyright 1970 by the American Psychological Association. Reprinted by permission.

There are some interesting differences between mice that have differing degrees of susceptibility to acoustic priming. Tepper and Schlesinger (1980) observed that acoustically priming their susceptible line produced damage to outer hair cells of the cochlea, whereas priming their resistant line produced no such damage; kanamycin priming damaged the cochleas of both lines. It would be interesting to know the cochlear functions of these different lines, both before and after priming.

Genetics, Auditory Development, and Acoustic Priming

Genetically determined differences of auditory development and degeneration are also strongly related to the sensitive or critical period for inducing susceptibility to audiogenic seizures by either acoustic or nonacoustic priming. In the next few paragraphs, several inbred strains are compared in terms of early auditory structure and function, and this is related to their abilities to be acoustically primed.

Priming has been studied most extensively in the C57BL/6 strain, in which the sensitive or "critical" period peaks at about 15–21 days of age. Lieff, Permut, Schlesinger, and Sharpless (1975) recorded inferior collicular evoked potentials (EPs) from awake C57BL/6 mice, noting that the EP amplitudes increased greatly between 15 and 22 days of age. They suggested that the critical period of this mouse is related to damage to auditory pathways that are developing at this time. Henry and Lepkowski (1978) noted degenerative changes of the cochlear microphonics and summating potentials of the 50-day-old C57BL/6 mouse, with functional losses increasing progressively with age. These changes were especially pronounced in the brainstem. Subsequent studies showed an early progressive loss of behavioral and electrophysiological auditory thresholds (Henry & Chole, 1980; Shnerson & Pujol, 1982).

The results of an earlier study (Saunders & Hirsch, 1976) suggested that there might be a relationship between the critical period of the C57BL/6 for acoustic priming and an age-related critical period for cochlear damage, but this correlation did not hold up in our laboratory. Both AP and CM thresholds were most susceptible to noise-induced losses at or near the time of puberty (30–40 days), although priming was barely effective at that age (Fig. 3.6). But the critical period for acoustic priming did agree with the developmental period when this strain has its most sensitive cochlear high frequency responses (Fig. 3.7).

In order to determine whether this lack of agreement with the ages of sensitivity to cochlear damage and to priming was related to the early auditory degeneration of the C57BL/6, it would be necessary to see how these factors relate in a mouse with normal audition, such as the CBA (Henry & Chole, 1980). The CBA mouse, which had not previously been characterized in terms of priming, was also most sensitive to noise-induced elevations of the CM and AP thresholds at about the age of puberty, becoming less sensitive thereafter (Fig. 3.8) But this strain, unlike the C57BL/6, had a close correspondence of the period of maximal sensitivity for AP threshold increase and for priming for audiogenic seizures (Fig. 3.9).

A comparison of Figs. 3.7 and 3.9 shows that the C57BL/6 mouse was not as severely affected by noise: its audiogenic seizures were less severe, and the cochlear threshold increases were less than was the case with the CBA strain.

The duration of priming-induced susceptibility also differs between these strains and is correlated with their auditory function. In the C57BL/6, it disappears rather quickly: a few 16-day-old primed mice will remain susceptible at 30 days, but this sensitivity disappears within the next week. In the CBA that is primed at 35 days, susceptibility lasts for at least 2 months.

This study suggests the following:

1. Acoustic priming is maximally sensitive in inducing susceptibility to audiogenic seizures at a developmental period when the auditory thresholds are lowest (approximately 18 days in the C57BL/6 and 36 days in the CBA).

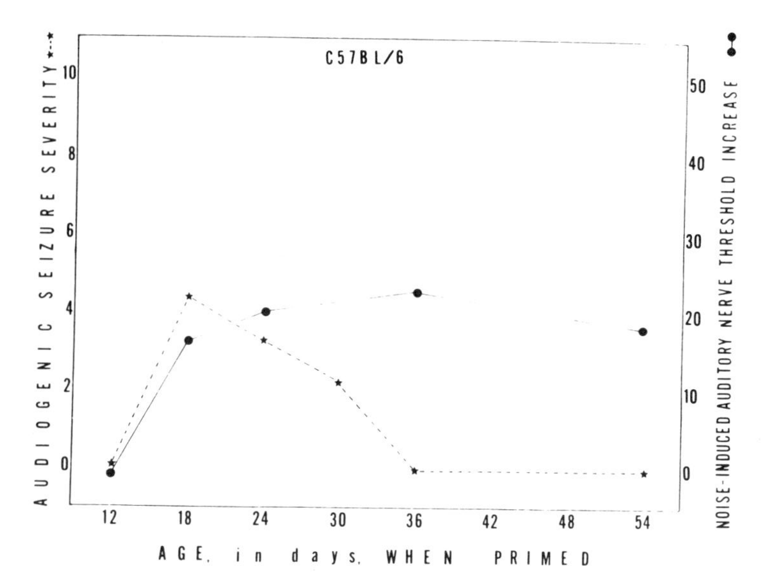

FIG. 3.6 Effects of age at priming on threshold elevation of the auditory nerve action potential and audiogenic seizure severity of the C57BL/6 mouse. The threshold measures were averaged across all frequencies from 2 to 64 kHz, and cochlear microphonic values showed the same age-related changes, which did not correlate highly ($r = .21$) with susceptibility to audiogenic seizures (unlike Fig. 3.5, in which a maximum score of 100 indicated death for all mice, in Figs. 3.6-9 a maximum score of 10 represents wild-running, clonic and tonic convulsions for all mice). (Redrawn from Henry, 1984b.)

2. This period for priming may correspond with the period at which the normal cochlea is most susceptible to noise damage, but genetically determined auditory abnormalities can alter this relationship.

3. The degree of susceptibility produced by acoustic priming is, within limits, greater in mice with higher noise-induced cochlear threshold increases (this is obviously not a monotonic function, because if noise were to completely destroy the cochlea, it could no longer provide CNS input that is necessary for audiogenic seizures to occur).

4. Progressive auditory degeneration reduces the ages at which the primed mouse will remain susceptible.

If these interpretations are correct, then these associations should also be seen in other strains of mice.

Fuller and Collins (1970) observed that susceptibility in the primed SJL mouse did not decline over a period of 25 weeks. This agrees with the observation (Henry, 1982a) that the SJL maintains good cochlear function for much of its life span. Fuller and Collins (1968a) did not trace the critical or sensitive period

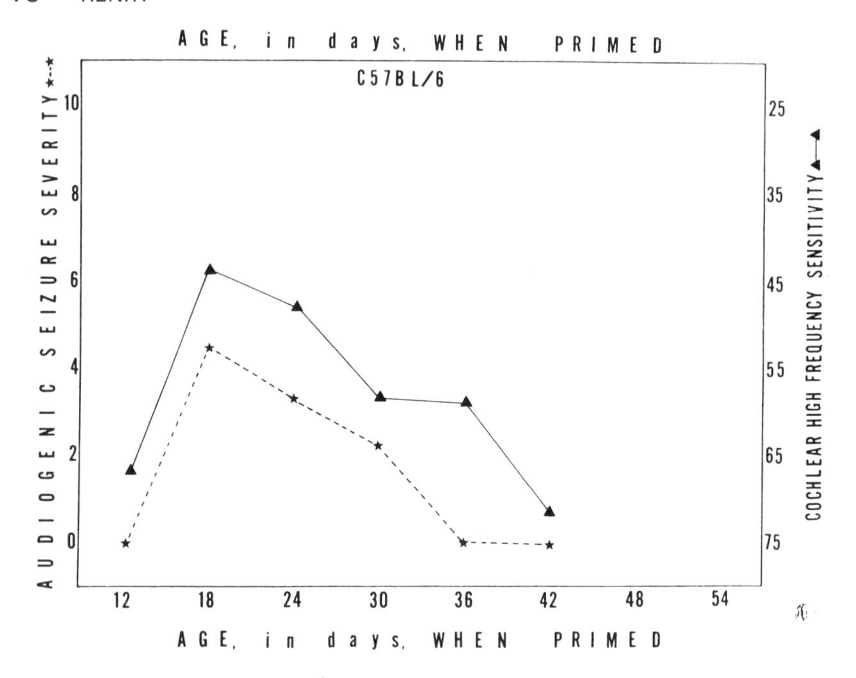

FIG. 3.7 Effects of age of the C57BL/6 on the effectiveness of priming and on cochlear high frequency sensitivity. The high-frequency thresholds are most sensitive at 18 days, and thereafter show the effects of early onset age-related hearing loss (presbycusis). These two measures correlate highly ($r = -.896$). (Redrawn from Henry, 1984a, 1984b.)

for priming, but its auditory profile suggests that it can be primed over a wide age span, from approximately 18 days to beyond 50 days. Twenty- and 60-day-old SJL and CBA mice also show similar high-frequency cochlear losses when exposed to a 12–24 kHz octave band noise (Fig. 3.10), suggesting that they might develop equally severe seizures when primed with this stimulus.

Only a few strains of mice have been characterized for cochlear function throughout a major portion of their life spans. Based upon this limited information, the preceding associations would predict the following:

1. The AU/Ss mouse also has good hearing in its youth, but loses its high-frequency cochlear function earlier than the CBA and SJL (Henry, 1982a; Fig. 3.11). Like the CBA, it has its lowest thresholds at approximately the time of puberty. Therefore, it should be susceptible to priming over a span that ranges from just prior to weaning until sometime after puberty, with maximal sensitivity occurring near puberty. Once primed, it should maintain its susceptibility for a matter of months, although not quite as long as the CBA and SJL mice.

2. The AKR strain, which has a moderate degree of innate susceptibility (Fuller & Sjursen, 1967), has slightly worse high-frequency cochlear sensitivity

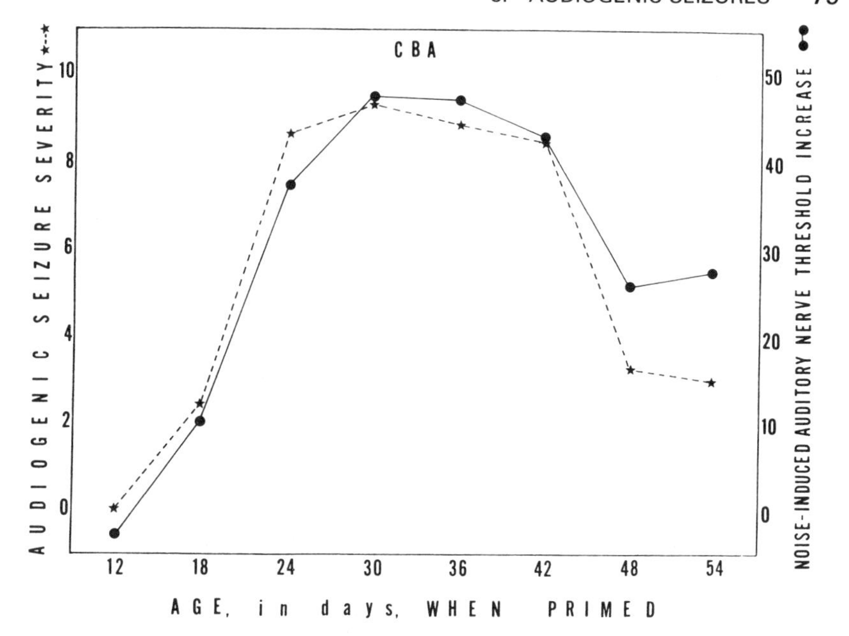

FIG. 3.8 Comparison of age when the CBA mouse was primed and its audiogenic seizure severity and auditory nerve threshold increase. Unlike the comparison for the C57BL/6 (Fig. 3.5), these two factors are highly correlated in the CBA (r = .946). (Redrawn from Henry, 1984b.)

than the C57BL/6, although its age-related cochlear losses are a bit slower than the C57BL/6 (Henry, 1982a; Fig. 3.12). Therefore, its critical period for priming would probably be of fairly short duration. It maintains some high-frequency cochlear function until just over 6 months of age, so it would probably retain susceptibility for longer than the C57BL/6, but not as long as the CBA, SJL, or AUSs.

These predictions should be correct if the hypotheses generated above are valid, and if no other genetic factors interact.

It might now be useful to return to an analysis of mice that are genetically susceptible to audiogenic seizures, and determine whether their susceptibility is correlated with cochlear functions.

COCHLEAR FUNCTIONS IN SUSCEPTIBLE STRAINS OF MICE: AN EVALUATION OF EXISTING DATA AND A PREDICTION OF FUTURE FINDINGS

This section only considers mice that have had a single exposure to the acoustic stimulus, so that genetic factors will not be confounded with possible priming effects. This strain comparison is made in reference to two hypotheses:

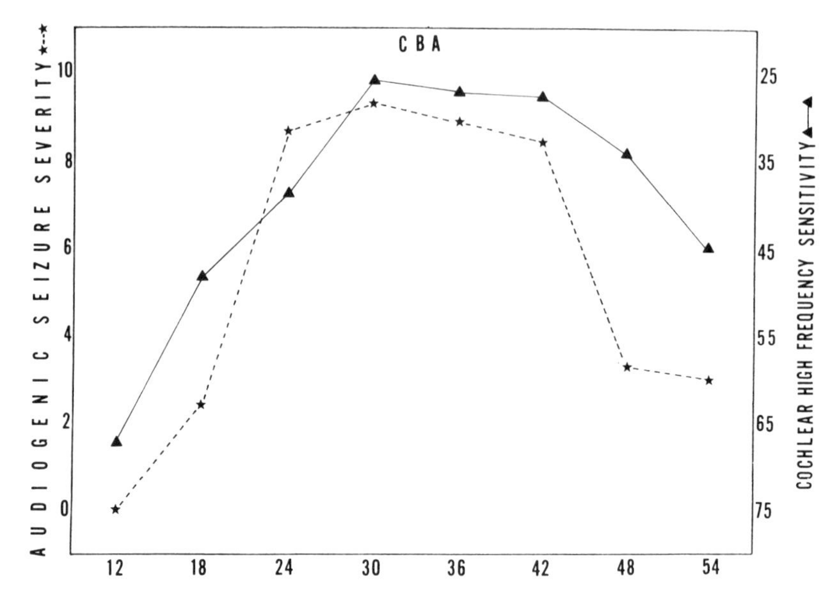

FIG. 3.9 Effects of age of the CBA on the effectiveness of priming and upon developmental changes of high frequency cochlear thresholds ($r = -.86$). (Redrawn from Henry, 1984a, 1984b.)

1. Strains of mice that are genetically susceptible to audiogenic seizures have, at a very early age, a marked increase of high-frequency cochlear thresholds, but middle- and low-frequency thresholds are relatively normal.
2. Susceptibility will be maintained so long as their cochlea continues to function in the manner described above, but will decline with further degeneration.

Niaussat (1969) compared cochlear auditory thresholds in susceptible and nonsusceptible sublines of the Swiss RB mouse. The susceptible line had poorer high-frequency responses early in life. Subsequent decline of susceptibility was associated with gross cochlear degeneration (Portmann et al., 1971).

Susceptibility in the DBA/2 mouse is very high at approximately 2–5 weeks of age, with 80% to 100% of the animals showing tonic seizures on their most susceptible day. The precise day of maximal susceptibility varies from study to study, and is probably influenced by environmental factors that affect maturation, as well as by the manner of testing (Schlesinger et al., 1965; Schreiber, 1978; Seyfried, 1981; Vicari, 1951). At its most sensitive age, the DBA/2 has a pronounced loss of high-frequency cochlear function (Henry et al., 1981; Ralls, 1967). Figure 3.2, shown earlier, compares this function with that of the acoustically and nonacoustically primed C57BL/6 mice. The rapid onset of age-related hearing loss in the DBA/2 mouse is probably responsible for the end of its susceptibility (Ralls, 1967).

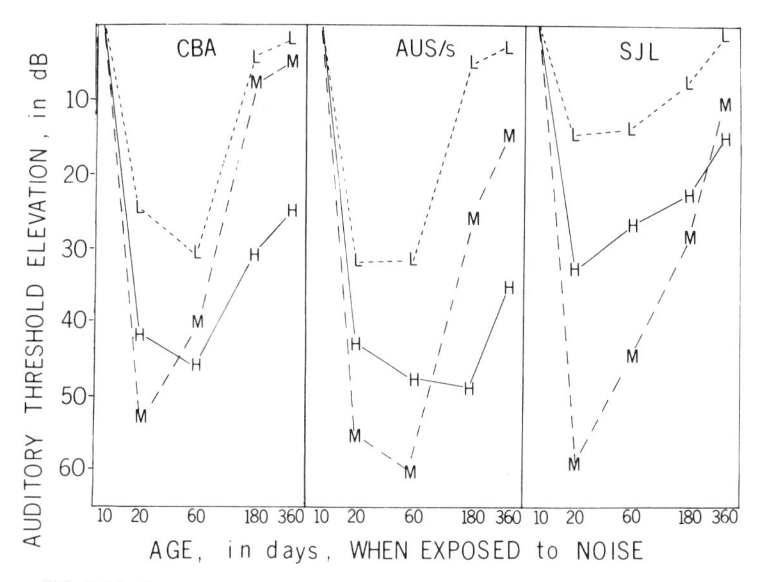

FIG. 3.10 Comparison of CBA, AUSs and SJL mice in their age-related changes of susceptibility to auditory threshold increases. These values might be useful in predicting the limits of the sensitive periods for acoustic priming. (*L, M, & H* represent, low, middle, and high frequencies, respectively). (Redrawn from Henry, 1984b, and unpublished data.)

The LP strain has been described as being as susceptible as the DBA/2 at 3 weeks of age, but its subsequent innate susceptibility is not known (Fuller & Sjursen, 1967). At this early age, its high-frequency functions are practically nonexistent, and all auditory function disappears in the LP before adulthood (Henry, 1982a; Fig. 3.13). If its susceptibility is determined by peripheral auditory function, then one would predict that it would also have a very circumscribed period of susceptibility, ending before adulthood. A study in progress (Henry & Buzzone) verifies this prediction, with susceptibility being highly correlated ($r = -.925$) with high-frequency cochlear thresholds.

Three-week-old mice of the AKR line are moderately susceptible (Collins, 1972; Fuller & Sjursen, 1967). At this age, they have moderate high-frequency cochlear losses, but maintain their hearing for longer than the DBA/2 and LP (Henry, 1982a; Fig. 3.12). Therefore, one might predict that they would maintain their innate susceptibility until sometime between 150 and 200 days of age.

SJL, CBA and C57BL/6 inbred strains are described as being resistant to innately determined audiogenic seizures at 21 days of age (Fuller, 1949; Fuller & Collins, 1968a; Fuller & Sjursen, 1967; Schlesinger et al., 1965). At this age, all these genotypes have normal peripheral auditory function (Henry, 1982a; Henry & Chole, 1980). These comparisons are consistent with the hypothesis

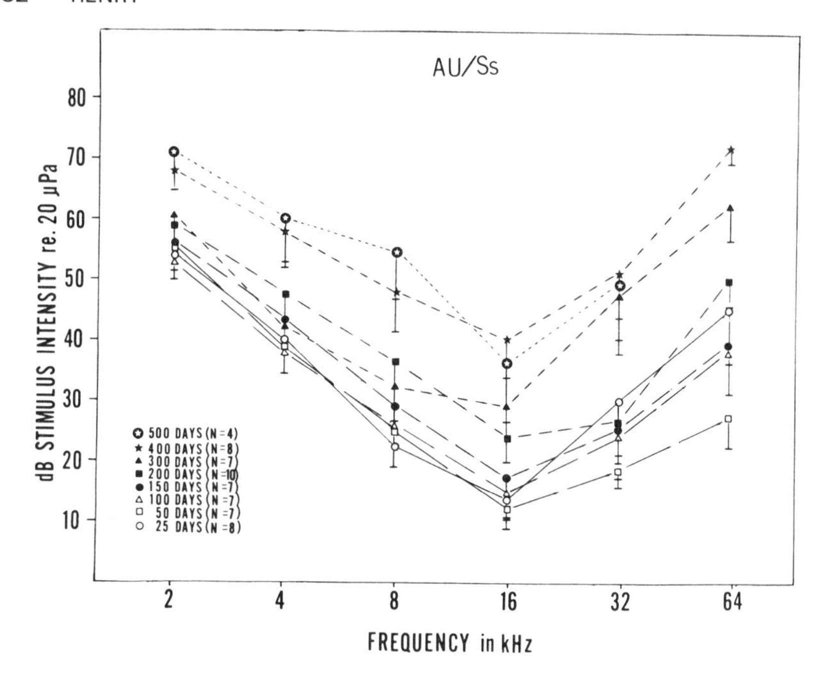

FIG. 3.11 Age-related cochlear nerve threshold changes in the AU/Ss mouse. (From Henry, 1982a, with permission of the *Journal of Gerontology*.)

that the normal development peripheral auditory function is associated with protection from audiogenic seizures.

Susceptibility in the A strain persists for most of its life span (Vicari, 1957). AP thresholds are elevated in the weanling A mouse, being especially poor in the high-frequency region (Henry, 1982a; see Fig. 3.14). This is the same pattern as that expressed by the highly susceptible DBA/2 and LP mice (Figs. 3.2 & 3.13). But, unlike these strains that are only susceptible for a few days, there is little sign of further degeneration until much later in life (Henry, 1982a). The span over which the A mouse maintains its aberrant cochlear function corresponds well with Vicari's description of its span for audiogenic seizures.

One report does not appear to fit this pattern. Gates and Chen (1975) described the BALB/c mouse as developing a mild susceptibility to audiogenic seizures late in life, although Fuller and Sjursen (1967) describe it as being mildly susceptible when young. This genotype has a progressive age-related hearing loss (Ralls, 1967), which has not yet been examined in mice as old as those tested for seizures by Gates and Chen. Therefore, the possibility exists that peripheral high-frequency auditory losses at any age can induce susceptibility to audiogenic seizures in the mouse. If this explanation is true, then the aged SJL, AU/Ss, and NMRI strains, which develop progressive high-frequency losses late in life (Ehret, 1974; Henry, 1982a), should also develop susceptibility at an advanced age. Alternatively, the older mice that Gates and Chen tested may

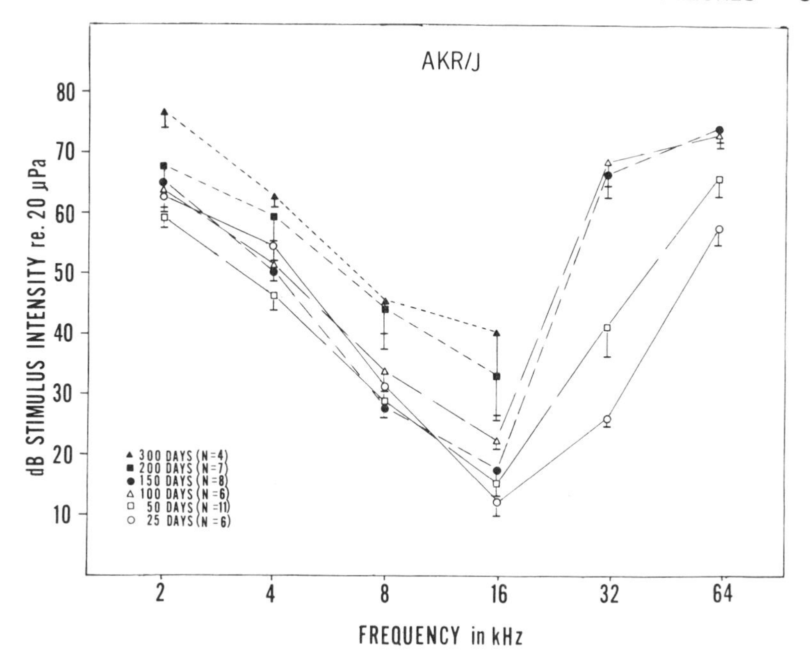

FIG. 3.12 Age-related changes of cochlear sensitivity in the AKR mouse. (Henry, 1982a, with permission of the *Journal of Gerontology*.)

have maintained sensitivity to audiogenic seizures since a much younger age. But this would mean that the age-related losses seen by Ralls either did not occur in the mice of Gates and Chen, or did not progress beyond a certain age. In any event, the report of Gates and Chen (1975) is potentially important, because it may be an exception to the generalizations described above.

If the development, degree, and duration of susceptibility can be predicted in every case from a knowledge of peripheral auditory function, then one might conclude that the genes that result in audiogenic seizures are possessed by all these mice; the only additional requirement would be that certain types of peripheral auditory dysfunctions must occur before these genes are phenotypically expressed. If a few exceptions are found, those strains could be examined for genes that modify CNS functions and directly affect the audiogenic seizure, itself.

CONCLUSIONS

Many investigator years have been spent over the past decades in the search for the genetic mechanisms of audiogenic seizures in rodents. It was concluded that few of the interpretations and conclusions of these investigators have withstood

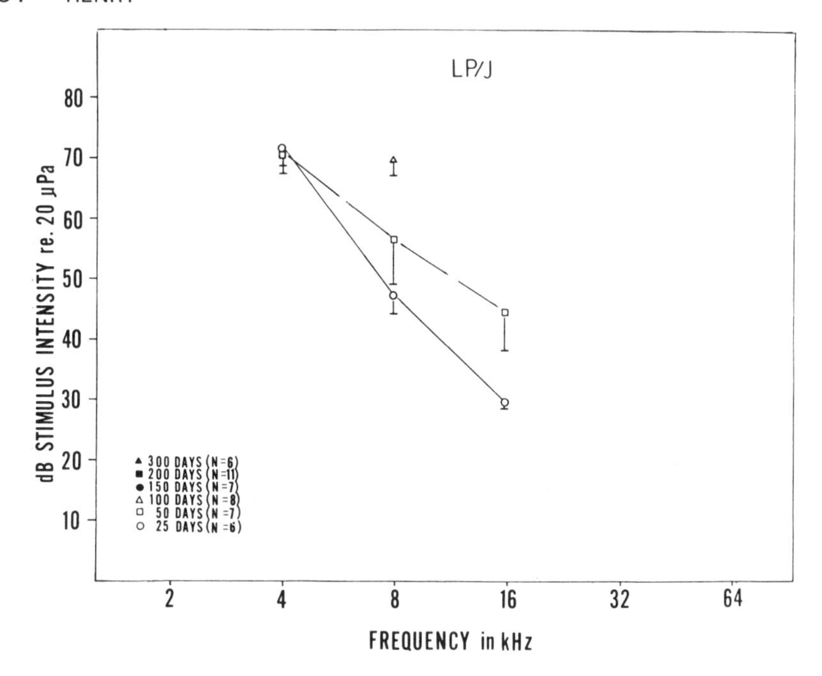

FIG. 3.13 Age-related changes of cochlear sensitivity in the LP mouse. (Henry, 1982a, with permission of the *Journal of Gerontology*.)

the test of time. Some of the problems that I believe exist in the majority of experimental designs are:

1. Having too little genetic variability in the parental strains, so that correlations derived from the crosses do not generalize well across other genotypes.
2. Using the DBA/2 and C57BL/6 strains as parental genotypes in many of the studies; both these mice have cochlear dysfunctions that interfere with the expression of the audiogenic seizure.
3. Restricting the analysis to a single age. This was probably duc to thc use of the DBA/2 and a few other genotypes whose rapidly developing age-related hearing losses greatly restrict the period of susceptibility.
4. Ignoring the auditory characteristics of the mice when considering the inheritance of susceptibility to audiogenic seizures. Of all the factors that have thus far been investigated, this is the one that seems to have the greatest predictive value. It is even possible that the search for "audiogenic seizure genes" is a will-of-the-wisp; perhaps most mice possess this characteristic, the only additional requirement being that genetic or experimental factors produce certain cochlear dysfunctions. A better approach might be to look for genes that, in mice with predisposing cochleas, still do not allow the expression of susceptibility to audiogenic seizures.

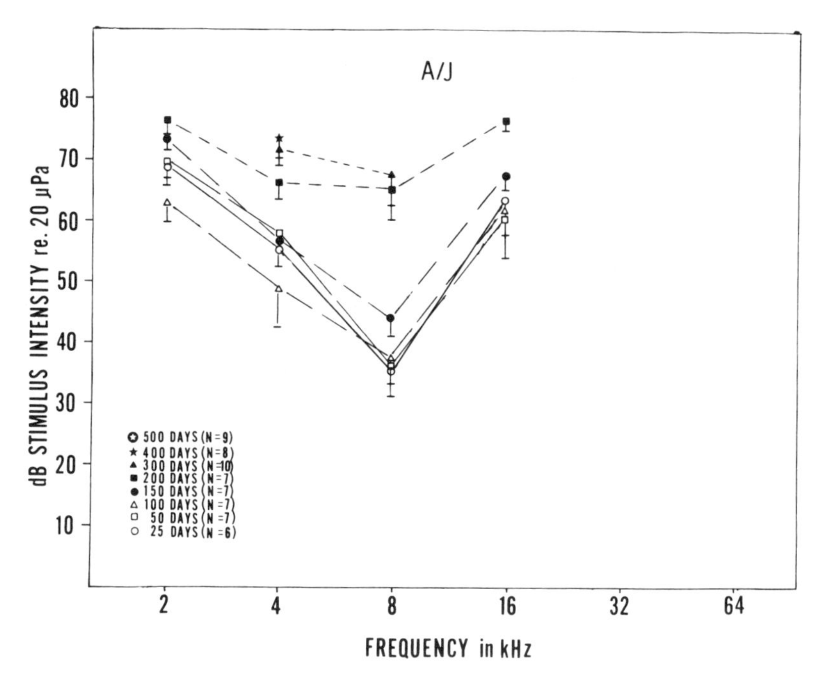

FIG. 3.14 Age-related changes of cochlear sensitivity in the A strain mouse. (Henry, 1982a, with permission of the *Journal of Gerontology*.)

5. Using a correlational approach, where a great many variables are uncontrolled. If an experimental method were used in conjunction with these genetic studies, it would be used to predict or verify results from genetic correlations. Acoustic priming appears to produce a phenocopy for producing susceptibility to audiogenic seizures. As an experimental method, in which a mouse can have one half of its auditory system primed while the other half is a nonprimed control, it is an extremely powerful technique for investigating the physiological and anatomical factors underlying audiogenic seizures. These factors could be identified, and used in evaluating genetic difference of audiogenic seizures in mice.

Studies of emotionality, intelligence and other behaviors that are influenced by a variety of genetic and environmental factors could also profit from the lessons learned from these years of studying the audiogenic seizures. Stated in a more general fashion, they are:

1. Use subjects that reflect a wide range of genetic variability, in order that the ultimate results may be more likely to be generalized to other populations.
2. Be aware that any "susceptible" and "nonsusceptible" genotype for the behavior under question may express this phenotype for reasons not closely related to the hypotheses of the investigator.

3. Test subjects over a wide range of ages, using both cross-sectional and longitudinal studies; if these two methods give different reasons (Fuller & Sjursen, 1967; Henry, 1967), it may indicate an influence of the test upon the behavior being examined.

4. Select the proper phenotypes to study, carefully considering those essential to the behavior under question. The auditory system is necessary for audiogenic seizures and the cerebral cortex for intelligence; the endocrine and limbic systems are involved in experience and expression of emotion. The behavior may be most fascinating and easier to study, but behaviors are not produced by genes alone: a great many intervening factors must be considered.

5. Behavior genetic studies, especially those involving humans, tend to be correlational in nature. The use of an experimental method to produce a phenocopy could help validate or reject hypotheses derived from genetic correlational studies.

REFERENCES

Alexander, G. J., & Gray, R. (1972). Induction of convulsive seizures in sound sensitive albino mice: Response to various signal frequencies. *Proceedings of the Society for Experimental Biology and Medicine, 140,* 1284–1288.

Alexander, G. J., & Kopeloff, L. (1976). Audiogenic seizures in mice: Influence of agents affecting brain serotonin. *Research Communications in Chemical Pathology and Pharmacology, 14,* 437–448.

Bakhit, C., Shenoy, A. K., Swinyard, E. A., & Gibb, J. (1982). Altered neurochemical parameters in the brains of mice (Frings) susceptible to audiogenic seizures. *Brain Research, 244,* 45–52.

Bevan, W. (1955). Sound-precipitated convulsions: 1947–1954. *Psychological Bulletin, 52,* 473–504.

Bock, G. R., & Chen, C. -S. (1972). Frequency-modulated tones as priming stimuli for audiogenic seizures in mice. *Experimental Neurology, 37,* 124–130.

Bock, G. R., & Saunders, J. C. (1977). A critical period for acoustic trauma in the hamster and its relation to cochlear development. *Science, 197,* 396–398.

Boggan, W. O., Freedman, D. X., Lovell, R. A., & Schlesinger, K. (1971). Studies in audiogenic seizure susceptibility. *Psychopharmacologica* (Berlin), *20,* 48–56.

Brown, R. D., Jobe, P. C., Bairnsfather, S., Mims, M. E., & Woods, T. W. (1980). Effects of multiple exposures to intense acoustic stimulation on audiogenic seizure susceptibility and intensity in rats (II). *Neuroscience Abstracts, 6,* 825.

Bures, J. (1963). Electrophysiological and functional analysis of the audiogenic seizure. In *Colloques Internationaux du Centre National de la Recherche Scientifique, No. 112: Psychophysiologie, Neuropharmacologie et Biochemie de la Crise Audiogene* (pp. 165–197). Paris: CNRS.

Buresova, O., Fures, J., & Fifkova, E. (1962). Analysis of the effect of spreading cortical depression on the activity of reticular neurons. *Physiologica Bohemoslovenica, 11,* 375–381.

Castellion, A. W., Swinyard, E. A., & Goodman, L. S. (1965). Effect of maturation on the development and reproducibility of audiogenic and electroshock seizures in mice. *Experimental Neurology, 13,* 206–217.

Chen, C. -S. (1973). Sensitization for audiogenic seizures in two strains of mice and their F1 hybrids. *Developmental Psychobiology, 6,* 131–138.

Chen, C. -S. (1980). Rapid development of acoustic trauma-induced risk in 3 strains of seizure-resistant mice. *Experientia, 36,* 1194–1196.

Chen, C. -S., & Fuller, J. L. (1976). Selection for spontaneous or priming-induced audiogenic seizure susceptibility in mice. *Journal of Comparative and Physiological Psychology, 90,* 765–772.

Chen, C. -S., Gates, G. R., & Bock, G. R. (1973). Effect of priming and tympanic membrane destruction on development of audiogenic seizure susceptibility in BALB/c mice. *Experimental Neurology, 39,* 277–284.

Chocholova, L. (1962). Role of the cerebral cortex in audiogenic seizures in the rat. *Physiologica Bohemoslovenica, 11,* 452–457.

Coleman, D. L., & Schlesinger, K. (1965). Effects of pyridoxine deficiency on audiogenic seizure susceptibility in inbred mice. *Proceedings of the Society for Experimental Biology and Medicine, 119,* 264–265.

Collins, R. L. (1970). A new genetic locus mapped from behavioral variations in mice: Audiogenic seizure prone (asp). *Behavior Genetics, 1,* 99–109.

Collins, R. L. (1972). Audiogenic seizures. In D. P. Purpura (Ed.), *Experimental models of epilepsy—A manual for the laboratory worker* (pp. 347–372). New York: Raven.

Collins, R. L., & Fuller, J. L. (1968). Audiogenic seizure prone (asp): A gene affecting behavior in linkage group VIII of the mouse. *Science, 162,* 1137–1138.

Consroe, P., Picchioni, A., & Chin, L. (1979). Audiogenic seizure susceptible rats. *Federation Proceedings, 38,* 2411–2416.

Darrouzet, J., & Guilhaume, A. (1967). A propos d'une absence de potentiel microphonique: Etude histologique de l'organe de Corti de la souris. *Revue Laryngologie* (Bordeaux), *88,* 813–833.

Darrouzet, J., Niaussat, M. -M., & Legouix, J. P. (1968). Etude histologilque de l'organe de Corti de souris d'une lignee presentant des crises convulsives au son. *Comptes Rendus de l'Academie des Sciences* (Paris), *266,* 1163–1165.

Deckard, B. S., Lieff, B., Schlesinger, K., & Defries, J. C. (1976). Developmental patterns of seizure susceptibility in inbred strains of mice. *Developmental Psychobiology, 9,* 17–24.

Deckard, B. S. (1977). *Genetic, biochemical and pharmacological correlates of responses to priming in mice. Behavior Genetics* (Abstract), *7,* 52–53.

Deckard, B. S., Teppar, J. M., & Schlesinger, K. (1976). Selective breeding for acoustic priming. *Behavior Genetics, 6,* 375–383.

Dice, L. R., & Barto, E. (1952). Ability of mice of the genus *Peromyscus* to hear ultrasonic sounds. *Science, 116,* 110–111.

Donaldson, H. H. (1924). *The rat: Data and reference tables* (2nd ed.). Philadelphia: Wistar Institute.

Duplisse, B. R. (1976). *Mechanisms of susceptibility of rats to audiogenic seizure.* Unpublished doctoral dissertation, University of Arizona, Tucson.

Ebel, A., Stefannovic, V., Simler, S., Randrainarisoa, H., & Mandel, P. (1974). Activity of cholinergic system enzymes in the cochlea of audiogenic seizure susceptible mice. *Experientia, 30,* 48–49.

Ehret, G. (1974). Age-dependent hearing loss in normal hearing mice. *Die Naturwissenschaften, 61,* 506.

Finger, F. W. (1947). Convulsive behavior in the rat. *Psychological Bulletin, 44,* 201–248.

Frings, H., & Frings, M. (1952). Acoustical determinants of audiogenic seizures in laboratory mice. *Journal of the Acoustical Society of America, 24,* 163–169.

Fuller, J. L. (1949). Genetic control of audiogenic seizures in hybrids between DBA subline 2 and C57BL subline 6. *Records of the Genetic Society of America, 18,* 86–87.

Fuller, J. L. (1975). Independence of inherited susceptibility to spontaneous and primed audiogenic seizures in mice. *Behavior Genetics, 5,* 1–8.

Fuller, J. L., & Collins, R. L. (1968a). Temporal patterns of sensitization for audiogenic seizures in SJL/J mice. *Developmental Psychobiology, 1,* 185–188.

Fuller, J. L., & Collins, R. L. (1968b). Mice unilaterally sensitized for audiogenic seizures. *Science, 162,* 1295.

Fuller, J. L., & Collins, R. L. (1970). Genetic and temporal characteristics of audiogenic seizures in mice. In B. Welch & A. Welch (Eds.), *Physiological effects of noise* (pp. 203–210). New York: Plenum Press.

Fuller, J. L., Easler, C., & Smith, M. E. (1950). Inheritance of audiogenic seizure susceptibility in the mouse. *Genetics, 35,* 622–632.

Fuller, J. L., & Sjursen, F. H. (1967). Audiogenic seizures in eleven mouse strains. *Journal of Heredity, 58,* 135–140.

Fuller, J. L., & Thompson, W. R. (1960). *Behavior genetics.* New York: Wiley.

Fuller, J. C., & Thompson, W. R. (1978). *Foundations of behavior genetics.* St. Louis: C. V. Mosby.

Fuller, J. L., & Williams, E. (1951). Gene controlled time constants in convulsive behavior. *Proceedings of the National Academy of Science* (U.S.A.), *37,* 349–356.

Furlow, T. W. (1981). A comparison of short-latency auditory-evoked potentials in two strains of mice: Possible neurophysiological correlates of susceptibility to audiogenic seizures. *Brain Research, 220,* 378–385.

Gates, G. R., & Chen, C. -S. (1975). Development of audiogenic seizure susceptibility in aged BALB/c mice. *Experimental Neurology, 47,* 360–363.

Gates, G. R., Chen, C. -S., & Bock, G. R. (1973). Effects of monaural and binaural auditory deprivation on audiogenic seizure susceptibility in BALB/c mice. *Experimental Neurology, 38,* 488–493.

Glenn, D. W., Brown, R. D., Jobe, P., Penny, J. E., & Bourn, W. M. (1980). A comparison of cochlear microphonics and N1 in audiogenic seizure susceptible and control rats. *Neurological Research, 2,* 85–100.

Gusic, B., Femenic, B., & Konic-Carnellutti, V. (1970). Mittelohrschleilmhautveranderungen bei experimenteller dysfunktion der schilddruse. *Acta Otolaryngologica, 69,* 281–285.

Guttman, R., & Lieblich, I. (1964). Apparent lack of a relation between dilute locus and audiogenic seizures in crosses between DBA/1J and C57BL/6J mice (abstract). *Genetics, 50,* 253.

Hall, C. S. (1947). Genetic differences in audiogenic seizures. *Journal of Heredity, 38,* 3–6.

Hamburgh, M., & Vicari, E. (1960). A study of some physiological mechanisms underlying susceptibility to audiogenic seizures in mice. *Journal of Neuropathology and Experimental Neurology, 19,* 461–472.

Haythorn, M. M. (1979). *Development of the auditory system and its relationship to audiogenic seizures in mice with different genetic backgrounds.* Unpublished doctoral dissertation, University of California, Davis.

Henry, K. R. (1966). *Audiogenic seizures.* Unpublished honors thesis, University of North Carolina, Chapel Hill.

Henry, K. R. (1967). Audiogenic seizure susceptibility induced in C57BL/6 mice by prior auditory exposure. *Science, 158,* 938–940

Henry, K. R. (1972a). Pinna reflex thresholds and audiogenic seizures: Developmental changes after acoustic priming. *Journal of Comparative and Physiological Psychology, 79,* 77–81.

Henry, K. R. (1972b). Unilateral increases of auditory sensitivity following early auditory exposure. *Science, 176,* 689–690.

Henry, K. R. (1980). Effects of noise, hypothermia and barbiturate on cochlear electrical activity. *Audiology, 19,* 44–56.

Henry, K. R. (1982a). Age-related auditory loss and genetics: An electrocochleographic comparison of six inbred strains of mice. *Journal of Gerontology, 37,* 275–282.

Henry, K. R. (1984a). Cochlear microphonics and action potentials mature and decline at different rates in the normal and pathologic mouse cochlea. *Developmental Psychobiology, 17,* 493–504.

Henry, K. R. (1984b). Noise and the young mouse: Genotype modifies sensitive period for effects on cochlear physiology and audiogenic seizures. *Behavioral Neuroscience, 98,* 1073–1082.

Henry, K. R., & Bowman, R. E. (1969). Effects of acoustic priming on audiogenic, electroconvulsive and chemoconvulsive seizures. *Journal of Comparative and Physiological Psychology, 67,* 401–406.

Henry, K. R., & Bowman, R. E. (1970a). Behavior-genetic analysis of the ontogeny of acoustically primed audiogenic seizures in mice. *Journal of Comparative and Physiological Psychology, 70,* 235–241.

Henry, K. R., & Bowman, R. E. (1970b). Acoustic priming of audiogenic seizures in mice. In B. L. Welch & A. S. Welch (Eds.), *Physiological effects of noise* (pp. 185–201). New York: Plenum Press.

Henry, K. R., Bowman, R. E., English, V. P., Thompson, K. A., & LeFever, M. (1971). Unilateral and bilateral effects of acoustic priming of audiogenic seizures. *Experimental Neurology, 32,* 331–340.

Henry, K. R., & Chole, R. A. (1980). Genotypic differences in behavioral, physiological and anatomical expressions of age-related hearing loss in the laboratory mouse. *Audiology, 19,* 369–383.

Henry, K. R., & Chole, R. A. (1984). *Hypothermia elevates cochlear action potential thresholds and protects the cochlea from noise damage.* In preparation.

Henry, K. R., & Haythorn, M. M. (1975). Auditory similarities associated with genetic and experimental acoustic deprivation. *Journal of Comparative and Physiological Psychology, 89,* 213–218.

Henry, K. R., & Lepkowski, C. M. (1978). Evoked potential correlates of genetic progressive hearing loss. *Acta Otolaryngologica, 86,* 366–374.

Henry, K. R., McGinn, M. D., Berard, D. R, & Chole, R. A. (1981). Effects of neonatal thyroxine, genotype, and noise on the ear and audiogenic seizures. *Journal of Comparative and Physiological Psychology, 95,* 418–424.

Henry, K. R., & Saleh, M. (1973). Recruitment deafness: Functional effect of priming-induced audiogenic seizures in mice. *Journal of Comparative and Physiological Psychology, 84,* 430–435.

Henry, K. R., Thompson, K. A., & Bowman, R. E. (1971). Frequency characteristics of acoustic priming of audiogenic seizures in mice. *Experimental Neurology, 31,* 402–407.

Henry, K. R., Wallick, M., & Davis, M. E. (1972). Inferior collicular lesions: Effects on audiogenic seizures and Preyer reflex. *Physiology and Behavior, 9,* 885–887.

Huff, S. D., & Fuller, J. L. (1964). Audiogenic seizures, the dilute locus, and phenylalanine hydroxylase in DBA/1 mice. *Science, 144,* 304–305.

Huff, S. D.,& Huff, R. L. (1962). Dilute locus and audiogenic seizures in mice. *Science, 136,* 318–319.

Huxtable, R. J., & Laird, H. E. (1978). The prolonged anticonvulsant activity of taurine on genetically determined seizure susceptibility. *Canadian Journal of Neurological Sciences, 5,* 215–221.

Infantellina, F., Sanseverino, E. R., & Urbano, A. (1964). Experimental epilepsy produced in cats by auditory stimulation after strychninization of the cerebellar auditory area. *Electroencephalography and Clinical Neurophysiology, 17,* 582.

Iturrian, W. B., & Fink, G. B. (1967). Conditioned convulsive reaction. [Abstract]. *Federation Proceedings, 26,* 736.

Iturrian, W. B., & Fink, G. B. (1968). Effects of age and condition-test interval (days) on an audioconditioned convulsive response in CF#1 mice. *Developmental Psychobiology, 1,* 230–235.

Iturrian, W. B., & Johnson, H. D. (1971). Conditioned seizure susceptibility in the hamster induced by prior auditory exposure. *Experentia, 27,* 1193–1194.

Jobe, P. C. (1981). Pharmacology of audiogenic seizures. In R. D. Brown & E. A. Daigneault (Eds.), *Pharmacology of hearing: Experimental and clinical bases* (pp. 271–304). New York: Wiley Interscience.

Koenig, E. (1957). The effects of auditory pathway interruption on the incidence of sound-induced seizures in rats. *Journal of Comparative Neurology, 108,* 383–392.

Kesner, R. P. (1966). Subcortical mechanisms of audiogenic seizures. *Experimental Neurology, 15,* 195–205.

Kesner, R. P., O'Kelly, L. I., & Thomas, G. J. (1965). Effects of cortical spreading depression and drugs upon audiogenic seizures in rats. *Journal of Comparative and Physiological Psychology, 50,* 280–282.

Kim, C. U. (1961). Effect of hippocampal ablation on the audiogenic seizure in rats. *Seoul Journal of Medicine, 2,* 29–32.

Kornfield, M., Geller, L. M., Cowen, D., Wolf, A., & Altman, F. (1970). Pathological changes in the inner ears of audiogenic seizure-susceptible mice treated with 6-aminonicitanamide. *Experimental Neurology, 26,* 17–35.

Kristt, D. A., Shirley, M. S., & Kasper, E. K. (1980). Monoaminergic synapses in infant mouse neocortex: Comparison of cortical fields in seizure-prone and resistant mice. *Neuroscience, 5,* 883–891.

Kruschinsky, L. V., Molodkina, L., Fless, D., Debrokhotova, L., Steshenko, A., Semiokino, A., Semiokina, A., Zorina, A., & Romonova, L. (1970). The functional side of the brain during sonic stimulation. In B. L. Welch & A. S. Welch (Eds.), *Physiological effects of noise* (pp. 159–183). New York: Plenum Press.

Laird, H. E., & Huxtable, R. (1978). Taurine and audiogenic epilepsy. In A. Barbeau & R. J. Huxtable (Eds.), *Taurine and neurological disorders* (pp. 339–357). New York: Raven.

Lehmann, A. (1977). Mechanisms underlying modifications in the severity of audiogenic convulsions. *Life Sciences, 20,* 2047–2060.

Lehmann, A., & Busnell, R. -G. (1963). A study of the audiogenic seizure. In R. G. Busnell (Ed.), *Acoustic behavior of animals.* (pp. 244–274). Amsterdam: Elsevier.

Lieff, B. D., Permut, A., Schlesinger, K., & Sharpless, S. K. (1975). Developmental changes in auditory evoked potentials in the inferior colliculi of mice during periods of susceptibility to priming. *Experimental Neurology, 46,* 534–541.

Lints, C. E., Willott, J. F., Sze, P. Y., & Nenja, L. H. (1980). Inverse relationship between whole brain monoamine levels and audiogenic seizure susceptibility in mice: Failure to replicate. *Pharmacology, Biochemistry and Behavior, 12,* 385–388.

Maxson, S. C. (1978). Strain differences in the development of susceptibility to audiogenic seizures after acoustic priming but not after hearing loss. *Experimental Neurology, 62,* 482–488.

Maxson, S. C., & Cowen, J. S. (1976). Electroencephalographic correlations of the audiogenic seizure response of inbred mice. *Physiology and Behavior, 16,* 623–629.

Maxson, S. C., & Sze, P. Y. (1976). Electroencephalographic correlations of audiogenic seizures during ethanol withdrawal in mice. *Psychopharmacology, 47,* 17–20.

McGinn, M. D., & Henry, K. R. (1975). Acute versus chronic acoustic deprivation: Effects on auditory evoked potentials and seizures in mice. *Developmental Psychobiology, 8,* 223–232.

McGinn, M. D., Willott, J. F., & Henry, K. R. (1973). Effects of conductive hearing loss on auditory evoked potentials and audiogenic seizures in mice. *Nature New Biology, 244,* 255–256.

Minami, Y. (1981). A case of audiogenic seizures. *Otolaryngology* (Tokyo), *53,* 657–661.

Morgan, C. T., & Morgan, J. D. (1939). Auditory induction of an abnormal pattern of behavior in rats. *Journal of Comparative Psychology, 27,* 505–508.

Niaussat, M. -M. (1969). Audiometric comparison of the evoked potential and Preyer reflex in normal mice and mice susceptible to audiogenic seizures. *Comptes Rendus Societe de Biologie, 163,* 2503–2508.

Niaussat, M. -M., & Laget, P. (1963). Electroencephalographic study of the audiogenic seizure. In R. G. Busnell (Ed.), *Psychophysiology, neuropharmacology and biochemistry of the audiogenic seizure* Paris.

Niaussat, M. -M., & Legouix, J. P. (1967). Abnormalities of the cochlear microphonic in a line of mice susceptible to audiogenic seizures. *Comptes Rendus, de l'Academie des Sciences, Paris, 264,* 103–105.

Penny, J. E., Brown, R. D., Hodges, K. B., Kupetz, S. A., Glenn, D. W., & Jobe, P. C. (1983). Cochlear morphology of the audiogenic seizure susceptible or genetically epilepsy prone rat. *Acta Otolaryngologica, 95,* 1–12.

Plotnikoff, N. (1960). Ataractics and strain differences in audiogenic seizures in mice. *Psychopharmacologica, 1,* 429–432.

Portmann, M., Darrouzet, J., & Niaussat, M. -M. (1971). The organ of Corti in the Swiss/RB mouse. *Audiology, 10,* 298–314.

Ralls, K. (1967). Auditory sensitivity in mice: *Peromyscus* and *Mus musculus. Animal Behavior, 15,* 123–128.

Reid, H. M., Bowler, K. J., & Weiss, C. (1983). Hippocampal lesions increase the severity of unilaterally induced audiogenic seizures and decrease their latency. *Experimental Neurology, 81,* 240–244.

Reid, H. M., Mamott, B. D., & Bowler, K. J. (1983). Hippocampal lesions render SJL/J mice susceptible to audiogenic seizures. *Experimental Neurology, 82,* 237–240.

Ross, S., Sawin, P. B., Denenberg, V. H., & Volow, M. (1963). Effects of previous experience and age on sound induced seizures in rabbits. *International Journal of Neuropharmacology, 2,* 255–258.

Saunders, J. C., Bock, G. R., Chen, C.-S., & Gates, G. R. (1972). The effects of priming for audiogenic seizures on cochlear and behavioral responses in BALB/c mice. *Experimental Neurology, 36,* 426–436.

Saunders, J. C., Bock, G. R., James. R., & Chen, C. -S. (1972). Effects of priming for audiogenic seizures on auditory evoked responses in the cochlear nucleus and inferior colliculus of BALB/c mice. *Experimental Neurology, 37,* 388–394.

Saunders, J. C., & Hirsch, K. A. (1976). Changes in cochlear microphonic sensitivity after priming C57BL/6 mice at various ages for audiogenic seizures. *Journal of Comparative and Physiological Psychology, 90,* 212–220.

Schlesinger, K., Boggan, W. O., & Freedman, D. X. (1965). Genetics of audiogenic seizures. I. Relation of brain serotonin and norepinepherine in mice. *Life Sciences, 4,* 2345–2351.

Schlesinger, K., & Griek, B. J. (1970). The genetics and biochemistry of audiogenic seizures. In G. Lindsay & D. Thiessen (Eds.), *Contributions to behavior-genetic analysis* (pp. 219–257). New York: Appleton-Century-Crofts.

Schreiber, R. A. (1975). Effects of stimulus intensity and stimulus duration during acoustic priming on audiogenic seizures in C57BL/6J mice. *Developmental Psychobiology, 10,* 77–85.

Schreiber, R. A. (1978). Stimulus frequency and audiogenic seizures in DBA/2J mice. *Behavior Genetics, 8,* 341–347.

Schreiber, R. A., & Graham, J. M. (1976). Audiogenic priming in DBA/2J and C57BL/6J mice: Interactions among age, prime-to-test interval and index of seizure. *Developmental Psychobiology, 9,* 57–66.

Schreiber, R. A., Lehmann, A., Ginsburg, B. E., & Fuller, J. L. (1980). Development of susceptibility to audiogenic seizures in DBA/2J and Rb mice: Toward a systematic nomenclature of audiogenic seizure levels. *Behavior Genetics, 10,* 537–543.

Servit, Z. (1960). The role of subcortical acoustic centres in seizure susceptibility to an acoustic stimulus and in symptomatology of audiogenic seizures in the rat. *Physiologia Bohemoslovencia, 9,* 42–47.

Servit, Z., & Sterc, J. (1958). Audiogenic epileptic seizures evoked in rats by artifical epileptogenic foci. *Nature, 181,* 1475–1476.

Seyfried, T. N. (1981). Developmental genetics of audiogenic seizures in mice. In V. E. Anderson (Ed.), *Genetics and epilepsy* (pp. 199–210). New York: Raven.

Seyfried, T. N. (1983). Genetic heterogeneity for the development of audiogenic seizures in mice. *Brain Research, 27,* 325–329.

Seyfried, T. N., & Glaser, G. H. (1981). Genetic linkage between the Ah locus and a major gene that inhibits susceptibility to audiogenic seizures in mice. *Genetics, 99,* 117–126.

Seyfried, T. N., Glaser, G. H., & Yu, R. K. (1978). Cerebral, cerebellar, and brain stem gangliosides in mice susceptible to audiogenic seizures. *Journal of Neurochemistry, 31,* 21–27.

Seyfried, T. N., Glaser, G. H., & Yu, R. K. (1979a). Genetic variability for regional brain gangliosides in five strains of young mice. *Biochemical Genetics, 17,* 43–55.

Seyfried, T. N., Glaser, G. H., & Yu, R. K. (1979b). Thyroid hormone influences on the susceptibility of mice to audiogenic seizures. *Science, 205,* 598–600.

Seyfried, T. N., Glaser, G. H., & Yu, R. K. (1981). Thyroid hormone can restore the audiogenic seizure susceptibility of hypothyroid DBA/2J mice. *Experimental Neurology, 71,* 220–225.

Seyfried, T. N., Glaser, G. H., & Yu, R. K. (1984). Genetic analysis of serum thyroxine content and audiogenic seizures in recombinant inbred and congenic strains of mice. *Experimental Neurology, 83,* 423–428.

Seyfried, T. N., & Yu, R. K. (1980). Heterosis for brain myelin content in mice. *Biochemical Genetics, 18,* 1229–1238.

Seyfried, T. N., Yu, R. K., & Glaser, G. H. (1980). Genetic analysis of audiogenic seizure susceptibility in C57BL/6J × DBA/2J recombinant inbred strains of mice. *Genetics, 94,* 701–718.

Shnerson, A., & Pujol, R. (1982). Age-related changes in the C57BL/6J mouse cochlea: I. Physiological findings. *Developmental Brain Research, 2,* 65–75.

Stanek, R., Bock, G. R., Goran, M. L., & Saunders, J. C. (1977). Age dependent susceptibility to auditory trauma in the hamster: Behavioral and electrophysiological consequences. *Transactions of the American Academy of Ophthalmology and Otolaryngology, 84,* 465–472.

Sterc, J. (1963). Experimental reflex epilepsy (audiogenic epilepsy). In Z. Servit (Ed.), *Reflex mechanisms in the genesis of epilepsy* (pp. 44–65). Amsterdam: Elsevier.

Tepper, J. M., & Schlesinger, K. (1980). Acoustic priming and kanamycin-induced cochlear damage. *Brain Research, 187,* 81–95.

Urban, G. P., & Willott, J. R. (1979). Response properties of neurons in the inferior colliculi of mice made susceptible to audiogenic seizures by acoustic priming. *Experimental Neurology, 63,* 229–243.

Van Middlesworth, L. (1977). Audiogenic seizures in rats after severe prenatal and perinatal iodine depletion. *Endocrinology, 100,* 242–245.

Van Middlesworth, L., & Norris, D. H. (1980). Audiogenic seizures and cochlear damage in rats after perinatal antithyroid treatment. *Endocrinology, 106,* 1686–1690.

Vicari, E. M. (1947). Establishment of differences in susceptibility to sound-induced seizure in various endocrinic types of mice. *Anatomical Record, 97,* 407.

Vicari, E. M. (1951). Fatal convulsive seizures in the DBA mouse strain. *Journal of Psychology, 32,* 79–97.

Vicari, E. M. (1953). Effect of 6n-propylthiouricil on lethal seizures in mice. *Proceedings of the Society of Experimental Biology and Medicine, 78,* 744–746.

Vicari, E. M. (1957). Audiogenic seizures and the A/Jax mouse. *Journal of Psychology, 43,* 111–116.

Wada, J. A., Terao, A., White, B., & Jung, E. (1970). Inferior colliculus lesion and audiogenic seizure susceptibility. *Experimental Neurology, 28,* 326–332.

Ward, R. (1971). Unilateral susceptibility to audiogenic seizure impaired by contralateral lesions in the inferior colliculus. *Experimental Neurology, 32,* 313–316.

Ward, R., & Sinnett, E. E. (1971). Spreading cortical depression and audiogenic seizures in mice. *Experimental Neurology, 31,* 437–443.

Willott, J. F. (1976). Effects of unilateral spinal cordotomy and outer ear occlusion on audiogenic seizures in mice. *Experimental Neurology, 50,* 30–55.

Willott, J. F. (1981). Comparison of response properties of inferior colliculus neurons of two inbred mouse strains differing in susceptibility to audiogenic seizures. *Journal of Neurophysiology, 45,* 35–47.

Willott, J. F., & Henry, K. R. (1974). Auditory evoked potentials: Developmental changes of threshold and amplitude following early acoustic trauma. *Journal of Comparative and Physiological Psychology, 86,* 1–7.

Willott, J. F., Henry, K. R., & George, F. (1975). Noise-induced hearing loss, auditory evoked potentials, and protection from audiogenic seizures in mice. *Experimental Neurology, 46,* 542–553.

Willott, J. F., & Lu, S. -M. (1980). Midbrain pathways of audiogenic seizures in DBA/2 mice. *Experimental Neurology, 70,* 288–299.

Willott, J. F., & Lu, S. -M. (1982). Noise-induced hearing loss can alter neural coding and increase excitability in the central nervous system. *Science, 216,* 1331–1332.

Willott, J. F., & Urban, G. P. (1978). Paleocerebellar lesions enhance audiogenic seizures in mice. *Experimental Neurology, 58,* 575–577.

Witt, G., & Hall, C. S. (1949). The genetics of audiogenic seizures in the house mouse. *Journal of Comparative and Physiological Psychology, 42,* 58–63.

4 Toward the Genetics of an Engram: The Role of Heredity in Visual Preferences and Perceptual Imprinting

Joseph K. Kovach
The Menninger Foundation, Topeka, Kansas

In the perspectives of many overstated generalizations from man to animal, or from animal to man, we again and again encounter the curious problem of enormous changes in the human brain in the brief moment, geologically speaking, since man passed the manlike, or hominid, level and became man. Against the backdrop of geological time it is extraordinary that a mere two million years have tripled the size of human brain, altered human posture, developed the usable hand with good eye-hand coordination, altered relations to tools of creative and aggressive capacity, and catapulted man into the world of symbols and language. We seem to be caught here between two equally unserviceable ideas: (1) the learning ability or thought capacity makes no difference to a basic human nature; man will be just as wild and mean as before but more clever in the implementation of his jungle drives; and (2) that man will soon creep away from his animal inheritance and live in a symbolic world filled with its own imaginative possibilities. The truth or, better still, the empirical reality is somewhere between these two extreme formulations. It is in unraveling the myriad of factors in this zone of in-between that the ultimate contributions of ethology and comparative psychology will lie.

(Murphy & Kovach, 1972, p. 365)

INTRODUCTION

The rediscovery of Mendel's work at the turn of the century set the stage for one of the most concerted efforts in all science. It specified a new subject matter (the study of heredity), the most appropriate subjects for experimentation (sexually reproductive organisms, preferably of short generation time), and a method of investigation (controlled breeding experiments). The new field of genetics became organized by the following questions: What are the units of heredity and where are they located? How do they operate? Do they change in time? What is their role in evolution? Does environment influence their expression, and if so how? What are they made of? And ultimately, how do they control development? These questions have become fundamentally important in the study of behavior as well, yet they alone do not define behavioral genetics as a field. The purpose of this chapter is to examine how these questions relate to the study of behavior, with special emphasis on the place of genetic procedures, and of the genetic code, in the study of the neural coding and processing of stimulus information.

At its very core the study of behavior is guided by questions such as the following: How do particular factors of experience, culture, and constitution interact in the development of behavior? What are the causes for behavioral variations in different persons or in the same person at different points of time? What are the continuities and discontinuities between variations of human and animal behaviors? The history of science teaches us that answers to even so pointedly anthropocentric questions about behavior are best sought through concepts and experimentation on species-general mechanisms. The mechanisms of learning and memory have been traditionally considered as such. It is generally accepted that some fundamental mechanisms of learning and memory cut across the evolutionary progression and various species-typical manifestations of behavior. As the Mendelian postulates about universal mechanisms of inheritance pointed the way to experimentation that led to the discovery of the genetic code, so this belief about universal mechanisms of learning fuels the hope for the ultimate discovery of a neural code or engram of information. In this chapter I emphasize and illustrate with data on early preference behaviors in the Japanese quail (*C. coturnix japonica*) that (1) the search for an engram has a distinctly behavioral genetic component, and that (2) the behavioral genetic search for an engram now calls for an inferential phase and synthesis of the molar and molecular concepts of genetics, neurobiology, and the study of learning.

Genetics and the Study of Behavior

The importance of the genetic method in the study of learning has been recognized ever since Tryon's (1940) massive genetic study of maze learning in rats. After 7 generations of selection Tryon obtained two distinctly different strains, which he called "maze dull" and "maze bright" rats, characterized by marginally overlapping distributions of low and high numbers of entries into blind alleys while learning to negotiate a maze. The two strains were different also in a variety of other measures in the maze (speed of running, hesitation at choice points, etc.), but they were not especially bright or dull in learning situations other than maze learning (Searle, 1949). Comparing the F_1 and F_2 crosses of the two strains indicated no segregation, indicating highly polygenic control of the manipulated behavior. In other words, the data implicted no universal learning mechanisms and the task specific influences of selection did not lend themselves to easy genetic analysis. This pioneering work nonetheless demonstrated that the genotype is a source of behavioral variations, and it made the field aware of the genotype as a factor to reckon with in psychological theories of learning (Caspari, 1977; Fuller & Thompson, 1960; Henderson, 1979; Thiessen, 1979).

The investigations that quickly followed in Tryon's trail alerted the field to the shortcomings of a long-standing typological mode of thinking in psychology (see Mayr, 1965) and pointed to the need to replace it with more appropriate concepts of hereditary and environmental individuality. The rapid creation of selected and inbred strains of various species exhibiting a variety of genetic influences in behavior (Fuller & Thompson, 1960; Ginsburg, 1958; Hirsch, 1962; Hirsch & Tryon, 1956; Manosevitz, Lindzey, & Thiessen, 1969) and some newly refined quantitative genetic procedures (Falconer, 1960; Mather & Jinks, 1971) expanded the scope of behavioral genetics. Animal models for various human behaviors and behavioral disorders were developed, and the involvement of heredity in mental illness became identified (by direct investigation of human populations; see Kety, Rosenthal, Wender, Schulsinger, & Jacobsen, 1978; Rosenthal, 1970; Winokur, 1981). These accomplishments changed the prevalent views about the sources of behavioral variation; they demonstrated that genes are involved in the development of behavior and paved the way for the next generation of more mechanistic and mediation oriented studies that are now in progress.

Still another source of the new research directions has been the study of gene expression in species-typical "instinctive" behaviors in simple organisms (see Manning, 1976). An interesting example of this work deals with song production and song recognition in crickets (Bentley 1971; Bentley & Hoy, 1972; Hoy, Hahn, & Paul, 1977). The results of this study indicate that male hybrids of closely related cricket species, each of which exhibit a different courtship song, produce intermediate songs, and hybrid females prefer the hybrid male song over the male song of either parental species. Song development in males and song recognition in females appear to be organized as a single package of inherited perceptual, motor, and informational factors. The results of another interesting study on the hygienic behavior of the honeybee (Rothenbuhler, 1964) also suggest modular packaging of interrelated and seemingly complex sensory and informational processes and motor patterns by simple Mendelian factors. Together, the various investigations of the genetic sources of variation among species-typical behaviors have assigned pivotal roles to genes in the development of various behaviors, including signal generation, signal recognition, and the production of appropriate responses. But these studies could not say much about how genes organize behavior, nor what the genetic packaging of stimulus-bound and stereotyped behaviors of insects may explain about genetic influences in such plastic processes as learning and memory.

This book contains a chapter about the genetic study of fly learning (chapter 5), so I mention here only one example of this work, that of the "dunce" fly. I do so to highlight the promise further and pinpoint some problems with an increasingly reductionistic and molecular orientation in the genetic study of learning and memory.

The "dunce" mutant of *Drosophila* (Dudai, Jan, Byers, Quinn, & Benzer, 1976; Quinn, Harris, & Benzer, 1974) does not exhibit a rather weak yet statistically reliable chemotactic learning that is detectable in the behavior of nonmutant flies. Whether the mutant gene affects sensory or motor or central neural mechanisms, and whether it is involved directly or only indirectly in the fly's capacity to learn, flies that have the gene in double dose do not learn. Working with such preparations is desirable for two major reasons, both arising from primarily genetic considerations: because the speed with which flies multiply is much valued in genetic studies, and because of the great wealth of genetic information available on the fly. Yet when viewed from the larger perspectives of behavioral genetics, there are also some drawbacks and disadvantages in working with such a lowly preparation.

An effective behavioral genetic investigation of learning and memory must deal with the complexities of both the genotype and the learning process. The weak acquisition in the fly has a lot to be desired as a preparation for the study of learning. The matter is complicated further by uncertainties about the degree and nature of participation in learning by the affected metabolic sites. Nonetheless, the path to the mechanisms of a neural engram is by necessity a reductionist path. Reaching its destination seems to require the type of molecular biological knowledge sought by fly geneticists, but preferably as one of a variety of equally important and parallel approaches. The need for complementary series of approaches is indicated by one outstanding question: How may the molecular biological tradition of genetics be combined with the molar tradition of the study of overt behavior?

The Molecular Biological Approach in Behavioral Genetics

Investigators pursuing what might be called a molecular approach in behavioral genetics (for example, see Benzer, 1973, and Brenner, 1974) usually concentrate on mechanisms of gene expression in simple behaviors. The simpler the organism and the better its genetics and physiology are understood, the more productive this approach is expected to be. The work on bacterial chemotaxis by Adler is a good example (Adler, 1973; Adler, Hazelbauer, & Dahl, 1973).

Adler identified different chemoreceptors in the outer membrane of *E. coli,* each receptor responding to a particular sugar in the medium. Different structural genes were shown to be responsible for building the different sugar receptors, and each in turn was being controlled by a regulatory gene activated by a sugar. The entire process agrees well with what is known about regulatory gene action (Jacob & Monod, 1961) and is readily relatable to behavioral problems of stimulus response association and coding of stimulus information. Adler's data raise the possibility that similarly environment-sensitive regulatory genes are involved in manufacture of synaptic and nuclear receptors and binding of neurotransmitters and/or hormones in neural tissues. A step down the road looms the question about possible involvement of regulatory gene action in learning and memory.

Other outstanding examples of the molecular approach in behavioral genetics are Brenner's (1974) experimentation on the nematode *Caenorhabditis elegans* and Benzer's (1973) study of the *Drosophila.* These studies seek to identify the neurobiological expression of particular mutant genes. They stress the genetic and physiological simplicity of studied organisms and the belief that it facilitates finding the sites and defining the mechanisms of behavior-relevant gene action. An interesting procedural variant is the use of genetic mosaics and chimeras (see Goldowitz & Mullen, 1980; Hotta & Benzer, 1970; Mullen & Whitten, 1971), which seems especially promising for the study of the structural organization of neurons and brain regions during embryonic and early postnatal development. Although the availability of particular mutant genes dictates the behavioral content of such studies, and the lack of predictable order in the mosaic and chimeric distribution of polymorphous tissue complicates them, this line of research points to resolving fundamental puzzles about the developmental organization of the central nervous system. Its potential for solving the problem of an engram, however, seems at best indirect.

Still another interesting development in this domain is the use of monoclonal antibodies that may identify particular classes of neurons and differentiate cell regions in the brains of such widely different creatures as fly and man (Miller & Benzer, 1984). The procedure seems to have opened a way for chemical mapping of the brain, and for identifying the structural-neurobiological expression of behaviorally significant genetic mutations.

In general, the molecular biological approaches to behavior seem most promising when it comes to questions about the roles of heredity in the developmental organization of the central nervous system. However, the critical procedures and data that would specify the direction of this molecular thrust in the search of an engram are not yet in. We do not yet have the genetic data that would be comparable in impact on neurobiology to the impact of "Avery's bombshell" in molecular genetics (see Watson, 1970). We do not know, for example, whether or not the regulatory processes of the genetic code are pertinent to understanding the neural engram. RNA and protein syntheses have been considered repeatedly (see McConnell, 1962; Ungar, 1973; Uphouse, MacInnes, & Schlesinger, 1974), but no lasting impact on current neurobiological thinking is discernible from this line of research. Do RNA and protein synthesis have no direct roles to play in the specific processes of learning, or are the available data about gene effects and gene-environment interactions not yet sufficiently detailed for generating the relevant inferences? We do not know. However, one thing seems reasonably certain. The hypothesis, inferences, and interpretations that are still missing will have to rely on the combined data and complementary investigations of behavioral genetics, comparative psychology, and neurobiology.

The Neural Engram and its Information

Whether molar or molecular and whether directly or only indirectly focused on particular mechanisms of mediation, the dominant approaches in the study of behavior are all anchored in questions about how the brain codes, stores, and processes stimulus information. These questions dominate modern neurobiology in ways that are comparable only to the dominant status of the code of inheritance in molecular biology. Comparable to cracking the genetic code, the work of cracking the neural information code now demands a multilayered set of approaches ranging from the inferential phase exemplified by Morgan's early work on the *Drosophila* to the recent algorithmic approaches of cognitive sciences (Fodor, 1983).

Calling for multilayered considerations in the study of perception, Marr (1982) argued that, no matter how exhaustive, the data about neuronal activity cannot alone explain overt behavior. The argument applies equally well to the associative jrocesses of learning and memory. From Pavlov to Hull and Skinner or Kandel the study of learning has proceeded on the grounds of an empiricism of behavioral and neural associations. The Gibsons' call for a "new empiricism based on discrimination" (Gibson & Gibson, 1955, p. 450) has only recently begun to gain some ground, with the advent and consideration of information theory and the study of artificial intelligence. Yet we are still not even sure whether the fundamental associations of an engram are perceptual or sensorimotor associations. We know that stimulus-response associations are essential in behavioral adaptation. However, it is not clear how the brain builds such associations; whether an engram is reducible to perceptual and/or motor, or only sensorimotor elements. The enormous effort invested in the study of associations has not yet resolved even such primary questions as whether the engram resides within the neuron or at the synapse or in the functional organization of specialized groups of neurons. The making of inferences that would pin it to particular places and processes seems to require a thorough understanding of the ways gene effects and environment effects interact and are combined as information that mediates the phenotypic expression of a behavior; and probing such information is not the sole matter of associations but also of discriminations.

Neurobiologists have long acknowledged the role of inheritance in the developmental organization of neuronal circuitry that makes behavior possible, but there has been no commensurate interest in the roles of genes and genetic processes that are involved in the phenotypic expression of perceptual and behavioral discriminations. The dominant outlook of neurobiology has effectively coordinated investigative effort in the study of associative processes in learning and memory for a good many years, but the more recent informational considerations and "new empiricism of discrimination" now call for a reorientation. The remainder of this chapter describes an inferential approach in the search for an engram that focuses on the mediation of separate and joint influences of genotypes and experiences in the processes of stimulus discrimination and stimulus preferences.

CONCEPTUAL AND METHODOLOGICAL CONSIDERATIONS

The project described here is committed to understanding the roles of genes in learning and memory in much the same way as the behavioral genetic studies discussed above, but in a reversed order: by inferences from the phenotypic expression and molar organization of the influences of genes and experiences in relatively complex perceptual and behavioral discriminations and preferences. Specifically, the project deals with how learning and memory are expressed in the overt variations of genetically and environmentally manipulated visual preferences in Japanese quail chicks (*C. coturnix japonica*). The primary questions that were posed for it are as follows: Do genes influence the coding of stimulus information that is implicit in the phenotypic expression of unlearned and learned stimulus discrimination and stimulus preferences? If genes are involved, what are their specific roles, and what is the relationship between genetic determination of a behavior and the learning of the same behavior? And, on the level of mechanisms, what similarities and differences might there be between the neural mediation and coding of stimulus information relatable to gene effects and the neural mediation and coding of overtly similar stimulus information relatable to learning?

A set of artificial selection experiments served as a point of departure for answering our questions. In one experiment, two genetic lines of quail were bidirectionally selected for approach preferences between two colors (*Blue* and *Red*). Another selection experiment focused on the genetic sources of variation in stimulus general processes of perceptual learning. Still another genetic experiment dealt with unconditional preferences between anchromatic patterns. Subjects from the variously selected genetic lines and their crosses, including crosses with a genetic control line that has been maintained without selection, have been used to investigate the following: (1) genetic influences in the visual preferences of newly hatched quail chicks; (2) genetic and environmental factors that contribute to differences in the outcome of perceptual learning; (3) interaction between genetic and environmental influences in the acquisition and manifest expression of particular stimulus preferences, and (4) behavioral and physiological mechanisms that may mediate the influences of genotypes, environments, and gene-environment interactions in the quail's initial perceptual preferences and learning.

Why the Japanese Quail?

The Japanese quail, a popular laboratory subject, is a small, marginally dimorphic gallinaceous bird (see Plate 4.1) that has been domesticated in Japan since at least the 12th century. The male has a dull, dark orange breast and darkly pigmented cheeks. Females are usually larger than males and their breast color is a light yellow, sprinkled with small black spots. The low unit cost of maintenance and short generation time make this bird an ideal subject for genetic, developmental, and joint genetic and developmental investigations.

The rate of maturation is about equal in the two sexes. Age at reaching sexual maturity is about 5 weeks. Fertile eggs are normally obtained from a mated hen kept in 16:8 LD cycles during the 6th posthatch week. Maximum fertility occurs shortly thereafter, and there is a gradual decline in fertility starting at about the 6th month of posthatch life. Life span is from 1,300 to 2,000 days. Average yearly egg production is well over 300. Among the general advantages of using birds as laboratory subjects is the fact that avian hatchlings are free of intrauterine and maternal environmental effects that must be considered in developmental research on mammalian neonates (Joffe, 1969; Vom Saal & Bronson, 1980; Ward, 1974). The highly precocial quail chick also exhibits a degree of self-reliance that makes experimentation with it easier than with mammalian neonates. Especially relevant is the fact that during the first 2 to 3 posthatch days the Japanese quail can live on the content of prenatally invaginated yolk sac and needs no food or water, which permits dark rearing, strict control of initial perceptual experiences, and behavior modifications by perceptual imprinting to well-controlled environmental stimuli.

As regards quail genetics, there are several known mutant conditions in this bird. Particularly well documented are genes responsible for plumage and egg coloration (Lauber, 1964; Lucotte, 1975; Shimakura, 1940; Sittman & Abplanalp, 1965). The normal quail egg is pale grey or green, overlaid with dark red-brown or greenish-brown speckles and blotches (Hachisuka, 1931; Jones, Maloney, & Gilbreath, 1964). Egg color and blotch distribution are strikingly uniform within and variable between families. When present, variations within families usually fall into 3:1 ratios, suggesting recessive gene effects. The physiology and genetics of egg pigmentation have been extensively studied in the quail by Poole (1964, 1965, 1967).

Despite a growing number of morphological genetic and physiological studies, the quail is nowhere close to the *Drosophila* or the *Mouse,* let alone *E coli,* in the thoroughness of its genetic and physiological understanding. In general, avian

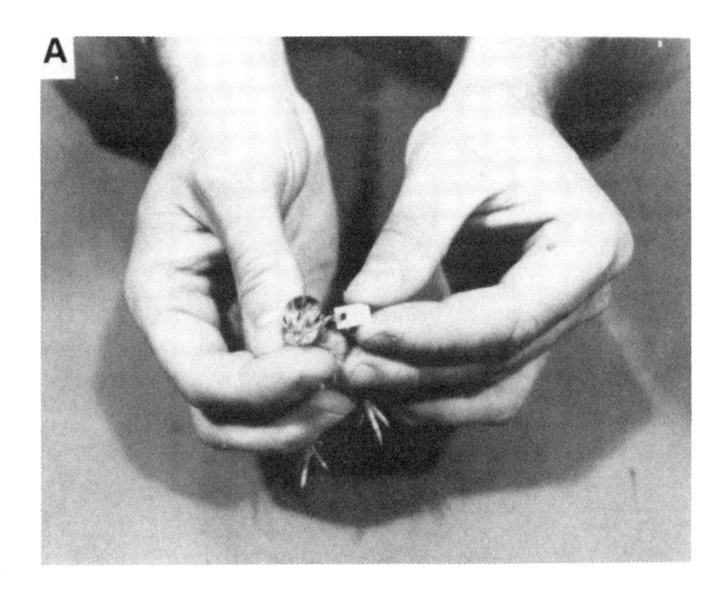

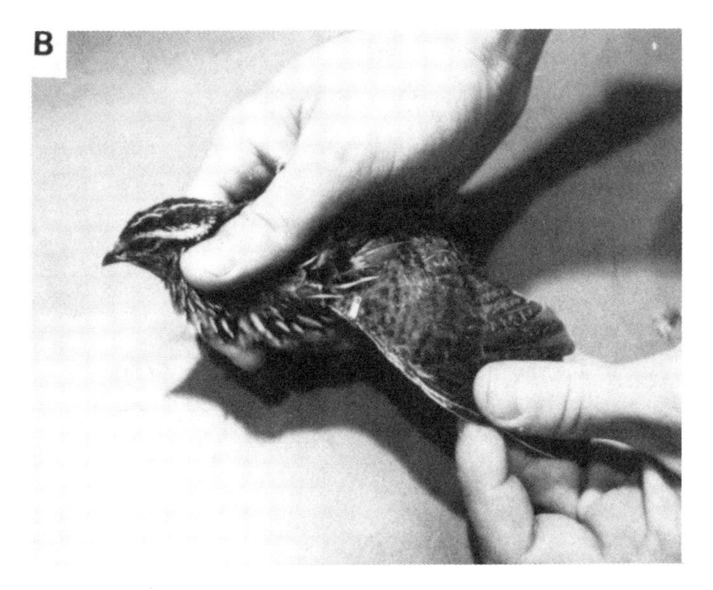

PLATE 4.1. One-day and 4-week-old Japanese quail. Note the temporary identification on the chick and the permanent wing tag on the mature hen. For an earlier review of behavioral research on quail see Kovach, 1974.

genetics and karyology are complex and neglected. Birds have unusually large numbers of chromosomes, most of them very small. Vegni-Talluri and Vegni (1965) identified a diploid complement of 76 chromosomes in the quail, plus the ZZ combination in the male and the ZW combination in the female. Interestingly, the endocrinologically neutral (anhormonal) sex seems also to be male, and estrogen is the organizing hormone in this bird (Adkins, 1979).

Studies by Bammi, Shoffner, and Haiden (1966) indicate that the Z and W chromosomes do not form direct end-to-end association during meiosis. Thus, exchange of genetic material on these chromosomes seems unlikely. The quail is also unusually sensitive to inbreeding depression. To date, all attempts at inbreeding quail have failed (Abplanalp, 1967; Iton, 1966; Kulenkamp, 1967; Maeda, Hiroma, Tsutomu, & Manjiro, 1978; Sittman, Abplanalp, & Fraser, 1966; see also discussion below of our attempt at inbreeding quail).

The quail has been used extensively in genetic studies of mating behaviors (Bernon & Siegel, 1981; Cunningham & Siegel, 1978; Sefton & Siegel, 1975). Results indicate strong responses to selection for and substantial genetic correlations between high mating frequency in males, aggressiveness, and the size of the cloacal gland. Mating preference studies have found that early social experiences influence adult sexual preferences by plumage morph (Gallagher, 1977, 1978), and may determine the balance between inbreeding and outbreeding in natural populations (Bateson, 1978a, 1978b). Although there has been traditional ethological concern with "innate" components of behaviors elicited by sign stimuli (Baerends, 1950; Hess, 1959; Lorenz, 1935; Tinbergen, 1951), there has been no genetic investigation of the sources of preference variations in the quail prior to the present project, nor, for that matter, in any other bird. Earlier genetic studies of avian imprinting (Fischer, 1969; Graves & Siegel 1968, 1969; Smith & Templeton, 1966) have dealt exclusively with heritabilities of high to low approach and following tendencies in the domestic chicken. In summary, the Japanese quail lends itself well to the control and experimental manipulation of particular genetic, environmental, and joint genetic-and-environmental influences in its early behaviors. Results of available studies indicate some distinct advantages and some drawbacks in using this bird in behavioral genetic studies. What recommends the quail as a laboratory subject is not the simplicity of its physiology nor the thoroughness of its genetic analysis, but the unusually short generation time, the ease of laboratory maintenance, an exceptionally high fecundity, and an early behavioral repertory that is uniquely suited for probing the nature of initial stimulus preferences, perceptual learning, and the related mediation of stimulus information.

Why Perceptual Preferences and Imprinting?

Newly hatched quail chicks readily approach, discriminate, and exhibit preferences between conspicuous visual stimuli and, after relatively short exposures, become attached to the stimuli. Because the natural object of attachment is the hen, and because the hen reciprocates the chicks' attachment with her own caretaking behaviors, quick learning is essential for the chicks' survival. The behavior is also believed to facilitate the development of perceptual templates by which conspecifics are identified in later life. The phenomenon has been thoroughly studied and discussed in the literature under the rubrics of filial and sexual imprinting (see Bateson, 1964; Hess, 1973; Immelmann, 1972; Lorenz, 1935; Rajecki, Lamb, & Obmascher, 1978; Sluckin, 1972).

Although much information is now available about sexual imprinting in the Japanese quail (Blohowiak & Siegel, 1983; Gallagher, 1977, 1978), early approach-avoidance tendencies and filial imprinting have been only sporadically studied. The earliest report (Schaller & Emlen, 1962) stated that avoidance responses in quail chicks are not influenced by the size, color, or shape of unfamiliar stimuli. A subsequent report (Rubel, 1970) concluded that there are no approach or following responses in this bird, and identified a sensitive period between 5 and 9 hours posthatch, during which time exposure to a stimulus was said to result in lasting diminution of avoidance responses. The data collected from many thousands of quail chicks in our laboratory indicate that approach and avoidance tendencies in the quail are exceptionally strong and are very much influenced by the color, pattern, shape, and intermittence of stimuli. Furthermore, our data indicate no narrowly delineated sensitive periods for the perceptual component of early learning in this bird (Kovach, 1974, 1979, 1983a, 1983b). However, the tendency to approach unfamiliar stimuli does diminish with age even in dark-reared birds, and it disappears altogether by about the 4th posthatch day.

In summary, Japanese quail possess the full complement of behaviors usually considered belonging to imprinting. The present project has concentrated on initial unconditional stimulus preferences in the quail (i.e., on preferences that appear at and shortly after hatching and are not conditional on experience with eliciting stimuli) because prior experimentation demonstrated that such preferences can be modified by both genetic selection and perceptual imprinting. It was anticipated that the quail's early preference behaviors would provide a key to the age-old problem of constitution-environment interaction in behavioral development, and would help deal with the related problem of gene-environment interaction in the coding and processing of stimulus information.

Procedures for Testing Visual Preferences

The apparatus developed for mass-screening visual preferences in the quail (see Plate 4.2) consists of 36 compartments arranged as a half-inverted hemipyramid, with a single compartment on top, two compartments in the second row, three compartments in the third row, and so on until there are eight collection boxes on the ground floor. Each of the 28 discrimination compartments offers an approach choice between two stimuli. Stimuli may be different in color, pattern, intermittence, and brightness, or a combination of two or more of these parameters. Subjects progress through the apparatus by approaching one in favor of

A

B

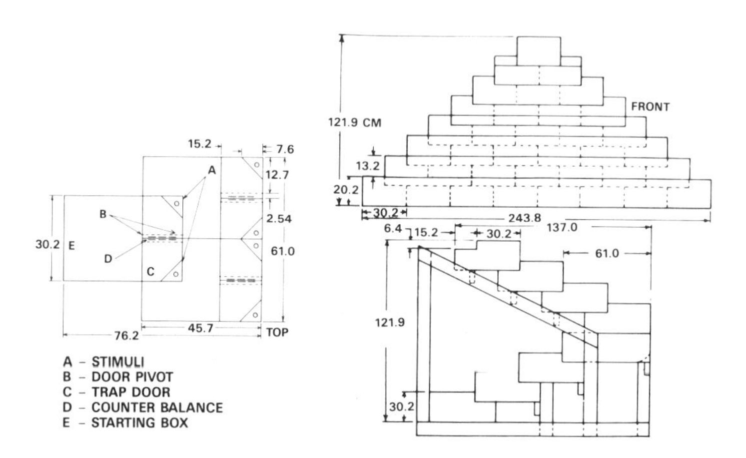

PLATE 4.2 Photograph and schematic representation of the apparatus used for mass-screening visual approach choices in quail chicks. Diagram B on left under the photograph shows a view from the top of starting box and two compartments in second row. Top and lower diagrams at right are schematic drawings of front and side views on a scale reduced from the scale of starting compartment and second row of boxes. All measurements are given in cm. Note location and size of trapdoors in front of each of the stimuli in each box. Note also that the direction of stimulus presentation is reversed in the fifth row, which was done to counteract position habit or extraneous stimulus influences and to make this large apparatus fit in a single room.

another stimulus in each compartment, and by falling from one to the next compartment through a trap door in front of the stimulus. Arrival in a collection box on the ground floor indicates the number of approaches made to one over another stimulus in 7 trials. Subjects may be tested individually or mass-screened in groups of up to 200 per testing session. When mass-screened, which is the usual procedure, subjects are placed into the starting compartment at 10 min intervals, in groups of about 25. They are identified and scored individually, and are tested in two consecutive runs through the apparatus (for 14 choice trials).

The mass-screening procedure permits rapid testing of large subject populations, which is especially needed for behavioral genetic studies (see Benzer, 1973; Dobzhansky & Spassky, 1967; Hay, 1973; Hirsch & Boudreau, 1958; Hirsch & Tryon, 1956). Another advantage of mass-screening is that it deals with stimulus discrimination and preference without requiring qualitative or quantitative differences in overt responses. Whatever variation there may be in the motor output of completing a series of trials in the mass-screening apparatus, it is not relevant for assessing stimulus preferences, because subjects have to travel the same distance through the apparatus regardless of preference, and a particular preference score does not depend on differentiating motor output. There are variable paths to all but the two extreme collection boxes, and the left-right position of stimuli vis à vis subject is reversed in the fifth row. Furthermore, the position of stimuli is routinely altered for halves of tested groups. The apparatus thus tests differences in perceptual capacities and stimulus information as revealed by discrimination and preferences, without the confounding variable of differential requirements in motor output. This consideration is important in experimenting with specific mechanisms of coding and processing stimulus information.

Figure 4.1 illustrates the distributions of mass-screened choices between two identical white stimuli and between each of four pairs of discriminable stimuli. As can be seen, the distribution of choices between nondiscriminable stimuli nicely conforms to the binomial distribution of $P = .5$, indicating that there were only negligible influences, or none at all, on performances from position habit, systematic alternation, and stimulus effects other than those of the testing stimuli. By contrast, choices between discriminable stimuli are distributed with means that are different from $P = .5$ and variances that are larger than *PON*.

Comparing repeated series of trials indicated that mass-screened performances are either free of or only marginally influenced by trial-to-trial variation in choice probabilities (see Kovach, 1977, 1979). Comparing individually tested and mass-screened performances in subjects belonging to the same genetic populations indicates statistically identical mean scores and individual variations (see Fig. 4.2). These data suggest no social interaction effects in mass-screened subjects. However, when subjects from two genetically different populations were mixed and mass-screened together in a single group, some social interaction effects were

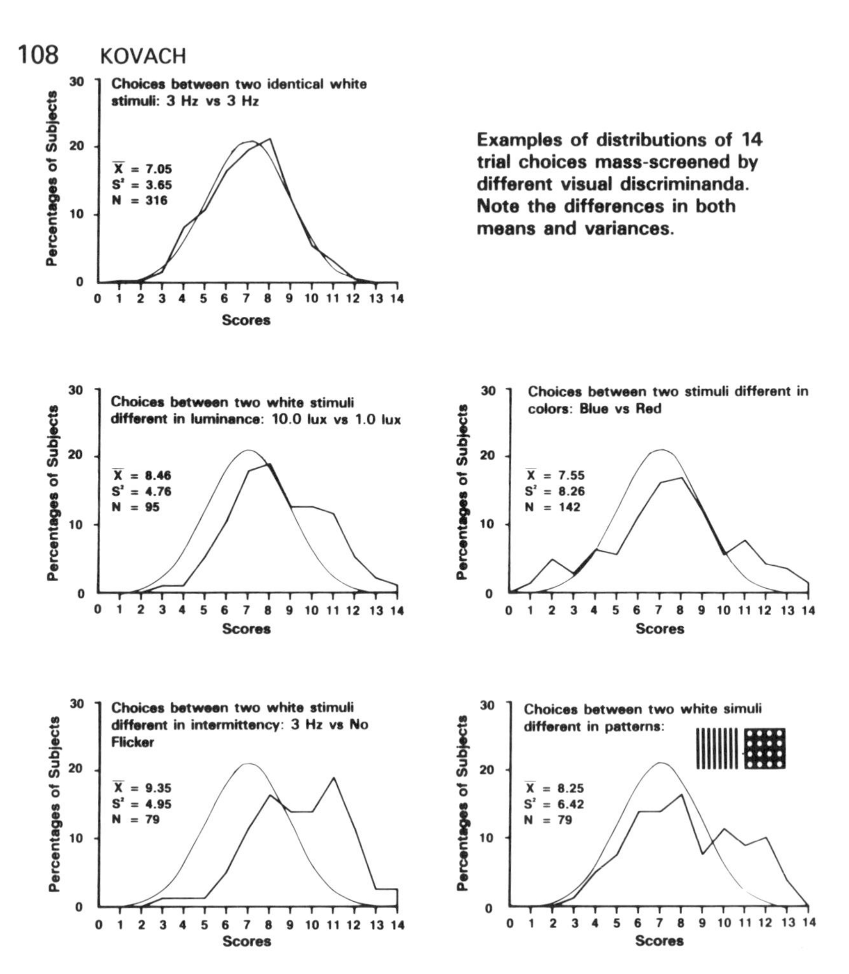

FIG. 4.1 Distribution of 14 trial choices assessed in the mass-screening apparatus by a pair of nondiscriminable white stimuli (upper diagram) and by four pairs of discriminable stimuli (four lower diagrams). Note that choices between the nondiscriminable white stimuli were distributed as a binomial of $P = .5$, whereas the distributions of choices between discriminable stimuli deviated from the binomial of $P = .5$ in both means and variances.

found (see Fig. 4.3). These influences are weak and need be monitored only in studies of mixed or segregating populations, where they may reduce variances.

Comparing the distributions illustrated in Figs. 4.2 and 4.3 also highlights the fact that the mass-screening apparatus in Plate 4.2 operates much like a Galton board. The variation of manifest choices tested in this apparatus with two identical stimuli, or with different stimuli presented to subjects that exhibit no preferences between them or exhibit no individual variation in preferences, is defined by binomial $\overline{PQN}$. Performances tested in a heterogeneous population by discriminable stimuli are distributed within the confines of a mixed binomial distributions. In both instances, means and variances are interdependent, and all

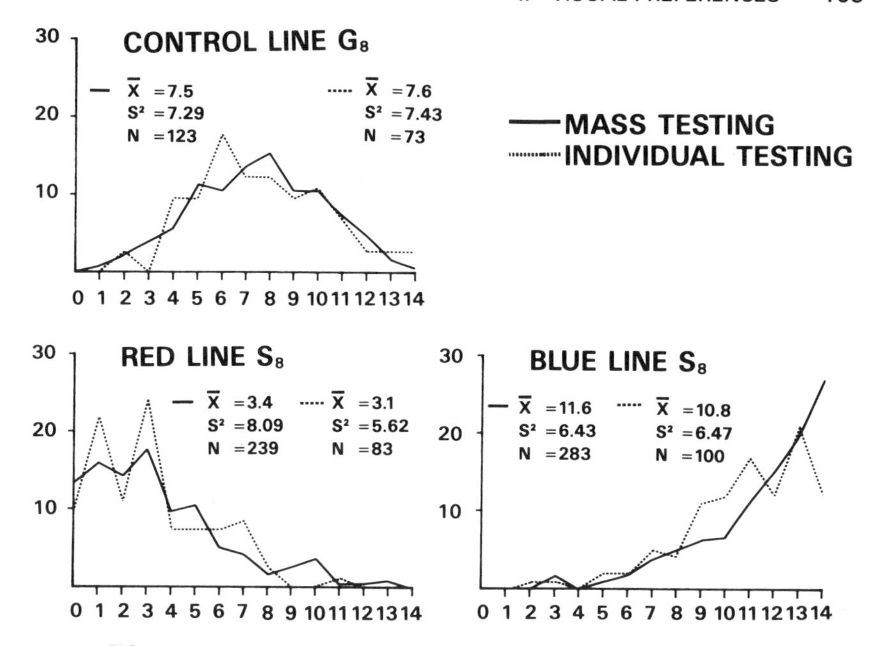

FIG. 4.2 The upper diagram illustrates distribution of individually and mass-tested B/R choices in unselected genetic control subjects. The lower diagrams illustrate similar performances in subjects taken from genetic lines that were genetically selected for preferences of Blue over Red and Red over Blue, respectively, for eight generations. Note the lack of reliable differences between individually tested and mass-screened performances in all comparisons.

(in cases of no individual variation) or some (in cases of individual variations) of manifest choice variation is due to the random error of the binomial. Nonrandom individual variations are indicated by $S_x^2 > \overline{PQ}N$, defined by the individual variations in the preference probabilities represented in the population. The mean probability ($\bar{P}$) is defined in such a distribution by $\bar{P} = \bar{X}/N$, and an index of nonrandom phenotypic variance may be calculated by $S_p^2 = S_x^2 - \overline{PQ}N \,/\, N(N\text{-}1)$, wherein mean and variance are no longer interdependent. A simple example of this situation is given in Fig. 4.4, and its implication for genetic selection of mass-screened preferences is illustrated in Fig. 4.5.

Figure 4.5 compares data from the first two generations of bidirectionally selecting quail for pattern choices (Achromatic Vertical Lines vs. Achromatic Circular Dots, henceforth designated by AVL/ACD) and color choices (Blue vs. Red, henceforth designated by B/R). Note that mean probabilities of preferring the Grated over the Dotted Pattern and Blue over Red were about the same in the respective foundation populations (for comparison see top diagrams and their replication in S_1 and S_2 illustrations by thin-line drawings), but the variance of the pattern choices was smaller. Responses to selection were likewise about equal in the two experiments (compare thin- and thick-line diagrams). However,

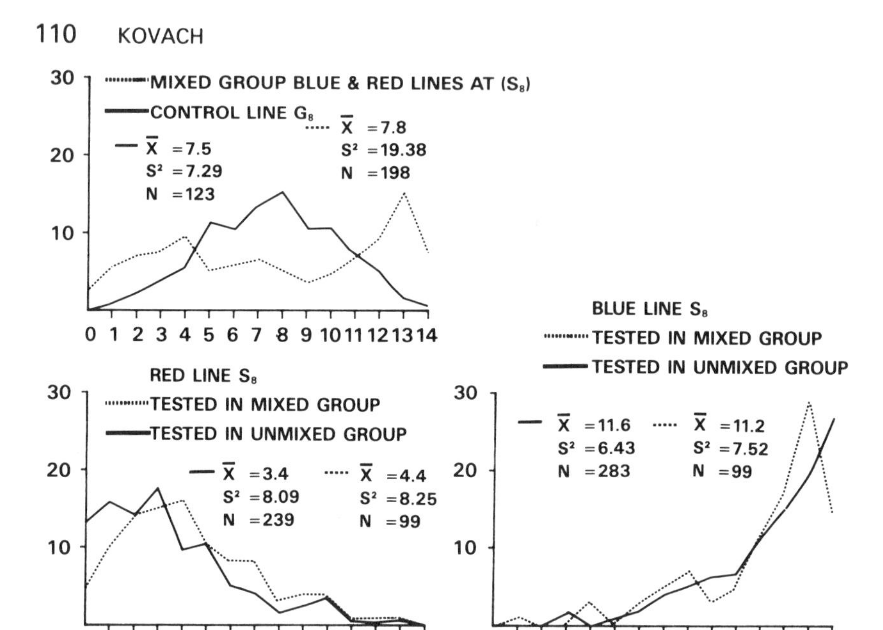

FIG. 4.3 In the top diagram the distribution of B/R choices in genetic controls is compared with a genetically mixed sample of subjects. Halves of the latter were drawn from the eighth generation of the genetic lines that were bidirectionally selected for blue and red preference. In the lower diagrams, the performances in the mixed groups distributed by membership in the two genetic lines are compared with the performances of independently tested samples from the same selected genetic lines. Note the small regressions of performances in the mixed tested samples. For more detail see Kovach, 1977.

since selection pressures were about the same, equal heritabilities are expected to result in weaker response to selection in the situation of pattern choices, because of their smaller foundation variance, greater binomial error, and therefore less reliable identification of extreme phenotypes. Accordingly, the illustrated selection results suggest a higher heritability of pattern than of color choices, which remains to be examined as pattern selection continues.

Sources of Error in the Expression of Genetic Influences

The measurement error described above (which is not a function of mass-screening but of binary choice alternatives) was dealt with by using large subject populations and strong selection pressures. There is, however, another type of potential error in gene expression that is both more difficult to pin down and more interesting in its implication for behavioral development.

Imagine the progression of choice events in the apparatus of Plate 4.2 as a developmental progression, wherein a choice made at a juncture cannot be retested

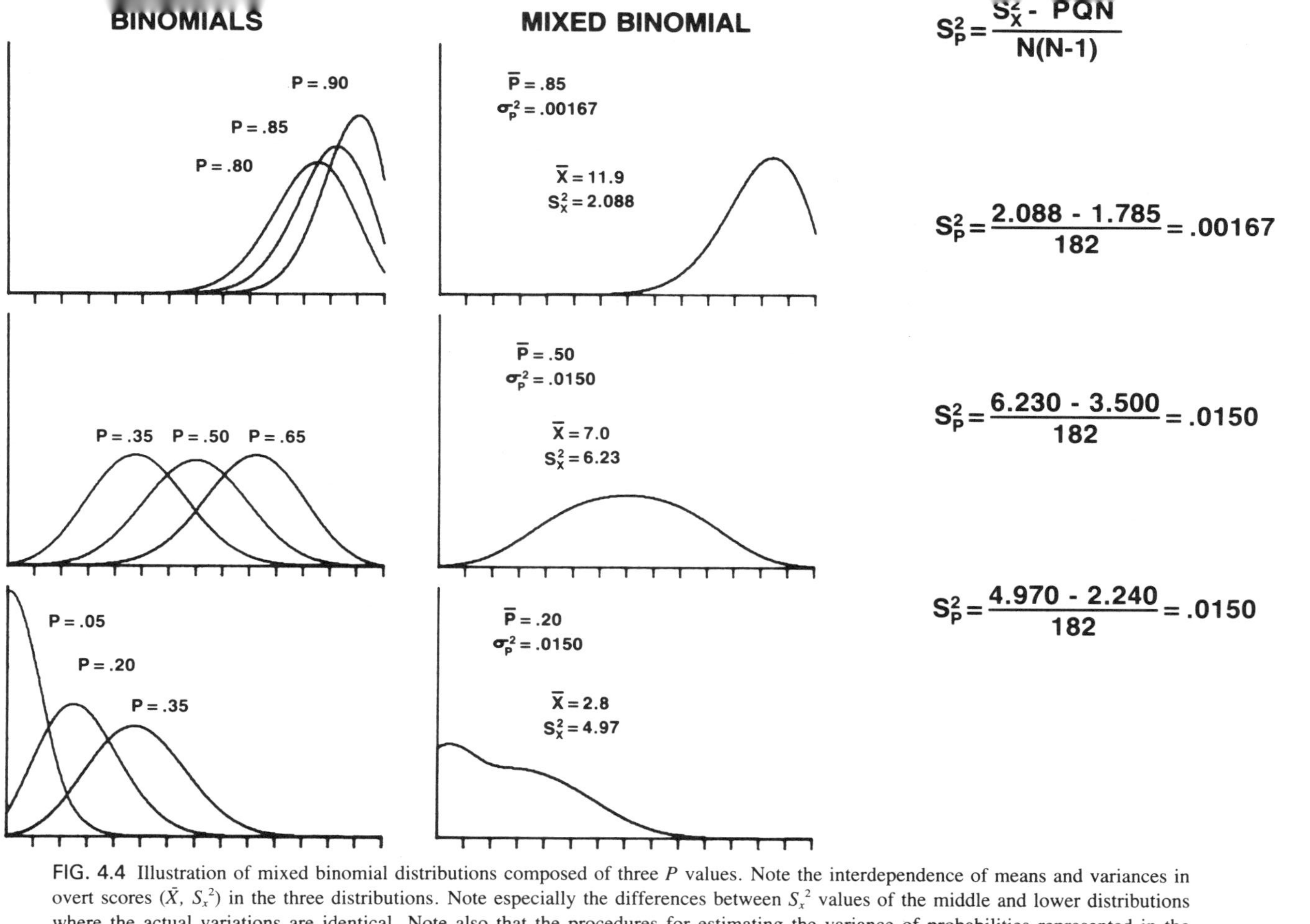

FIG. 4.4 Illustration of mixed binomial distributions composed of three P values. Note the interdependence of means and variances in overt scores ($\bar{X}$, S_x^2) in the three distributions. Note especially the differences between S_x^2 values of the middle and lower distributions where the actual variations are identical. Note also that the procedures for estimating the variance of probabilities represented in the distributions give true values of nonrandom variation.

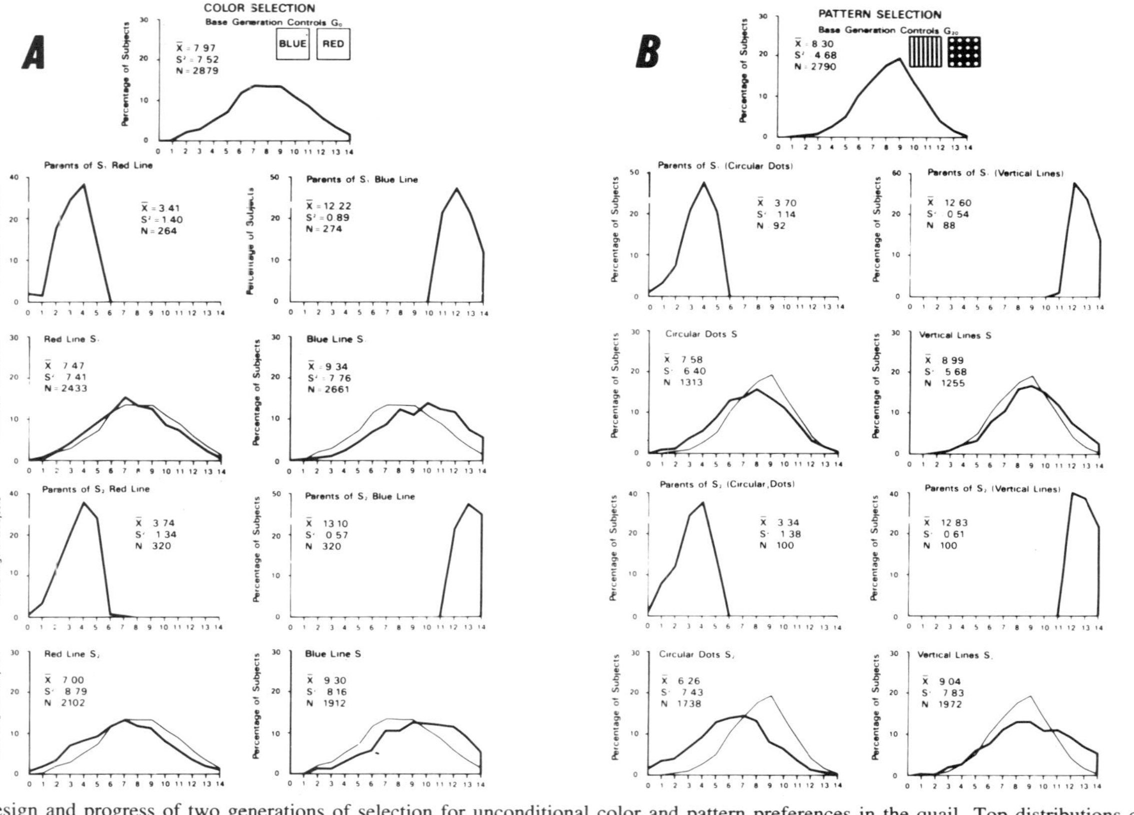

FIG. 4.5 Design and progress of two generations of selection for unconditional color and pattern preferences in the quail. Top distributions on left and right illustrate the performances of the respective foundation populations. The next two sets of rows illustrate, respectively, the choice distributions of the initially selected parents and compare performances of the corresponding foundation population (illustrated by thin lines) and first-selected generations (illustrated by thick lines). The two sets of rows at the bottom repeat the same for the second generation of parental and selected populations. Note the difference between variances of the two foundation populations, and the responses to selection in each experiment revealed by differences between foundation and selected performances. Means, variances, and numbers of subject ($\bar{X}$, S^2, and N) always refer to the thick-line distribution.

because the finality of development would not permit it. Measurement error in this situation could be related only to incorrect identification of subjects in particular end boxes. Imagine also that the manifest developmental probability is biased (it is not $P = .5$) and that each individual in a group possesses the same P value. Given a substantially large sample of subjects, the resulting distribution would be judged as a *Normal* or a *Poisson,* depending on the value of $\bar{P}$, and there could be no response to genetic selection of the trait. In this situation we would conclude that the trait is not heritable, even if the developmental P were completely gene dependent. Thus, a heritability estimate clearly reflects not on genetic determination but on trait variation that is subject to selective identification by a particular agent of natural or artificial selection.

Now imagine that the population contains considerable individual variation in its developmental P values, each P still being entirely gene dependent. The resulting "developmentally mixed binomial distribution" could still be judged as Normal or Poisson, and heritabilities as zero or very small, depending on the magnitudes, relative frequencies, and distribution characteristics of the P values represented in the population. Only a large variation of broadly distributed P values would dictate inferring nonparametric distribution and heritabilities larger than zero. In other words, standard procedures of estimating heritabilities may say little about genetic or environmental determination of a trait in situations where the expression of the trait is developmentally probabilistic.

An interesting example of possible accidental developmental variation in a genetic trait is human handedness. According to one view, based on data indicating that between 85% and 95% of humans exhibit right-hand preference, handedness is determined by a major Mendelian gene with variable penetrance (Annett, 1964, 1967; Chamberlain, 1928; Jordan, 1914; Levy & Nagylaki, 1972; Rife, 1940). According to another view, derived from data indicating that subtle individual variation in handedness is normally distributed, handedness is determined by polygenic effects (Annett, 1978). Still another interpretation claims that handedness and laterality are environmentally determined. Proponents of the last view draw support from family and twin data, especially from the fact that the pairwise right-right, right-left, and left-left combinations of handedness in twins confirm the random variation of the binomial of $P = .85$ to $.95$ for right-handedness. Observations that animals do not respond to genetic selection for laterality also support this point of view (Collins, 1968, 1977).

Annett (1978) has offered a resolution that effectively deals with each of these views about handedness and laterality. She proposed that handedness is due to "accidential variation, modified in man by a genetic factor inducing dextral bias" (p. 227). The possibility of "developmental error" discussed above suggests that the genetic factor inducing dextral bias need not be invariable in human populations, although if it varies it probably does so within narrow limits and with low frequencies of extreme variants. This amended version of Annett's interpretation is attractively parsimonious. It accommodates all that is known about human handedness, and may even explain the lack of response to selection of mice for laterality. In this case, true genetic variation may be obscured by both the developmental error in gene expression and the episodic binomial error in identifying its selectable variants. However, the question whether Annett's interpretation of handedness is or is not correct is of secondary concern for the present discussion. The issue of developmental error is raised here simply because it highlights how tenuous the concepts of heritability, genetic variation, and even response to genetic selection really are when applied in the study of behavior.

Environment Effects and Gene Expression

It could be argued that trait variation not subject to selective identification in a given situation is irrelevant for evolution or developmental molding of the trait. However, as demonstrated by studies of Waddington (1940, 1942) and Becker and Bearse (1973), this argument is weak. Waddington exposed a stock of pupal flies to heat shock and found the posterior wing vein missing in about 30% of adult flies. Normally reared flies did not exhibit the trait. Artificial selection for high incidence of missing posterior vein in heat-shocked flies led to spontaneous occurrence of the trait. Continued selection of flies resulted in 95% of them spontaneously exhibiting the trait, without prior heat shock. Obviously the heat shock did not cause the trait. Rather, it brought its phenotypic expression above a variable and normally very high threshold of penetrance. Once identified, selection could latch onto the gene or genes responsible for the normally rare, extreme low threshold variant and increase its frequency in the population.

Becker and Bearse (1973) have conducted a similar experiment on the occurrence of blood spots in the egg yolk of White Leghorn hens that were raised on a normal or vitamin A deficient diet. As in the Waddington experiment, environmental treatment (here vitamin A deficiency) made the trait detectable, and genetic selection resulted in the spontaneous prevalence of the trait in an environmentally untreated population.

These experiments illustrate that gene expression is developmentally probabilistic, and that the developmental penetrance of gene effects need not be constant but is influenced by environment. In other words, a genetic trait that is not subject to selective identification and genetic selection under one set of developmental circumstances may become identifiable and naturally or artificially selectable under another set of conditions. Heritabilities and genetic responses to natural or artificial selection are thus dependent on environmental factors that facilitate or inhibit the phenotypic expression of a trait.

In the sections that follow, examples are given of changes in behavioral variation resulting from either episodic or developmental influences in the quail's preference behaviors, or both. The crux of the discussion in this section, however, is that the selective identifiability of a genetically influenced trait may be a function of either the episodic conditions that identify the trait or the environmental factors that influence its development, or both. These considerations are relevant to the genetic study of behavior because, unlike a typical morphological phenotype, manifest behavioral phenotypes are always variable in both developmental and episodic terms.

THE GENETIC CONTEXT OF UNCONDITIONAL PREFERENCES

Artificial Selection for Color Preferences

Artificial selection of quail for initial unconditional color preferences has now reached the 22nd generation. It has resulted in nearly perfect preferences of Blue over Red in one (BL) and Red over Blue in another (RL) selected line. Selection also at first increased then drastically reduced individual variation (Kovach, 1979, 1980). Fluctuation of mean and variance from generation to generation was small and mostly insignificant in the unselected Genetic Control Line (CL). These data are illustrated in Fig. 4.6.

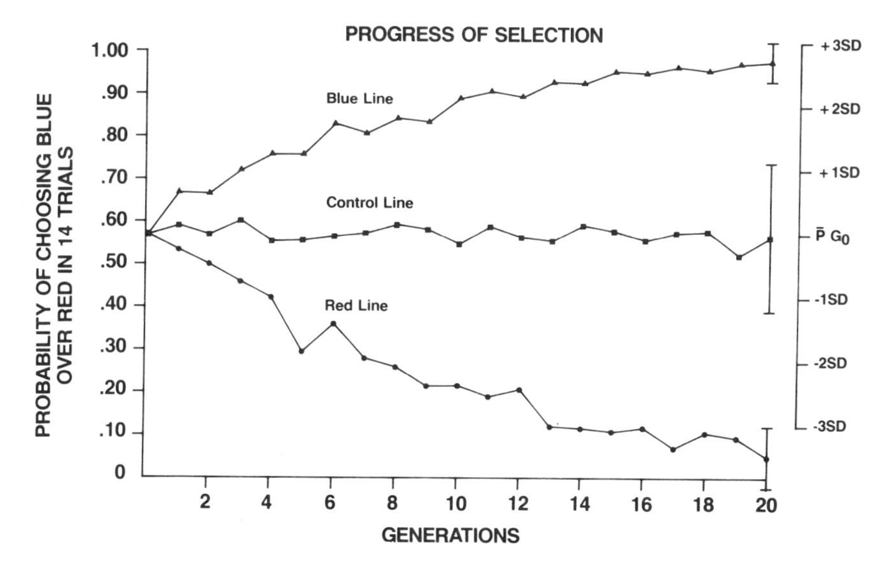

FIG. 4.6 Results of long-term bidirectional genetic selection of quail for unconditional approach preferences between Blue and Red. Performances are expressed by mean probabilities of choosing Blue over Red. Scale on the right indicates standard deviation units of the foundation population (G_0). Vertical lines cutting through means at the 20th generation indicate ± 1 *SD* values of the respective distributions. The latter indices estimate true phenotypic variations of choice probabilities. They are free of truncation effects and binomial error.

As the selection procedures have been described elsewhere (Kovach, 1979, 1983a), I only note here that very large samples were tested at each generation of each genetic line (usually close to or over 1,000) to ensure sufficiently strong selection pressures on distributions that contained a good deal of binomial error, and to maintain maximum genetic heterogeneities. Mating was organized at each generation by protocol, which permitted no shared ancestry in a mated pair for three preceding generations. Each subject used in the study was individually identified, which now gives us access to pedigree data on well over 50,000 subjects.

The Genetics of Color Preferences

Artificial selection of behavior usually is done to identify genetic influences and to investigate the "genetic architecture" of a behavior. Identifying the genes responsible for Blue and Red preferences in the quail has been a task for the present project as well. However, understanding genetic architecture is considered secondary to the task of understanding the roles and interactions of genetic and environmental influences and related coding of preference information.

The data on hybrid and backcross populations suggest simple genetic determination (Kovach, 1980). Table 4.1 compares the outcome of F_1 and F_2 crosses at S_{12}, S_{18} and S_{20} of the color-selected lines. Performances of unselected control subjects are also included for comparison. The indicated k values refer to estimates of segregating Mendelian factors. They were calculated by $k = (2a^2 + d^2)/4({S_{F2}}^2 - {S_{F2}}^2)$, where a is half the distance between parental phenotypic values, d is dominance effect estimated from deviation of F_1 mean from the midpoint between parental values, and variances refer to the respective phenotypic values (${S_p}^2$). Assumptions were (1) that all "blue genes" were in one and all "red genes" were in another parental line, (2) that parental lines were selected long enough so that genes segregating in F_2 were at frequencies of one-half, (3) that there were no linkages, and (4) that genes interacted additively with equal magnitudes of effects on thc trait.

There was no independent proof for these assumptions. But even if none were met, the obtained estimates of k could define the minimum number of segregating factors; the maximum number could not be far above, because that would obscure segregation. However, the implicit assumption of shared loci for Blue and Red preferences is not supported by the data. The F_1 variances were invariably larger (by a factor of about 2) than the variances of foundation and genetic control populations, which suggests independent "blue" and "red" loci. Other related and relatively consistent indicators were the nearly uniform regression of means in F_2 generations, which suggests some directional dominance for the blue genes and some reciprocal cross effects.

Our attempt to identify pertinent genes and processes of gene expression would be greatly simplified by having available inbred lines of quail, preferably

TABLE 4.1
Data from Mendelian crosses (reciprocal and total) of the two genetic lines that were selected for color preferences, at S_{12}, S_{18}, and S_{20}. Unselected genetic control samples, which were tested at corresponding generations and times, are included for comparison.

		N	X	S_x^2	P	S_p^2	F	df	$p <$	Estimates of Segregating units (k)
S_{12}										
Blue Males × Red Females	F_1	802	8.92	12.34	.64	.050	1.32	1	.01	4.35
	F_2	1472	8.04	16.25	.57	.070				
Red Males × Blue Females	F_1	882	7.60	15.14	.54	.064	1.26	1	.01	7.26
	F_2	1526	7.79	17.33	.56	.076				
Total	F_1	1684	8.26	13.74	.59	.057	1.28	1	.01	5.28
	F_2	2998	7.92	16.79	.57	.073				
Controls		140	7.89	10.62	.56	.039				
S_{18}										
Blue Males × Red Females	F_1	629	9.09	13.76	.65	.058	1.05	1	NS	—
	F_2	988	8.57	14.40	.61	.061				
Red Males × Blue Females:	F_1	629	9.27	13.51	.66	.057	1.24	1	.01	6.17
	F_2	78	7.27	16.42	.52	.071				
Total	F_1	1258	9.18	13.63	.66	.058	1.15	1	.05	9.60
	F_2	1771	8.00	15.70	.57	.067				
Controls		87	7.91	8.62	.56	.028				
S_{20}										
Blue Males × Red Females:	F_1	437	8.03	13.41	.57	.055	1.31	1	.01	4.78
	F_2	457	7.12	16.78	.51	.072				
Red Males × Blue Females:	F_1	339	6.63	15.61	.47	.067	1.07	1	NS	—
	F_2	482	6.17	16.48	.44	.072				
Total	F_1	776	7.42	14.84	.53	.062	1.17	1	.05	7.29
	F_2	939	6.63	16.83	.47	.073				
Controls		142	8.25	8.26	.59	.027				

congenic inbred lines. Unfortunately, earlier difficulties encountered by others attempting to inbreed quail (see discussion above, and Iton, 1966; Kulenkamp, 1967; Maeda et al., 1978; Sittman & Abplanalp, 1965) have surfaced in our laboratory as well. Of 36 families originally started (18 from each color line) only 4 remain after six generations of inbreeding, and their general viabilities are drastically reduced.

Correlated Responses to Selection: Behavioral Effects

The foremost question raised by the data described so far was whether selection modified color perception or color preferences, or both. Both behavioral and physiological experimentations have been used to answer this question. In this section, I present the behavioral data.

The color specificities of the selection effects were tested first at progressive generations of selection, with six pairs of four major colors (Blue-Green, B/G; Blue-Yellow, B/Y; Blue-Red, B/R; Green-Yellow, G/Y; Green-Red, G/R; and Yellow-Red, Y/R). Colors were presented through wide-band Wratten Gelatin filters and narrow-band Interference filters. Choice performances are illustrated in Fig. 4.7 (and in all comparable bar graphs below) by magnitudes of deviation from no preference (from $P = .50$). P values refer to preference of the first over the second stimulus in each color pair. Each performance was tested in a large sample (N at least 60, but usually over 100). As can be seen, standard errors rarely exceeded 2 points on the 100-point probability scale (see *T*s on bars), and selection effects were highly significant.

The data indicate progressive changes with selection in the choices between all colors, excepting the regions of smallest wavelength differences at the spectral extremes favored in selection (B/G in BL and Y/R in RL). In general, BL subjects exhibited progressively stronger preference for the shorter over the longer and RL subjects for the longer over the shorter wavelengths. Genetic controls exhibited only small and largely negligible generation effects in a comparatively weak yet statistically reliable tendency of preferring the middle over the extreme and the shorter over the longer wavelengths. To examine the wavelength specificities of these performances, subjects were retested at Generation 14, using wide-band and narrow-band color filters and a simplified testing procedure. The filter types were identical in mean wavelengths, interference filters having only 10 nm or less half-band widths.

Because interference filters are expensive, equipping each choice point in the mass-screening apparatus with such filters could not be done. A simplified screening was used instead, in which preferences were tested in 6 subgroups of 25 subjects drawn by replacement from population samples of 250 subjects, each subgroup yielding a single P value (see Kovach & Wilson, 1981). The results were in general agreement with other mass-screened data. Only marginally significant filter-type effects were found, and even then only in genetic control performances ($df = 1$, $F = 4.38$, $p < .05$; see Fig. 4.8). In the selected lines, the only detectable filter type effect was related to G/Y performances in RL subjects. In this situation, the wide-band filter appeared decidedly more orange to the human observer than the narrow-band filter. The RL birds chose the former. This behavior suggests that hue perception in the quail may be quite similar to human hue perception. In further tests of this possibility, subjects from each genetic line were tested and were found to choose according to their genetic preferences between hues differing by as little as 10 nm. Data of still another experiment negated the possibility that these discriminations are made and genetic preferences are exhibited by perceived brightness rather than color (see Kovach, Yeatman, & Wilson, 1981). Overall, these experiments demonstrate that artificial selection for unconditional preferences between Blue and Red modified responses to all colors. Blue-selected subjects tended to prefer the shorter over the longer, and red-selected subjects the longer over the shorter wavelengths. Exceptions

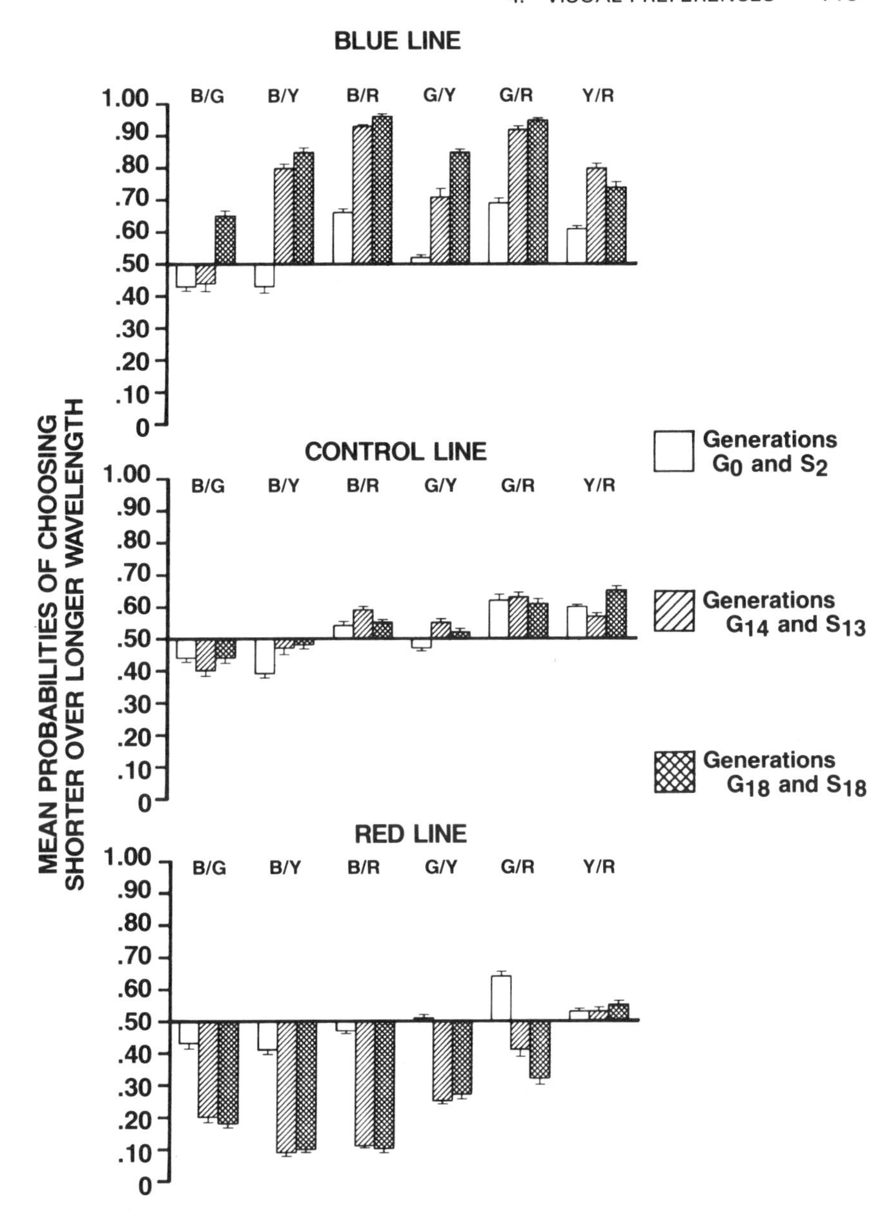

FIG. 4.7 Preferences between each of six color pairs in subjects drawn from progressive generations of selecting quail for blue and red preferences. In this and all other comparable bar graphs of this chapter, the small *T*s on bars indicate 1 standard error of the depicted mean probability values ($\bar{P}$), and $\bar{P}$ refers to choosing the first over second stimulus in each pair of stimuli. For more detail see Kovach and Wilson, 1981.

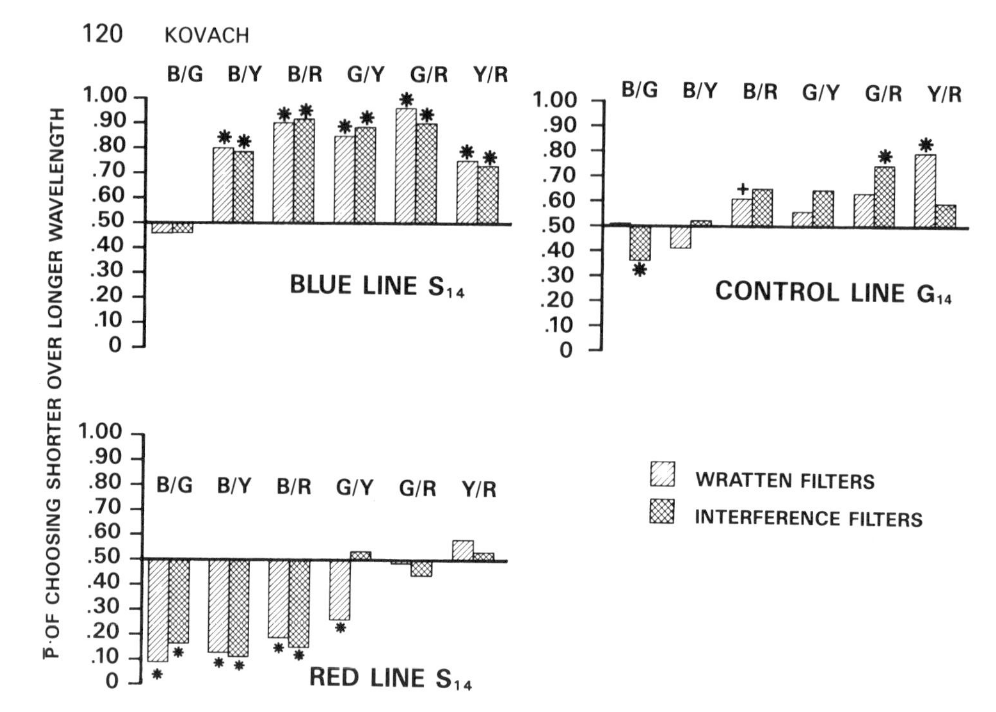

FIG. 4.8 Comparison of choices between each of six paired combinations of four major colors administered through wide-band and narrow-band color filters. Six trials were tested per filter type, each in 25 subjects drawn by replacement from a population sample of 250 subjects. Asterisks identify performances that were reliably different from $\bar{P} = .50$, at $p \leq .01$, and cross at $p \leq .05$.

were the colors at genetically favored extremes of the spectrum (B/G for BL and Y/R for RL).

Correlated Responses to Selection: Physiological Effects

A striking feature of avian retinas is the presence of brightly colored oil droplets between the outer and inner cone segments. These droplets are variable in color, size, and retinal distribution. Most birds possess primarily orange, red, and yellow droplets (Coulombre, 1955; Strother, 1963), but the quail retina contains a dominant amount of green droplets as well (Konishi, 1965; Kovach, Wilson, & O'Connor, 1976). Researchers have long maintained (King-Smith, 1969; Roaf, 1933; Wald & Zussman, 1938) that oil droplet coloration is responsible for avian color vision. Although this postulate is no longer acceptable (Bloch & Maturana, 1971; Bowmaker, 1979; Donner, 1960; Pedler & Boyle 1969), oil droplet pigmentation is still believed to be involved in both avian color perception and avian color preferences (see Bowmaker, 1979; Hailman, 1964; Wallman, 1972).

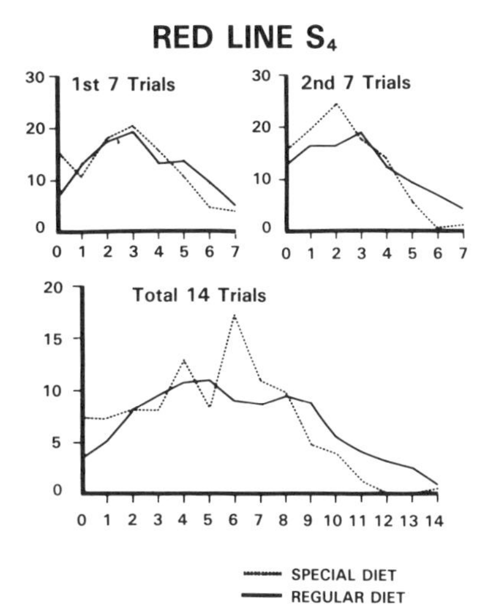

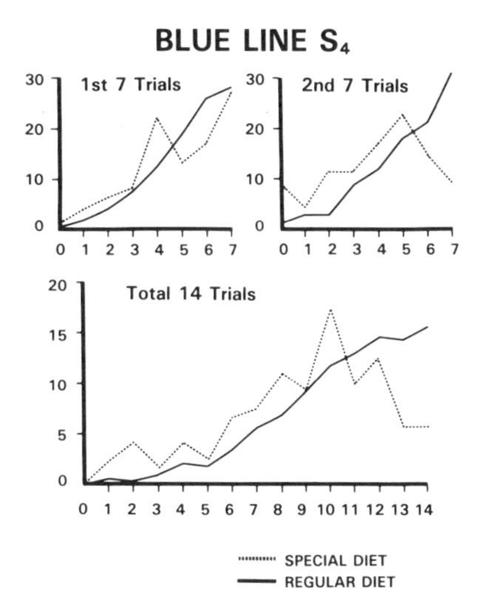

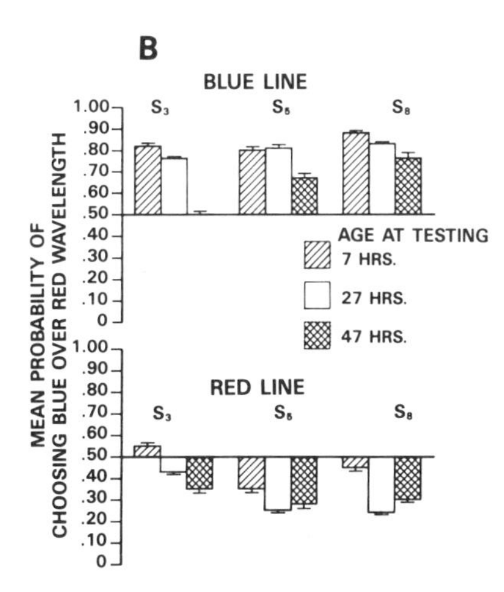

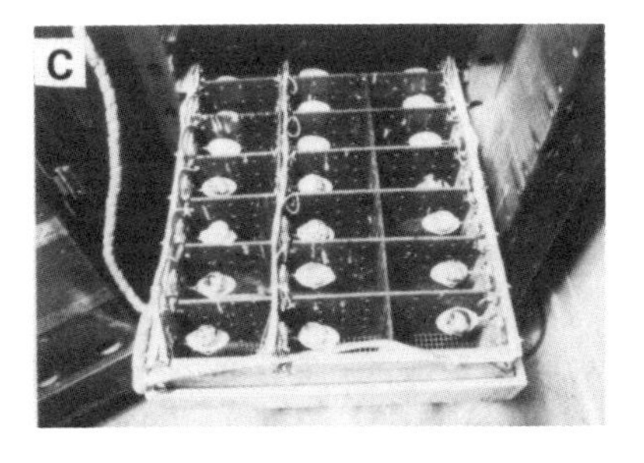

FIG. 4.9 Effects of age and carotenoid deprivation of B/R preferences. Diagram A shows a small yet statistically reliable red shift in carotenoid-deprived subjects from the fourth generation of color-preference selection. The upper row of the two sets of diagrams illustrates B/R choice distributions on first and second 7 trials. The lower row illustrates 14-choice distributions. Both compare choices in normally reared subjects (regular diet) and subjects hatched from eggs laid by hens raised on a carotenoid-free diet (special diet). All were tested at 27 hr posthatch (for more detail see Kovach, Wilson, & O'Connor, 1976).

Diagram B shows age effects in B/R preferences at progressive early generations (S_3, S_6, S_8) of selection. Note that selection reduced the magnitude of a red shift with age (for more detail see Kovach, 1977). Diagram C illustrates the hatching tray used to determine posthatch ages. The weight of the incubating egg closed an independent microswitch. At hatching, the chick fell off the spoon and the microswitch marked the event on a timer.

These considerations prompted us to test whether or not genetically selecting quail for color preferences may have modified pigmentation and relative distribution of retinal oil droplets (Kovach et al., 1976).

Microscopic examination of control and artificially selected birds (at Generation 4) revealed no differences in oil droplet coloration. Chicks obtained from carotenoid-deprived hens were expected and found to have no pigmentation in their retinal oil droplets, which agrees with similar observations by Meyer (1971) and Wallman (1972). These birds also exhibited a small yet consistent red shift in their color choices; however, the magnitude of differences between selected lines remained unchanged (see diagram A of Fig. 4.9). Overall, the data suggest that although oil droplet pigmentation most likely plays limited roles in color discrimination, it is not essential for color vision and does not mediate the observed genetic influences in the quail's color preferences.

A tendency to shift from Blue toward Red preference was also observed in relation to the age of subjects (see diagram B of Fig. 4.9). Therefore the red shift resulting from carotenoid deprivation possibly was not due to sensory effects of colorless oil droplets but to developmental influences from absence of carotenoid substances in the diet; conversely, the developmental changes in oil droplet coloration may have been responsible for age effects in color choices. These alternatives are yet to be examined.

The exact number of photopigments in avian retinae is still debated but the earlier view that there is only a single photopigment and that the pigmentation of retinal oil droplets is pivotal in avian color vision is no longer acceptable. Actually, there have been four or five classes of visual pigments identified in the rods and cones of avian retinas (with apparent sensitivities in the 400–415 nm, 460–480 nm, 515–540 nm, and 560–575 nm regions of the spectrum for cones, and 500–507 nm for rods; see Jacobs, 1981). However, it seems unlikely that genetic variation in the quail's color preferences is mediated by variable activities of photopigments. The data collected so far strongly implicate central rather than primary sensory mediation of the artificially selected color preferences (see data above and the discussion of learning experiments below). However, the retina is part of the central nervous system and the possible involvement of higher level retinal mechanisms cannot be excluded at this time. This too will be investigated as the project continues.

In still other experimentation visually evoked electrical potentials (VEP) were related to preferred and unpreferred colors at Generations 13 and 14 of selection (see Fig. 4.10). The most striking feature of these data has been the difference between two basic VEP types in each genetic line, one elicited by the Blue stimulus, the other by the Red. As can be seen, responses to a color were more similar between the selected lines, regardless of the selected preference, than between the control and the selected lines. Stepwise discriminant analysis (see Kovach et al., 1981) correctly classified color VEPs for each genetic line, with the exception of some red responses in RL subjects. But the procedure failed to

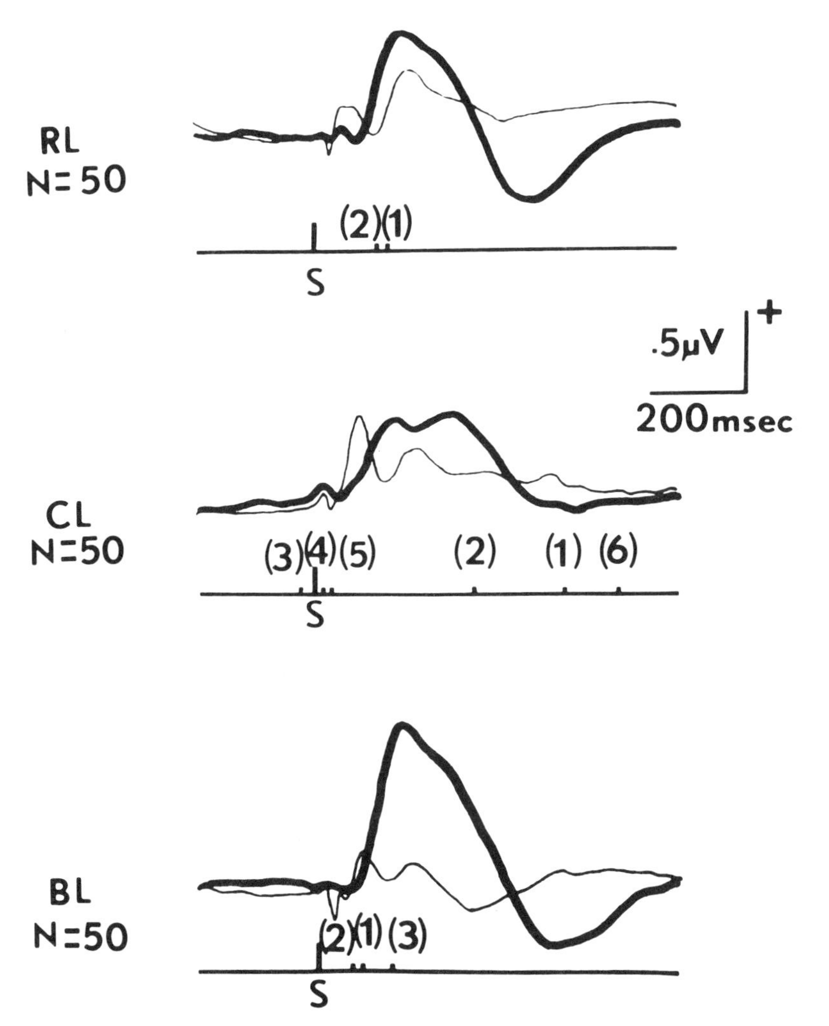

FIG. 4.10 Averaged visually evoked electrical potential waveforms collected from subjects at Generations 13 and 14 of the bidirectionally selected color preference lines (BL and RL) and unselected Control Line (CL). Evoked potentials were averaged for 50 subjects per sample, each subject contributing 400 responses to the Red and 400 responses to the Blue stimulus. Responses to Red are indicated by a thin line, responses to Blue by a bold line. Numbers on abscissas indicate points of significant differences between waveshapes within lines, as detected by stepwise discriminant analysis (for more detail see Kovach, Yeatman, & Wilson, 1981).

generate significant discriminant functions according to the preference or the avoidance of a color. There were reliable differences in the waveforms of evoked potentials according to the eliciting color, but not by the genetic preference of the color. As was expected from behavioral experimentation, these data also indicate that the primary processes of color vision were not influenced by selection. However, the lack of differentiation of VEP by preferences was unexpected.

There are several possible reasons for the failure to identify VEP differences relatable to genetically influenced preference or nonpreference of a color. Neuroelectrical activities relating to genetic differentiation of preferences may be too finely tuned for identification by the gross recordings employed in this experiment. The records obtained by placing electrodes on predominantly visual areas may reflect on dominant electrical activities of general visual responses or on activities relating to general rather than color specific preferences (for the latter see also Fig. 4.11). Identifying electrophysiological correlates of the latter may require filtering the former. Furthermore, the neural activities relating to selected preferences may be called into play only by simultaneous discrimination. Each of these possibilities will be examined in the continued search for electrophysiological correlates of the neural representation of genetically determined and acquired color preferences in the quail (see discussion relating to Fig. 4.22).

Overall, the data presented in this section support the conclusion from behavioral observations (Kovach & Wilson, 1981) that selection did *not* alter color vision. However, the employed evoked potential techniques were insensitive for detecting factors in centrally mediated gene effects in the preferences of particular colors.

Developmental Expression of Gene Effects and Episodic Interactions between Genetically Influenced Unconditional Preferences

The data collected in search of genetic correlation between preferences in different visual modalities indicate the existence of a stimulus general preference component. Both the BL and RL subjects preferred more than did CL subjects the achromatic grated over the achromatic dotted pattern, a brighter over a dimmer stimulus, a stimulus flickering at 3 Hz over a stimulus that did not flicker, and a stimulus that flickered at low rates (2–4 Hz) over stimuli flickering at higher rates (5–10 Hz). The data suggest genetic influences in a stimulus general preference component, as illustrated in the upper diagram of Fig. 4.11 in relation to preferences of colors and patterns.

For collecting the data of Fig. 4.11, the color-selected subjects (from BL and RL at S_{18}) were backcrossed to unselected genetic control subjects (from CL at G_{18}). The parental and hybrid populations were then tested for unconditional

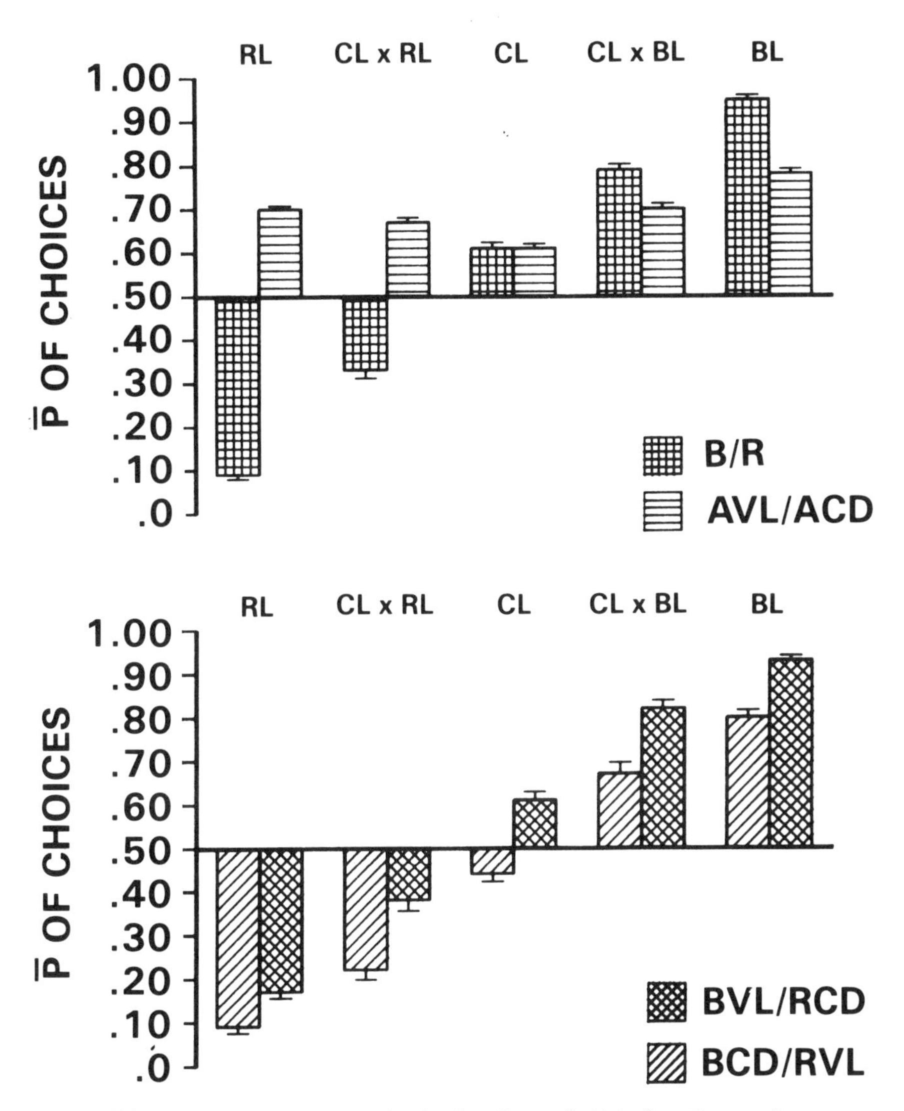

FIG. 4.11 Stimulus preferences in visually naive quail chicks from five genetic populations. Upper portion shows performances tested by two colors (B/R) and by two achromatic patterns (AVL/ACD). The lower portion shows performances tested with two pairs of stimuli that combined colors and patterns (BCD/RVL and BVL/RCD). Subjects were drawn from the Red (RL) and Blue (BL) selected lines, the unselected genetic control line (CL), and their respective genetic crosses (RL × CL and BL × CL), all at Generation 18. Performances are represented by mean probabilities of choosing Blue over Red, AVL over ACD, BVL over RCD, and BCD over RVL. For more detail see Kovach, 1983a.

preferences between colors (B/R) and patterns (AVL/ACD). Additional samples from each population also were tested with reciprocal combinations of the two colors and the two patterns (with BVL/RCD and BCD/RVL). The data confirm genetic correlation between color and pattern preferences.

In addition to the linear to U-shaped relationship between color and pattern preferences, the data in the upper diagram of Fig. 4.11 indicate simple additive gene effects in the choices between Blue and Red. Responses to stimuli combining colors and patterns (see lower diagram of Fig. 4.11) also confirm these additive gene effects. However, the latter data also indicate an episodic interaction between color and pattern effects, in which color effects are partially dominant over pattern effects. The reciprocal combinations of colors and patterns BCD/RVL and BVL/RCD altered performances, but the differences within genetic populations were not as large as would be expected from equal contribution of color and pattern effects.

The data in Fig. 4.12 illustrate similar episodic interactions between preferences of colors and flicker. In these data, flicker effects are completely dominant over color effects in the performances of genetic control subjects, and color effects partially dominate over flicker effects in the performances of the genetically selected subjects. The data suggest that the episodic dominance of one over another stimulus is related to the relative strengths of preferences that are genetically determined. Interestingly, conflicting preference combinations of composite stimuli also greatly increase individual variations of choice responses (see diagram B, Fig. 4.12).

The distributions illustrated in diagram B of Fig. 4.12 bring to mind the directional dominance effects and increased variances observed in segregating Mendelian populations. However, these data implicate not developmental but episodic processes of gene expression. They implicate interaction not of genes but of the genetically influenced elements of neurally mediated stimulus information. The data point back to considerations discussed above in relation to the Waddington (1940, 1942) and Becker and Bearse (1973) experiments. According to these data, selective identification of extreme behavioral phenotypes may be inhibited or facilitated by congruent or conflicting stimulus information that brings the behavior about. In other words, gene expression is influenced by factors of environment pertinent to not only the development but also the episodic expression of behavior.

THE GENETIC CONTEXT OF PERCEPTUAL IMPRINTING

Gene–Environment Interactions

Initial stimulus preferences of birds can be modified by imprinting (Bateson, 1964; Hess, 1959, 1973; Lorenz, 1935), which is the main reason for the above genetic manipulation of the quail's visual preferences. We anticipated that variation in the outcome of imprinting quail of various genotypes with variously

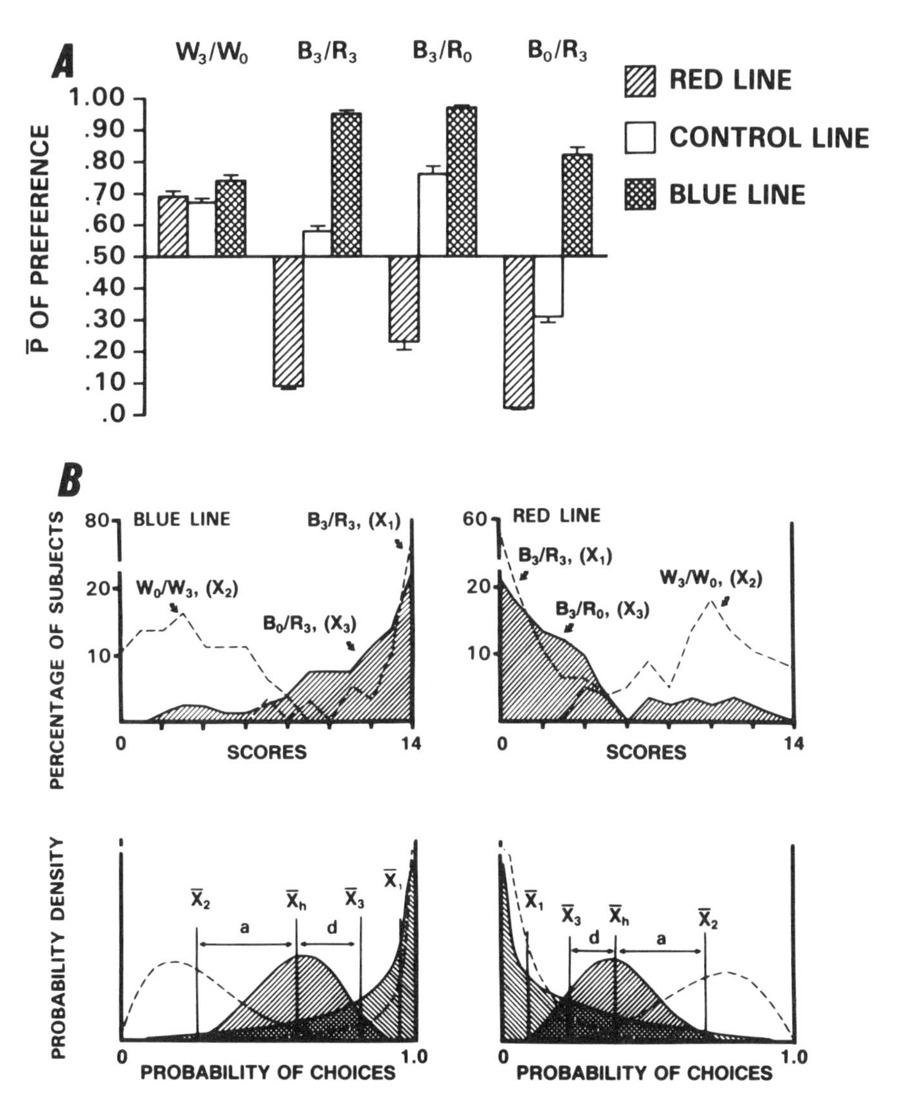

FIG. 4.12 Choice performances tested with equally flickering colors (B_3/R_3), flicker-no flicker (W_3/W_0), and composite discriminada of colors and flicker (B_3/R_0 and B_0/R_3) at Generation 18 of bidirectional selection for color preferences. Diagram A illustrates mean probabilities of choices, and diagram B the related choice distributions. Note that flicker effects were dominant over color effects in the performances of genetic controls and color effects were partially dominant in selected subjects. The lower portion of diagram B shows these relationships, including a hypothetical distribution (X_h) expected, were the effects of flicker, and color simply additive. *X*s in the diagrams indicate respective means, small *a*'s additive stimulus effects, and *d*'s stimulus dominance. Note that scoring of choices between achromatic and chromatic flicker discriminanda were matched by flicker for this illustration (W_3/W_0 with B_3/R_0 for RL and W_0/W_3 for BL). Note also that conflicting preference combinations resulted in variances that were larger than the variances of choices between either of the two colors tested alone or the flicker-no flicker alternatives. For more detail see Kovach, 1983b.

preferred stimuli would point the way to the mechanisms of genetic canalization and gene-environment interaction in behavioral development. The present section examines genotype-environment interactions in the quail's preferences, with emphasis on whether perceptual imprinting is stimulus general or stimulus specific and what the roles of the genotype might be in this learning.

Fig. 4.13 illustrates procedures developed for imprinting quail to various visual stimuli, and Fig. 4.14 some data on preference changes resulting from imprinting. The data in Fig. 4.14a illustrate (1) that 12 hr imprinting exposure to Blue or Red modified color choices in the quail, and (2) that the gain from imprinting to a genetically unpreferred color diminished somewhat with the progress of selection (see BL performances after exposure to Red and RL performances after exposure to blue). The data in Fig. 4.14b were collected to test whether genetic variation in unconditional preferences may cause variation in perceptual imprinting. Subjects were drawn at Generation 18 from the artificially selected color-preference lines and from the unselected control line, and from crosses of selected and control lines (see also Figs. 4.9 and 4.11). As before, subjects were exposed to Blue or Red for 12 hr in the apparatus shown in diagram A of Fig. 4.13. However, instead of an independent test for each condition, in this experiment the genetic samples receiving different exposures were randomly mixed and were tested in single mass-screened group to determine how, if at all, variation of the unconditional preferences may influence gain from imprinting exposures.

The data indicate that unselected CL subjects and CL × RL or CL × BL hybrids gained reliably more from exposures than did the selected subjects. Such an outcome is self-evident for exposures to genetically preferred colors, where the nearly perfect unconditional preferences left little room for further improvement by experience, which was the reason that RL and BL subjects were not exposed to preferred colors. However, learning was also found less pronounced in the selected than in the hybrid subjects, after respective exposures to genetically unpreferred colors. The phenomenon was especially pronounced in BL subjects. These data raise the following questions. Was it learning or the expression of what had been learned that was influenced by genetic variation in preferences? Was perceptual learning stimulus general or stimulus specific? Two paths were selected for answering these questions. First, an attempt was made to identify stimulus general processes in perceptual learning by genetic selection, and second, stimulus specific learning effects were probed in different genotypes.

Search for Stimulus General Processes of Perceptual Learning

In the initial genetic experiment (Kovach, 1979), quail were artificially selected for high gain from imprinting, one line selected for gain from 12 hr exposure to Blue, another for gain from 12 hr exposure to Red. Four generations of such

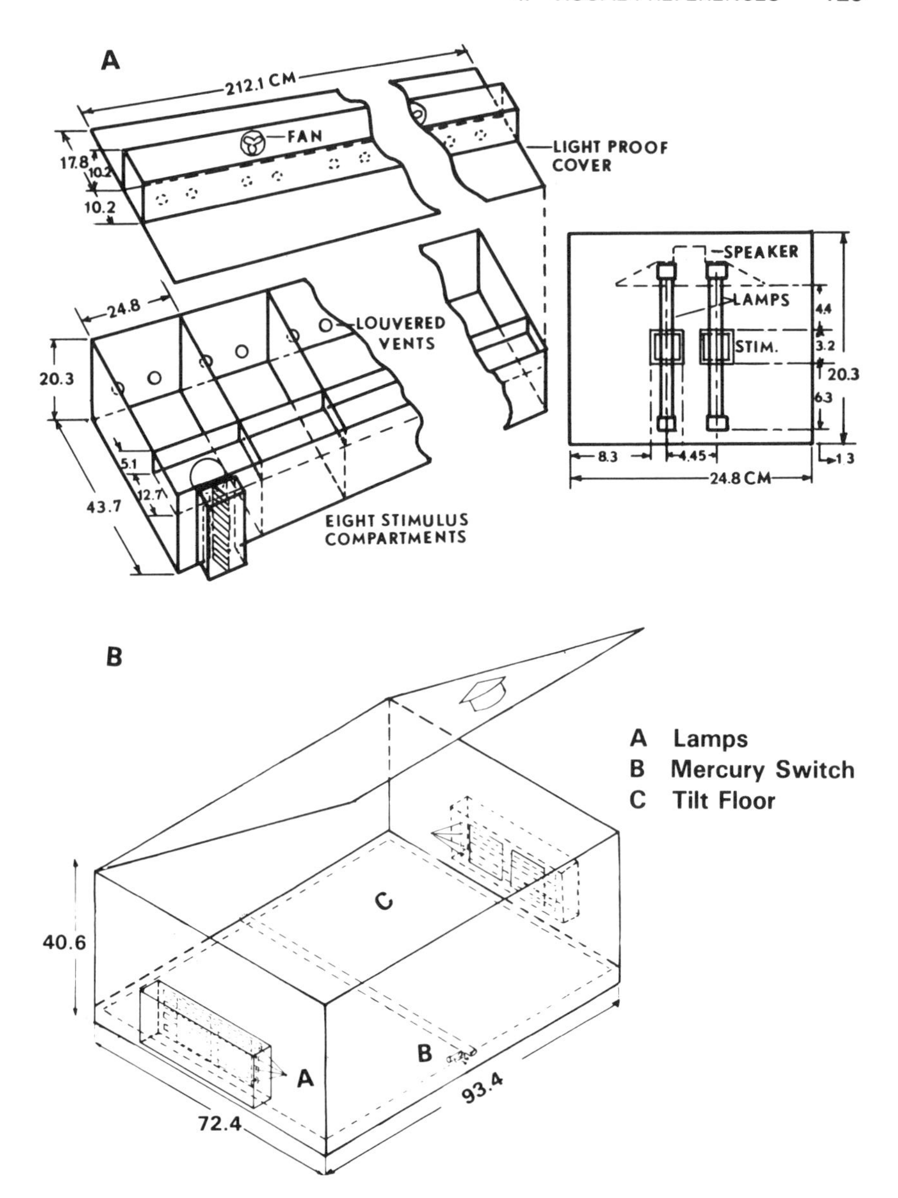

FIG. 4.13 Schematic representation of apparatuses used for mass-exposing quail chicks to stimuli. Diagram A illustrates an eight-compartment exposure; diagram B was used for longer exposures. In this apparatus, stimuli in the end-plates were "on" alternatingly during entire exposure period, for 10 min at each end at a time, which made the chicks approach stimuli from one end to another end of apparatus. When needed, chicks' movements were monitored by a microswitch activated by tilting the floor. All measurements of appratuses are in cm.

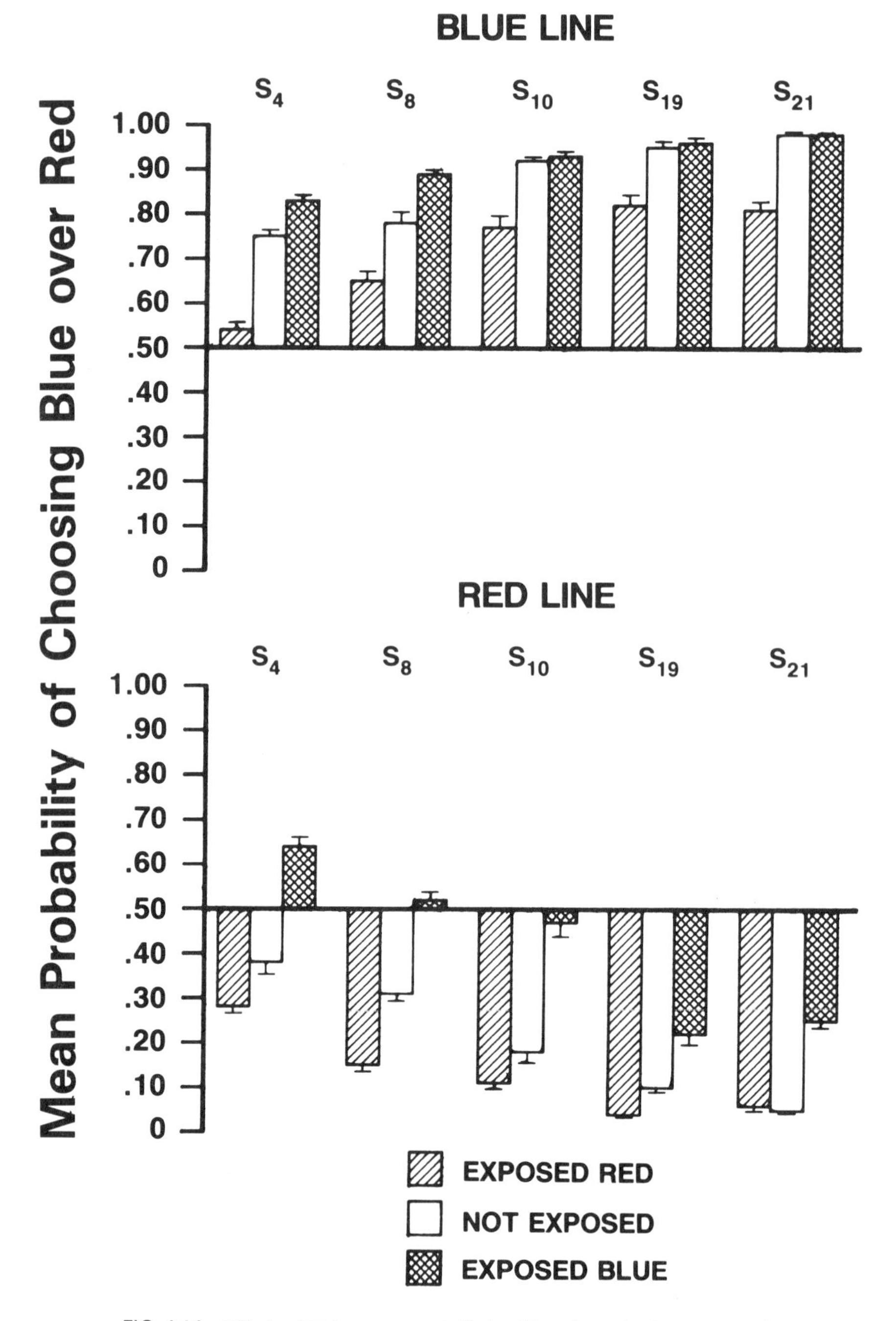

FIG. 4.14a Effects of 12 hr exposures to Red or Blue of samples from progressive generations of the genetically selected lines are shown. Note that selection reduced somewhat the gain from exposure to unpreferred colors.

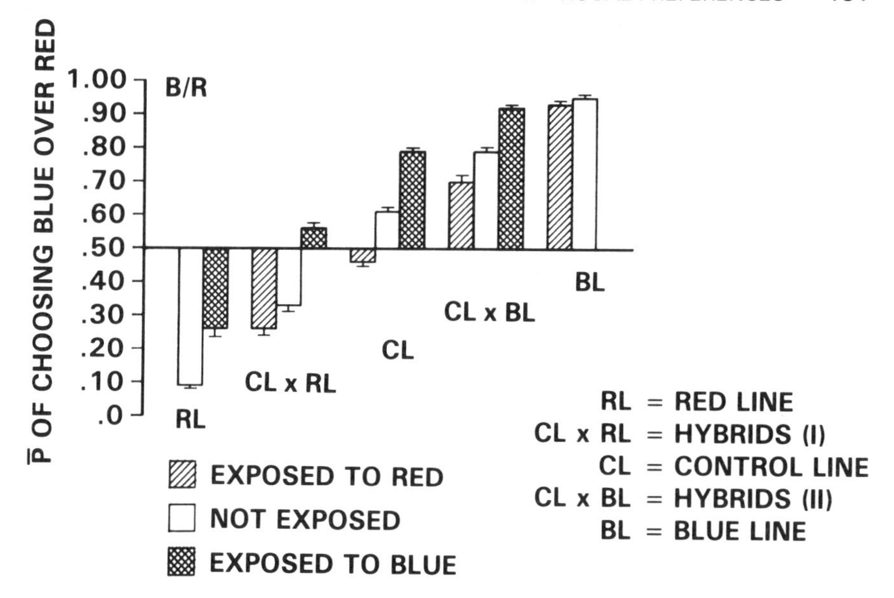

FIG. 4.14b Mean probabilities of choosing Blue over Red (B/R) in subjects at Generation 18 of five genetic populations are shown: the genetic line that was selected for Red preference (RL), the unselected control line (CL), the genetic line that was selected for Blue preference (BL), and their respective crosses (CL × RL and CL × BL). For more detail see Kovach, 1983a.

selection increased mean gains in both situations, and modified unconditional preferences as well. Crossing the two selected lines brought the latter back to foundation level and left the increased tendency to gain from imprinting intact. These data suggest stimulus general learning effects. For directly probing these effects, a selection experiment was initiated by the design illustrated in Fig. 4.15. Instead of selecting for high color-specific learning as before, the experimenters selected high and low lines for color-general learning. To eliminate color effects and manipulate stimulus-general learning only, the Blue- and Red-exposed subjects were crossed at each generation (blue-exposed males to red-exposed females, and vice versa; see Fig. 4.15).

Because hybridization greatly increased the individual variation of initial preferences (compare the foundation populations in the selection experiments illustrated in Figs. 4.5 and 4.15), giving large proportions of subjects with maximum potential for preference change by experience (from initially 0 score to 14 after Blue exposure, or from 14 to 0 after Red exposure), assortatively mated F_3 hybrids of the two color-preference lines (BL × RL) were used as foundation population in this experiment. As can be seen in Fig. 4.16, five generations of such selection resulted in very small differentiation of high and low performances. There was no response to low selection when learning was tested with Blue exposure, and testing with Red exposure indicated no genetic gain in the high selected lines. So far there have been no variance increases either, which would indicate breakages in an initially linked polygenic system.

The results illustrated in Fig. 4.16 were both interesting and unexpected. They were unexpected in the light of usually strong and immediate responses to artificial selection for learning (see McGuire & Hirsch, 1977; Oliverio, Eleftheriou, & Bailey, 1973; Tryon, 1940; van Abeelen, 1975). They were interesting because the generally higher average gain from exposure to Red (see Fig. 4.16) and the more pronounced response to low selection when tested after Red exposure may both be related to the red shift discussed above (see Fig. 4.9). If correct, even the small genetic responses observed in this selection study would be extraneous to processes of color general learning.

The lack of responses to selection in this experiment also may have been due to a strict stimulus specificity of individual variations in learning, and to the related recombination of selectable genetic variation at each generation of cross-breeding subjects by exposure color. Or variations in stimulus general processes of learning may depend on tightly knit polygenic systems, the breakage of which into selectable components may need longer selection and stronger selection pressures than employed so far. Finally, the fundamental mechanisms of neural coding and storing stimulus information derived from experiences may not be genetically variable, which would imply that the genetic variation in learning usually detected by artificial selection of classically or instrumentally conditioned behaviors are related to factors that influence but are not pivotal to the coding of stimulus information in learning. Although they are highly speculative at this stage, exploring these various possibilities justifies continuing the selection experiment discussed in this section.

Selective Learning from Joint Exposures to Different Visual Stimuli

Experimentation under this heading compared learning from exposure to stimuli of different visual parameters. Subjects from the color-selected and genetic control lines were exposed for 12 hr to the VL or CD patterns presented on Achromatic (AVL and ACD), Red (RVL and RCD), and Blue (BVL and BCD) backgrounds. Exposure effects were then tested with pairs of achromatic and chromatic patterns (AVL/ACD, BVL/RCD, BCD/RVL). The purpose was to probe the stimulus specificity of learning, and to see whether and how genetically influenced unconditional preferences may serve as reinforcers for learning by perceptual association. The results are illustrated in Fig. 4.17.

This experiment again failed to identify a stimulus general learning component. Exposures resulted in strong color learning, but there was no pattern learning, except possibly some marginal learning from CD exposures in unselected controls (see Fig. 4.17). Associating exposure patterns with preferred or

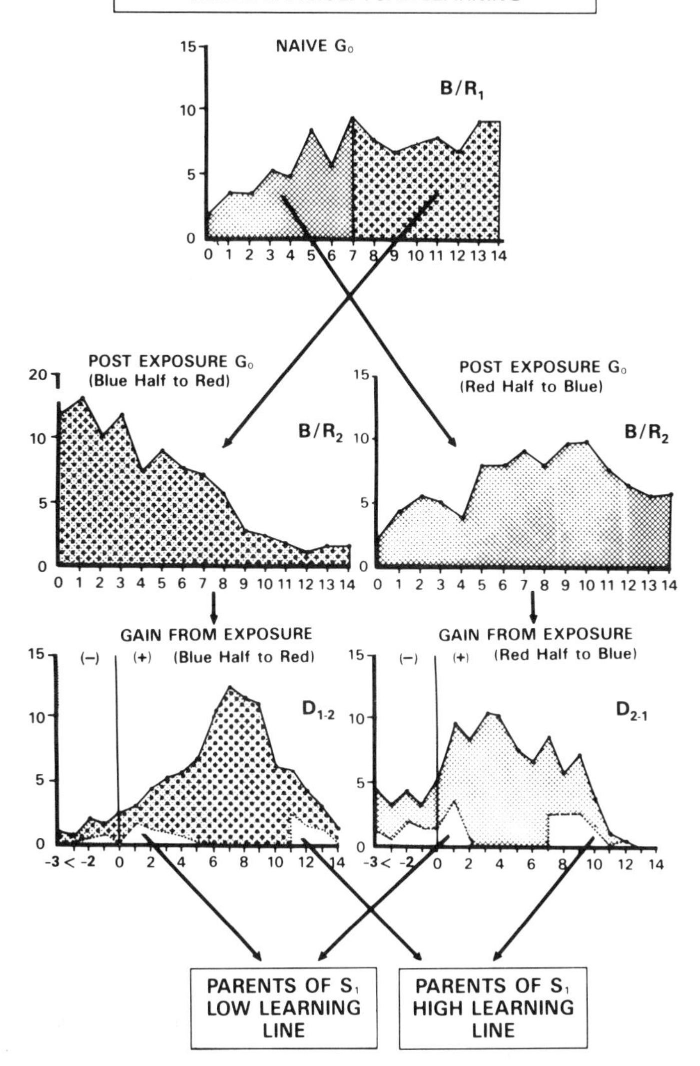

FIG. 4.15 Design and data of genetic selection for color-general learning effects. The naive foundation distribution was split. Subjects from Red half of the illustration (i.e., those preferring Red over Blue) were exposed 12 hr to Blue, and subjects from Blue half were exposed to Red. Differences between the pre- (B/R_1) and post- (B/R_2) exposure choices (D_{1-2} and D_{2-1}, respectively) were then calculated. Subjects with highest gain from the respective exposures were crossed to create Generation 1 of the high line, and subjects with lowest gain were crossed for the low line. Selection was continued by similar crossing of high performing Red-exposed with high performing Blue-exposed subjects in the high, and low-performing Red-exposed with low-performing Blue-exposed subjects in the low selected line.

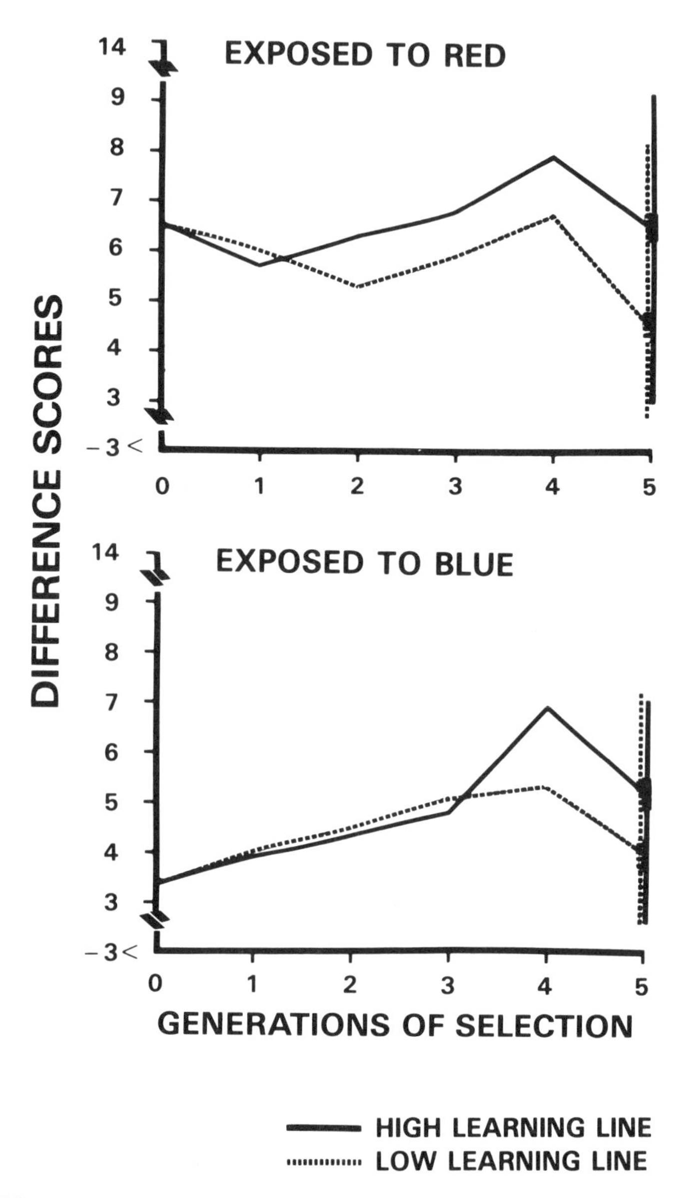

FIG. 4.16 Results of the first five generations of selection for color-general learning. Upper part of the figure illustrates progress of selection for pre- to post-exposure gains when tested after Red exposure, and the lower part illustrates progress of selection tested after Blue exposure. The thick portions of vertical lines at S_5 indicate *SE*s at Generation 5. The extended thinner portions illustrate *SD*s (note the related break in the scale at left, which is the reason for the apparent asymmetry of the values).

unpreferred colors made no difference (see A at top of Fig. 4.17). It remains to be determined whether pattern learning was absent in these data because the quail pay attention only to colors of exposure stimuli (including white) or because at the tested age they cannot as yet learn pattern.

Interaction between Unconditional and Acquired Preferences: Variable Gene Effects.

In addition to stimulus and apparently perceptual channel specific processes of learning, the data of Fig. 4.17 also indicate some interesting pattern effects in the expression of color learning. The congruent or conflicting combination of unconditional pattern with acquired color preferences in composite stimuli influenced the expression of learning and had done so by more than a mere additive shift of the preference base lines against which learning was expressed. Combining initially preferred color with preferred pattern (BVL/RCD for BL exposed to R, RVL, or RCD; and vice versa for RL) inhibited and, combining the initially unprefered exposure color with the preferred pattern (BVL/RCD for RL exposed to B, BVL or BCD; and vice versa for BL), facilitated the expression of acquired color effects. This relationship is illustrated further in Figs. 4.18 and 4.19.

Figure 4.18 compares performances of five genetic populations by exposure color and by testing stimuli that combined colors and patterns. These data illustrate again the straightforward developmental additivity of gene effects in color preferences, the linear to U-shaped genetic relationships between color and pattern preferences, and the inhibiting and facilitative effects of unconditional preferences in testing stimuli on the expression of learning effects. Figure 4.19 illustrates the inhibitory and facilitative effects of preference combinations in testing stimuli in relation to learning tested with composite stimuli that combined colors and flicker. These data also indicate that unconditional preferences (1) set the baseline against which learning effects were expressed, and (2) facilitate or inhibit the expression of learning effects. Both influences are matters of the relative strength and combination of unconditional and acquired preferences in the composite testing stimuli.

Interaction between Unconditional and Acquired Preferences: Variable Environment Effects

Preliminary data indicated asymptotic or near asymptotic preference changes in unselected quail chicks after about 12 hr exposure to colors. For this reason, 12 hr exposure periods were used in the above experimentation. However, we did not know whether asymptotic expression of learning implies exhaustion of learning potential in a given situation. The issue became important because 12 hr exposure to colors resulted in reliably less learning in the genetically selected than in the genetic control subjects (see Fig. 4.14). Was this fact a sign of a

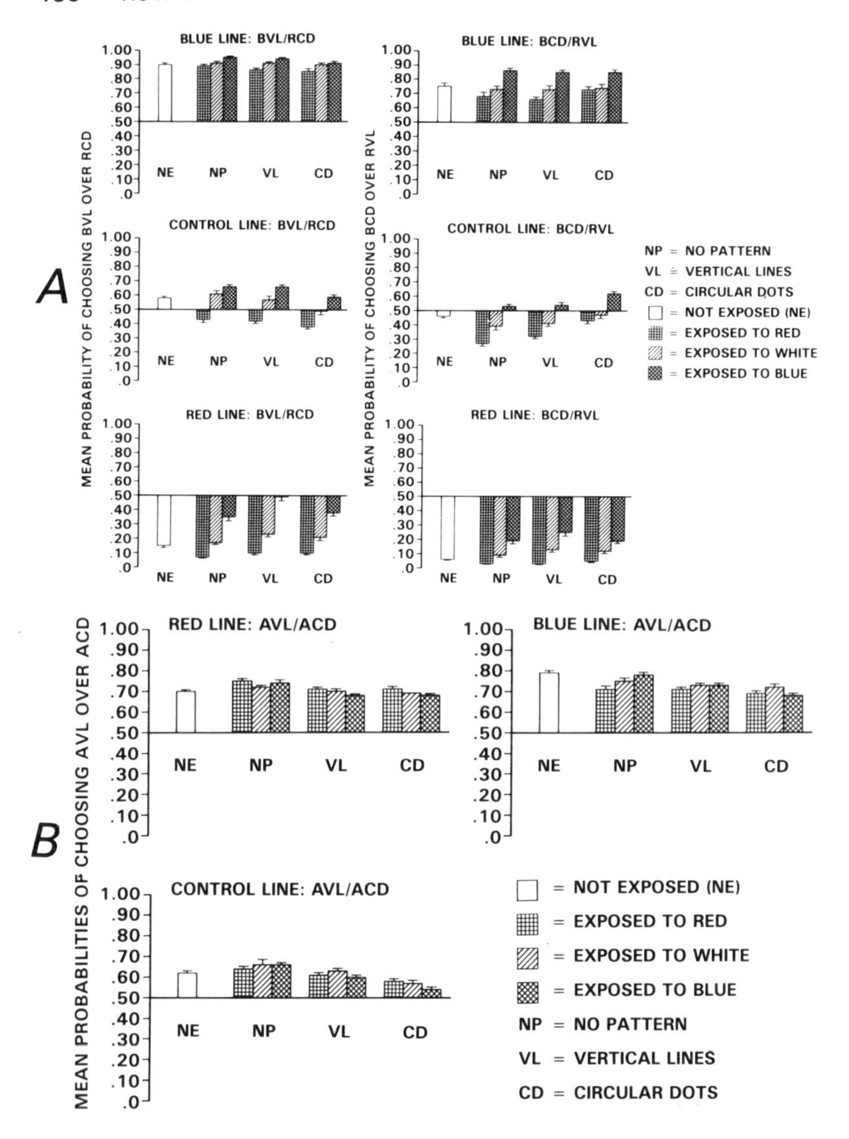

FIG. 4.17 Learning from separate and joint exposures to colors and patterns. Diagram A illustrates mean probabilities of choosing the Blue Vertical Lines over the Red Circular Dots (BVL/RCD) and the Blue Circular Dots over the Red Vertical Lines (BCD/RVL) in visually naive subjects (NE) from the three genetic populations (BL,CL,RL), in subjects that were first exposed to Red, White, or Blue without simultaneous exposure to associated pattern (NP), and in subjects that were also exposed to a pattern (VL or CD). Note that exposure effects were invariably greater when tested with congruent combinations of exposure color and preferred pattern (compare BVL/RCD with BCD/RVL choices within and across selected lines). Diagram B illustrates performances tested with achromatic patterns (AVL/ACD) after the same exposures as in diagram A. For more detail see Kovach, 1983d.

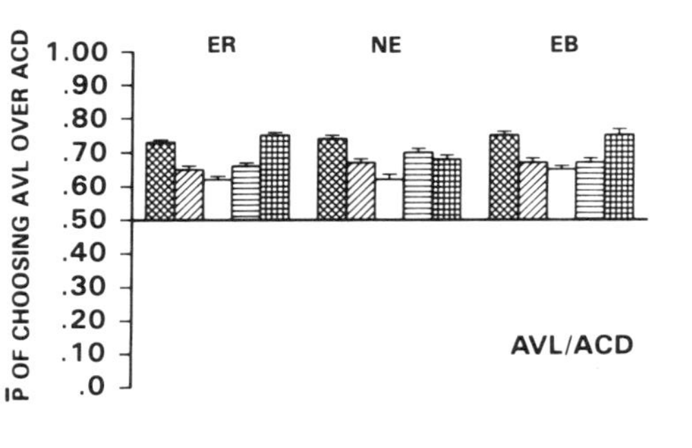

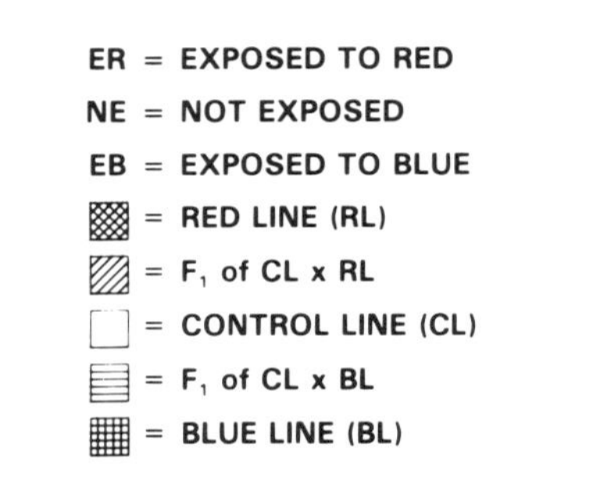

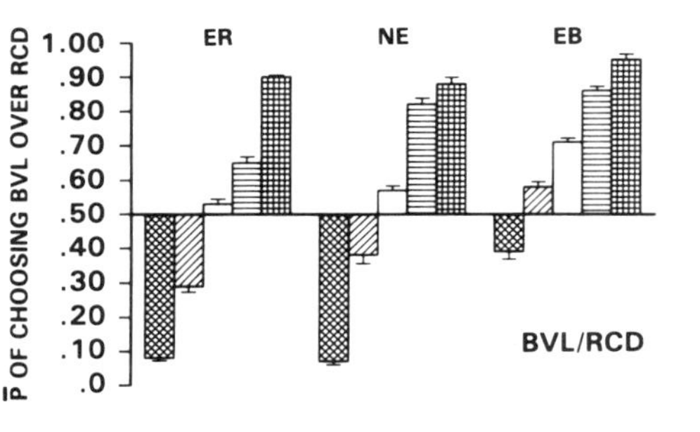

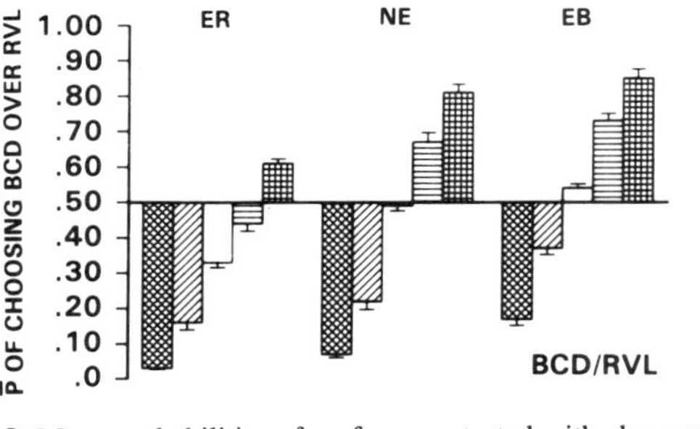

FIG. 4.18 Mean probabilities of preferences tested with chromatic and achromatic patterns in five genetic populations, with and without prior exposures to colors. Compare performances within and across populations, and note the linear to U-shaped relationship between performances elicited by colored and achromatic patterns. For more detail see Kovach, 1983d.

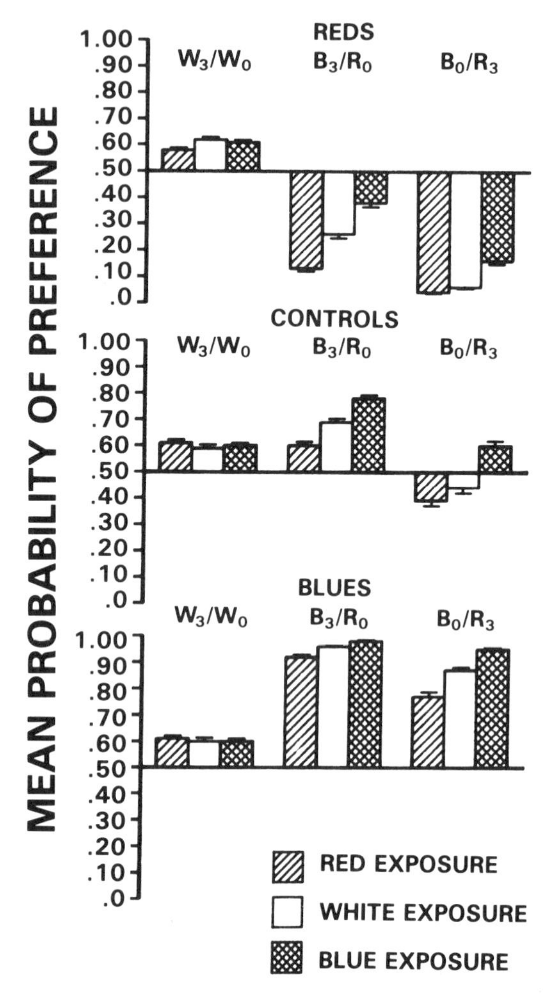

FIG. 4.19 Mean probabilities of choices between composite discriminanda of colors and flicker in samples from Generation 18 of three genetic populations after 12 hr exposures to Red, White, or Blue. The designations of "Reds," "Controls," and "Blues" refer, respectively, to blue-, control-, and red-selected genetic lines. Note that congruent flicker in testing stimuli facilitated and conflicting flicker inhibited the expression of learning from exposure to colors. For more detail see Kovach, 1983b.

lack of learning when stimuli were already maximally preferred for reasons of genotype, or a sign of reduced learning from exposure to genetically unpreferred colors? Or was the limited expression of learning in these situations also a matter of behavior control by combined preference values of testing stimuli? For probing these alternatives, first longer than 12 hr imprinting was introduced. Subjects were tested at 27 hr posthatch after 12 hr exposure as before, and were then

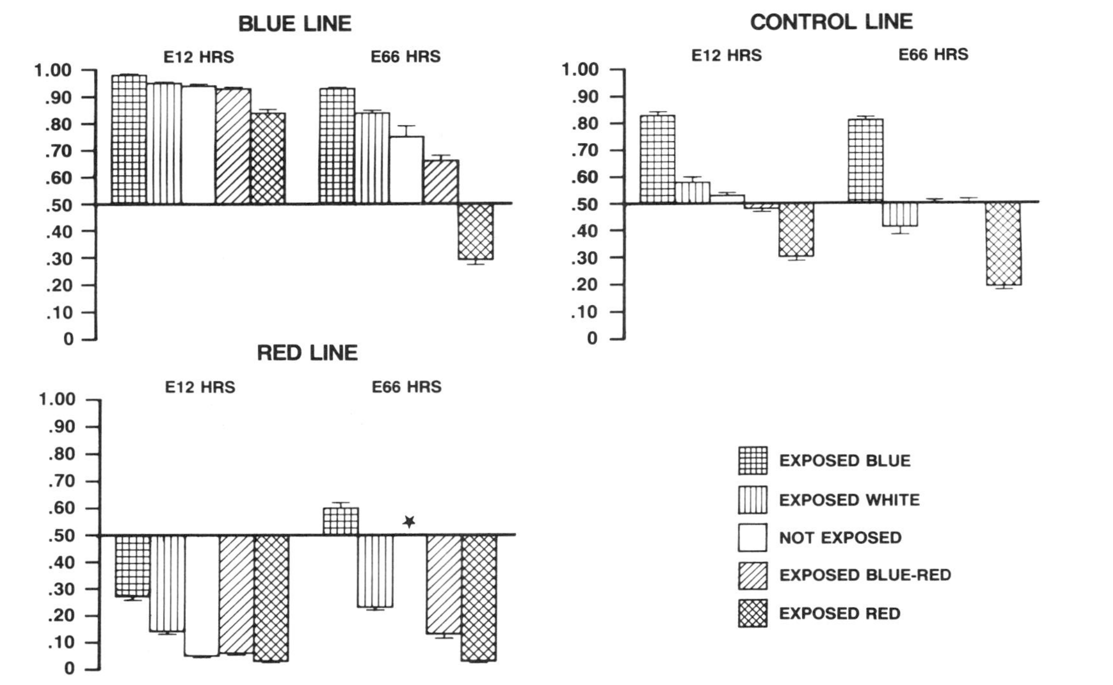

FIG. 4.20 Learning from 12 hr and 66 hr imprinting in genetically selected (RL and BL) and genetic control (CL) subjects at Generation 20. Conditions included exposures to Blue, White, Blue and Red, Red, and no exposure. All performances were tested with B/R at 27 hr and again at 100 hr of mean age. Note the differences between exposure length effects in selected and control subjects. Asterisk indicates that no subject completed 14 trials (in the not-exposed RL group). For more detail see Kovach, 1984a.

returned to the apparatus and tested again at 100 hr of age after an additional 54 hr (a total of 66 hr) exposure. Exposure was distributed evenly throughout the 4 days, 12 hr on 1st day and 18 hr on each of the next 3 days. Subjects were kept in darkness in the exposure apparatus during the remaining 12 hr on 1st and 6 hr on each of the next 3 days. Conditions included exposures to Blue, Red, or White, or to simultaneously presented Blue-and-Red. Control samples were tested from each genetic group without systematic exposure to stimuli. Food and water were available to all subjects from 2nd day on. Figure 4.20 summarizes the data.

As expected (see Fig. 4.14), selected subjects again gained less from 12 hr exposure than did the unselected controls. The latter exhibited no or only marginal gain from the prolonged exposure, whereas selected subjects continued to gain after the initial 12 hr of exposure, and their overall gain was at least as much as was the gain of the genetic controls. Clearly, the initially strong unconditional preferences were overcome in selected subjects by the prolonged experiences, and learning was not restricted to the initial 12 hr as it appeared to be from the control data alone.

The prolonged exposure to White (EW66) was not neutral as expected, but resulted in either regression of performances or shift toward red. Again, these effects may be related to the data summarized in Fig. 4.9. Also, the changes from simultaneous exposures to Blue and Red in selected subjects suggested no differentiation of attention and learning by genetic preference of colors. To test further whether the unconditional preference or avoidance of a color may be related to the ease with which it is learned, the Blue, Red, and Blue-and-Red exposures were redone in additional samples and tested with pairs of colors other than B/R (Fig. 4.21).

The data indicate that genetically selected subjects learn about as well from 12 hr exposure to colors as do unselected control subjects. Learning was detectable after both 12 hr and 66 hr exposures, and after exposure to initially preferred as well as initially unpreferred colors. Simultaneous exposures to Blue and Red resulted in about equal learning of both colors, regardless of differences in initial preferences of Blue or Red. When tested with Blue versus Yellow (B/Y) subjects exposed to Blue and Red exhibited as good or better learning than did the subjects exposed to Blue alone. When tested with Yellow versus Red (Y/R), subjects responded comparably to those exposed to Red alone. Clearly, subjects from both genetic lines learned both Blue and Red from the joint exposure to Blue and Red.

These data suggest that instead of serving as "templates" for learning, unconditional stimulus preferences may canalize early learning (1) by preferential responses and resulting selective exposures to the preferred stimuli, and (2) by facilitation and inhibition of the expression of congruent and conflicting

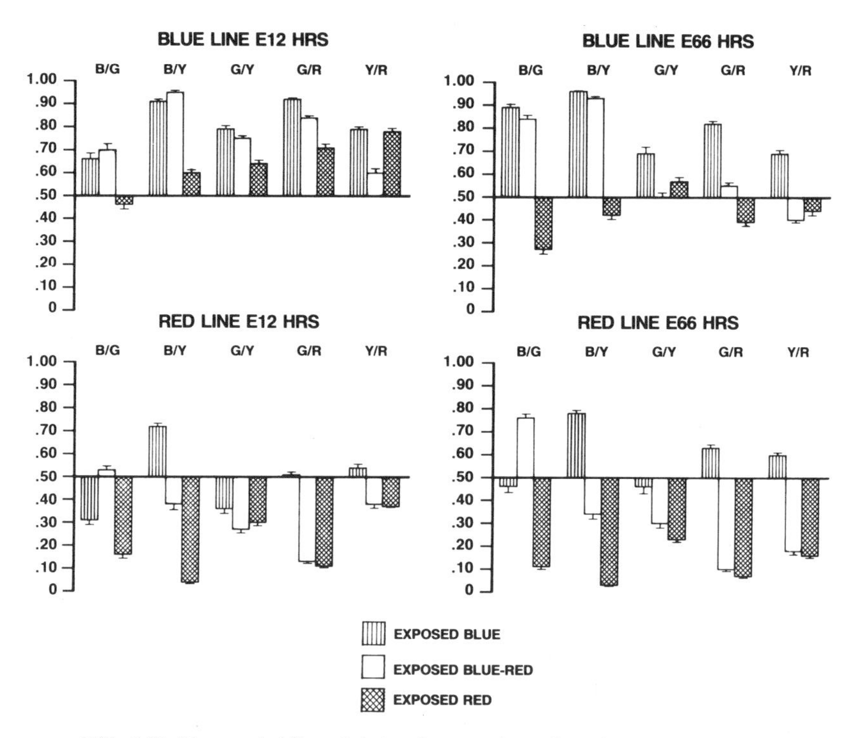

FIG. 4.21 Mean probability of choices between five color pairs in samples from the lines that were selected for Blue (BL) and Red (RL) preferences after 12 hr and 66 hr exposures to Blue, Blue and Red, or Red. For more detailed information see Kovach, 1985.

preferences acquired by exposures to stimuli. Neither process seems to impose rigid limitations on behavioral development, as revealed by the outcome of prolonged exposures to unpreferred stimuli that overcame initially very strong genetic biases in preferences. Rather, the process seems to add up to an intriguingly simple yet potentially very effective mechanism by which natural selection may canalize the ontogenetic development of behavior in ways that are genetically buffered yet highly plastic and adaptive to unpredictable environmental demands. The data suggest joint operation of genetic (Waddington, 1942, 1957) and behavioral (Holt, 1931; Janet, 1907; Murphy, 1947) canalization of development (for further discussion see Kovach, 1984a, 1984b).

The Neural Engram: Comparing the Mediation of Gene Effects and Environment Effects

The data described so far indicate that the quail responds exceptionally well to genetic selection by initial visual preferences, and that some of these preferences may be modified equally well by perceptual imprinting. With this information on hand, we now return to the major theme raised in the introduction—to the behavior genetic search for an engram. The following assumptions guide the related considerations in this section. Given appropriate analytic techniques, comparing two overtly identical but genetically and experientally different behavioral phenotypes (one unconditional and genetically determined, the other acquired due to prior experience administered to a genotype that otherwise would not exhibit the preference) should yield information about similarities and differences between the ways gene effects and environment effects are represented as information in the brain. The related model of comparisons appears in Fig. 4.22.

This model and its terminology bring to mind the traditional phenotypephenocopy distinction of physiological genetics. But the similarity is mostly heuristic. It implies neither simple genetic determination of the behavioral "phenotypes" nor necessarily shared processes in the mediation of the overtly identical but genetically and experientially different "phenocopies." Rather, the implication is that environment effects may mimic gene effects in a behavior, and that comparing the overtly identical "phenotypes and phenocopies" may teach us something about the separate and interactive mediation of gene effects and environment effects in the quail's preferences.

Figure 4.23 illustrates behavioral data that were collected with the help of the model in Fig. 4.22. The B/R data in the upper diagram indicate that imprinting genetic control subjects to Blue or Red results in color preferences that mimicked to a large extent, though by no means completely, the unconditional color preferences of selected subjects. The lower diagram of Fig. 4.23 shows that the acquired Blue and Red preferences transferred more to other colors yet generalize less readily than do the genetically determined preferences (compare especially the B/G, Y/R, and G/Y performances). Comparable, although somewhat less pronounced, differences in the transfer and generalization of preferences were observed in the naive and reciprocally imprinted performances subjects from the genetic lines that were selected for unconditional preferences (see the related E66 hr performances in Figs. 4.20 and 4.21).

Overall, these data indicate some interesting differences between the behavioral mediation of genetic and environmental influences that need further examination. But the most important implication of these data is that environment effects may copy gene effects in a behavior, and thus the model in Fig. 4.22 may be used in the behavioral and biological search for the engram that mediates the behavior. We hope to use this model in the search for similarities and

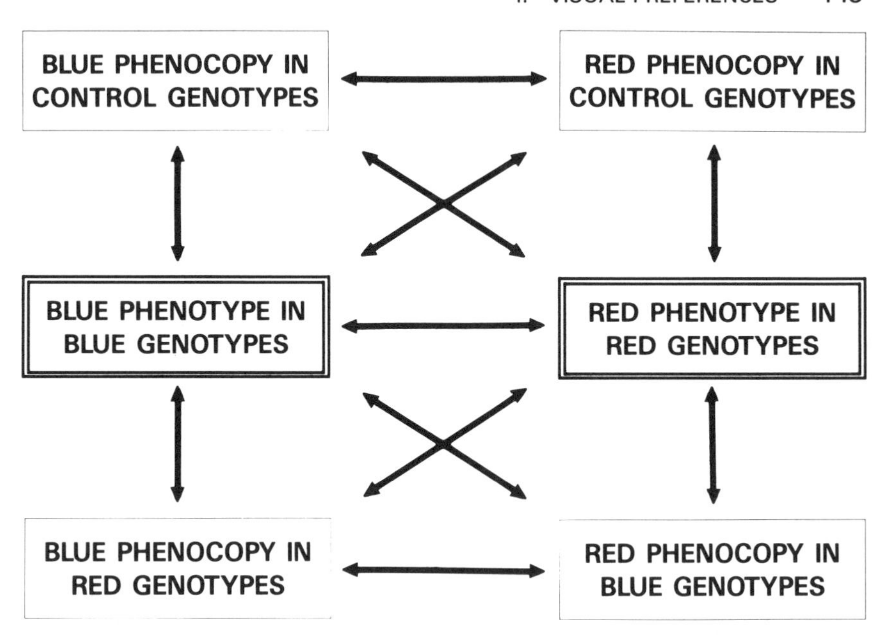

FIG. 4.22 An experimental paradigm for studying the behavioral and neural mediation of gene effects, environment effects, and gene-environment interaction in the quail's visual preferences. The term *phenotype* refers to preferences resulting from genetic selection in experimentally naive subjects tested at a specified age. The term *phenocopy* refers to similar preferences tested at the same age after exposure to colors in genotypes that would not otherwise exhibit the preference. Arrows indicate comparisons (for behavioral or neurobiological indicators).

differences between the neural representation of genetically determined and acquired stimulus preferences in the quail, and for probing specific interactions and the joint expression of gene effects and environment effects in the manifest preference behaviors.

CONCLUSION

Most researchers agree that the nature-nurture distinction is useless for explaining behavior. Unfortunately, this consensus is not based on a firm foundation of knowledge about the ways genes and environments interact in the development of behavior, and echoes of the traditional nature-nurture controversy continue to reverberate in persisting disagreements over the roles of inheritance and experience in behavior (see, for example, Eysenck, 1974; Hirsch, 1978; Jensen,

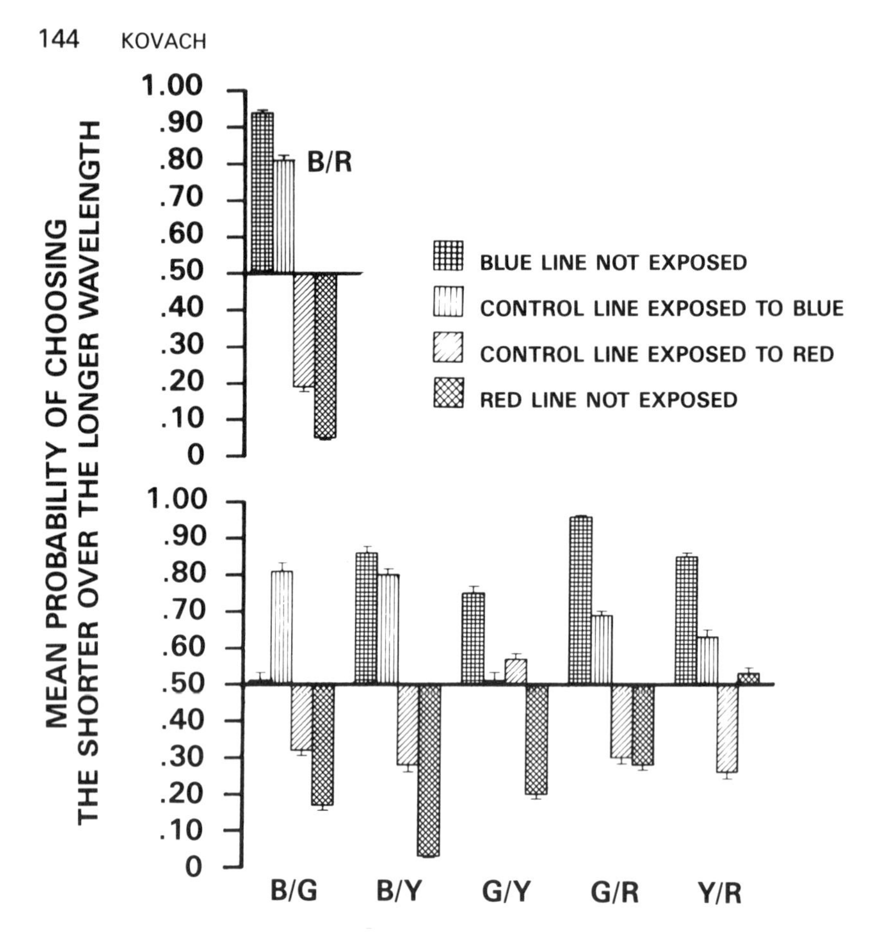

FIG. 4.23 Differences between the generalization and transfer of phenotypically similar but genotypically and environmentally different color preferences, in genetically selected naive and in imprinted genetic control subjects. Imprinting lasted for 66 hr and subjects were tested on the 5th posthatch day. Note the similar trends in the performances of genetically selected subjects in Figs. 4.20 and 4.21 that were reciprocally imprinted to the genetically unpreferred colors.

1970; Lewontin, Rose, & Kamin, 1984; Skinner, 1971; Wahlsten, 1979; Wilson, 1975). The many "overstated generalizations" emphasized in the introductory quotation of this chapter are still with us, and they still serve little more than the flights of fancy of ideologues and would-be engineers of human nature. Behavioral geneticists cannot overlook this fact, not in the apocalyptic shadows of evil perpetrated in our century in the name of the ideological pronouncements about behavior and inheritance. We should not overlook the fondness of modern ideologists for "scientific" justification of even the most outrageous claims and deeds. Only an uncompromising commitment to an incessant updating and continuous clarification of the meaning and limits of scientific concepts, facts, and procedures will cool this insalubrious attraction. In this regard, the gist and message of the work described in this book and chapter are these: Behavioral genetics is not a science of human nature, nor a way to biological or social engineering. The concepts and procedures of behavioral genetics are but tools in a fundamentally multilayered scientific search for processes and mechanisms that may explain behavior.

In the title and introduction of this chapter, I emphasized that (1) behavioral genetics is an integral part of behavioral sciences, and that (2) one of its coordinating paradigms is the postulate of the neural engram. In this concluding section, I overview the roundabout ways and threads that tie my work on imprinting to major issues of behavioral sciences, including the search for the basic mechanisms of coding, storing, and processing stimulus information by the brain.

The early learning associated with avian imprinting and song learning lends itself exceptionally well to experimentation in the search for universal mechanisms of learning and memory. For this reason, this behavior has attracted a sustained interest from ethologists, comparative psychologists, and developmental psychobiologists (see Hess, 1973; Hinde & Stevenson-Hinde, 1972; Immelmann, 1972; Marler & Peters, 1977; Nottebohm, 1970). The present data highlight that the unconditional preferences that influence this learning are genetically determined and may represent mechanisms of evolutionary preadaptation of behavioral development. The data examined in this chapter indicate that the stimulus preference characteristics of a population may readily and quickly be changed as a result of artificial selection. Preference data from crosses of different genetic lines indicated additive gene effects and suggested relatively simple genetic determination of centrally mediated color preferences. The stronger the genetically influenced preference of a color, regardless of wavelength, the stronger the unconditional preference of a grated over a dotted achromatic pattern, a brighter over a dimmer stimulus, and a flickering over a nonflickering stimulus. The implicit linear to U-shaped relationship suggests genetic influences in a small yet consistent stimulus-general preference component (see Kovach, 1983a, 1983b, 1983c, and Kovach & Wilson, 1981). Choices tested with composite stimuli of colors, patterns, and flicker indicate additive and dominance effects in the episodic interaction between stimuli of different unconditional preference values.

Although imprinting to colors modified color choices, imprinting to patterns did not modify pattern choices. Joint imprinting exposures to colors and patterns likewise modified only the choices between colors. Whether the lack of learning a pattern by perceptual exposure to it has an adaptive significance of its own or is but a matter of functional maturation of the quail's nervous system, or both, remains to be determined. To deal with this and related issues, an experiment has been initiated in which quail are bidirectionally selected for unconditional preferences of two achromatic patterns. Response to selection in this experiment suggests that, although the genetic control of stimulus preferences is most likely a widespread phenomenon that encompasses all visual channels, there may be large variation among the responses to selection for preferences of different stimulus parameters.

No firm conclusions can be made as yet about the outcome of the perceptual learning selection study, but data point to interesting research possibilities. These data suggest possible discontinuities between coding of perceptual information in learning on the one hand, and the formation of S-R connections that express such learning (and have been traditionally considered the fundamental component of learning) on the other hand.

As regards the behavior-canalizing roles of the observed genetic influences, the data suggest directive pull of attention and responses by genetically preadapted stimulus preferences. The resulting selective exposure to stimuli (which is to be expected in natural situations) implies strengthening of the initial preferences and learning of other associated stimuli.

The genetically preadapted unconditional preferences were found to inhibit the expression of conflicting and facilitated the expression of congruent learning. Interestingly, unconditional preferences did not facilitate or inhibit the learning process itself, at least within the perceptual channel of colors. The emerging picture does not fit the currently prevalent idea of perceptual templates for early learning. Rather, the developmental control of perceptual learning by unconditional preferences adds up to a set of simple yet potentially very effective mechanisms for evolutionary preadaptation and ontogenetic canalization of behavioral development (see Kovach, 1984b).

Does the observed genetic control of unconditional stimulus preferences and expression of learning effects represent species-general mechanisms of genetic and ontogenetic canalization of behavioral development? Might the behavior-canalizing roles of unconditional preferences in the quail help us in understanding the place of the genotype in psychological theory? And, more important, may the procedures and data described in this chapter help us to identify the neurobiological attributes of a species-general code of neural information that is relatable to the interactive influences of genotypes and learning? It is too early to tell. However, the separate and joint genetic and environmental manipulations of early preference behaviors and perceptual imprinting in the Japanese quail do seem to provide an effective tool for probing these and such issues.

ACKNOWLEDGMENTS

This work was supported by Grant 2 HD06770-10 from the National Institute of Child Health and Human Development, by Research Career Development Award 5-K02MH-20140-12 from the National Institute of Mental Health, and by the Menninger Foundation.

The author is grateful to Dr. Lolafaye Coyne for help in matters of statistical analysis, and to Mr. Gregory Wilson for assistance in experimentation and data processing.

REFERENCES

Abplanalp, H. (1967). Genetic studies with Japanese quail. *Zuechter, 37,* 99–104.

Adkins, E. K. (1979). Effect of embryonic treatment with estradiol or testosterone on sexual differentiation of the quail brain. Critical period and dose response relationships. *Neuroendocrinology, 29,* 178–185.

Adler, J. (1973). Chemotaxis in Escherichia coli. In J. Perez-Miravete (Publication Coordinator), *Behavior of microorganisms* (pp. 1–15). Oxford: Plenum.

Adler, J., Hazelbauer, G. L., & Dahl, M. M. (1973). Chemotaxis towards sugars in Escherichia coli. *Journal of Bacteriology, 115,* 824–847.

Abplanalp, H. (1967). Genetic studies with Japanese quail. *Zuechter, 37,* 99–104.

Annett, M. (1964). A model for the inheritance of handedness and cerebral dominance. *Nature, 204,* 59–60.

Annett, M. (1967). The binomial distribution of right, mixed, and left handedness. *Quarterly Journal of Experimental Psychology, 19,* 327–333.

Annett, M. (1978). Genetic and nongenetic influences on handedness. *Behavior Genetics, 8*(3), 227–249.

Baerends, G. P. (1950). Specialization of organs and movements with a releasing function. *Symposia of the Society for Experimental Biology, 4,* 337–360.

Bammi, R. K., Shoffner, R. N., & Haiden, G. J. (1966). Nonrandom association of somatic chromosomes in the chicken-coturnix quail hybrid and the parental species. *Canadian Journal of Genetics and Cytology, 8,* 537–543.

Bateson, P. P. G. (1964). Relation between conspicuousness of stimuli and their effectiveness in the imprinting situation. *Journal of Comparative and Physiological Psychology, 58,* 407–411.

Bateson, P. P. G. (1978a). Early experience and sexual preference. In J. B. Haldrin (Ed.), *Biological determinants of sexual behavior* (pp. 29–53). London: Wiley.

Bateson, P. P. G. (1978b). Sexual imprinting and optimal outbreeding. *Nature, 273,* 649–660.

Becker, W. A., & Bearse, G. E. (1973). Selection for high and low percentages of chicken eggs with blood spots. *British Poultry Science, 14,* 31–47.

Bentley, D. R. (1971). Genetic control of an insect neuronal network. *Science, 174,* 1139–1141.

Bentley, D. R., & Hoy, R. R. (1972). Genetic control of the neuronal network generating cricket (*Teleogryllus gryllus*) song patterns. *Animal Behaviour, 20,* 478–492.

Benzer, S. (1973). Genetic dissection of behavior. *Scientific American, 229*(6), 24–37.

Bernon, D. E., & Siegel, P. B. (1981). Fertility of chickens from lines divergently selected for mating frequency. *Poultry Sciences, 60,* 45–48.

Bloch, H. S., & Maturana, H. (1971). Oil droplet distribution and color discrimination in the pigeon. *Nature (London) New Biology, 234,* 284–285.

Blohowiak, C. C., & Siegel, P. B. (1983). Plumage phenotypes and mate preferences in Japanese quail 2. Sexual imprinting. *Behavioural Processes, 8,* 225–275.

Bowmaker, J. K. (1979). Visual pigments and oil droplets in the pigeon retina, as measured by microspectrophotometry, and their relationship to spectral sensitivity. In A. M. Granda & J. H. Maxwell (Eds.), *Neural mechanisms of behavior in the pigeon.* New York: Plenum.

Brenner, S. (1974). The genetics of *Caenorhabditis elegans*. *Genetics, 77,* 91–94.

Caspari, E. W. (1977). Genetic mechanisms and behavior. In A. Oliverio (Ed.), *Genetics, environment ad intelligence* (pp. 3–22). Elsevier/North-Holland: Biomedical Press.

Chamberlain, H. D. (1928). The inheritance of left handedness. *Journal of Heredity, 19,* 557–559.

Collins, R. L. (1968). On the inheritance of handedness. I. Laterality in inbred mice. *Journal of Heredity, 59*(1), 9–12.

Collins, R. L. (1977). Origins of the sense of asymmetry: Mendelian and nonMendelian models of inheritance. *Annals of the New York Academy of Sciences, 299,* 283–305.

Coulombre, A. J. (1955). Correlation of structural and biochemical changes in the developing retina of the chick. *American Journal of Anatomy, 96,* 153–187.

Cunningham, D. L., & Siegel, P. B. (1978). Response to bidirectional and reverse selection for mating behavior in Japanese quail (*Coturnix coturnix japonica). Behavior Genetics, 8,* 387–397.

Dobzhansky, T., & Spassky, B. (1967). Effects of selection and migration on geotactic and phototactic behaviour of *Drosophila* I. *Proceedings of the Royal Society of London Series B–Biological Sciences, 168,* 27.

Dobzhansky, T., & Spassky, B. (1967). Effects of selection and migration on geotactic and phototactic behavior of *Drosophila.* I. *Proceedings of the Royal Society of London, Series B.* 168, 27.

Donner, K. O. (1960). On the effect of the coloured oil droplets on the spectral sensitivity of the avian retina. In *Proceedings of the 12th Ornithological Congress* (Vol. 2, pp. 167–172). Helsinki: Akuteeminen Kirgakanpta Akademiska Bokhandeln.

Dudai, Y., Jan, Y. N., Byers, D., Quinn, W. G., & Benzer, S. (1976). Dunce, a mutant of Drosophila deficient in learning. *Proceedings of the National Academy of Sciences (USA), 73*(5), 1684–1688.

Eysenck, H. J. (1974). *The inequality of man.* London: Temple Smith.

Falconer, D. S. (1960). *Introduction to quantatitive genetics.* Edinburgh: Oliver & Boyd.

Fischer, G. J. (1969). Heritability in the following response of white leghorns. *The Journal of Genetic Psychology, 114,* 215–217.

Fodor, J. A. (1983). *Representations: Philosophical essays on the foundations of cognitive science.* Cambridge, MA: MIT Press.

Fuller, J. L., & Thompson, W. R. (1960). *Behavior genetics.* New York: Wiley.

Gallagher, J. E. (1977). Sexual imprinting: A sensitive period in the Japanese quail (*Coturnix coturnix japonica). Journal of Comparative and Physiological Psychology, 91,* 72–78.

Gallagher, J. E. (1978). Sexual imprinting: Variations in the persistence of mate preference due to difference in stimulus quality on Japanese quail (*Coturnix coturnix japonica). Behavioral Biology, 22*(4), 559–564.

Gibson, J. J., & Gibson, E. J. (1955). What is learned in perceptual learning? A reply to professor Postman. *Psychological Review, 62,* 447–450.

Ginsburg, B. E. (1958). Genetics as a tool in the study of behavior. *Perspectives in Biology and Medicine, 1,* 397–424.

Goldowitz, D., & Mullen, R. J. (1980). Weaver mutant granule cell defect expressed in chimeric mice. *Neuroscience Abstracts, 6,* 743.

Graves, H. B., & Siegel, P. P. (1968). Chicks' response to an imprinting stimulus: Heterosis and evolution. *Science, 160,* 329–330.

Graves, H. B., & Siegel, P. P. (1969). Bidirectional selection for response of *Gallus gallus* chicks to an imprinting situation. *Animal Behaviour, 17,* 683–691.

Hachisuka, M. (1931). On the eggs of Japanese quail. *Bulletin of British Ornithologists' Club, 52,* 37–38.

Hailman, J. P. (1964). Coding of the colour preference of the gull chick. *Nature (London), 204,* 710.

Hay, D. A. (1973). Genotype-environmental interaction in the activity and preening of *Drosophila melanogaster. Theoretical and Applied Genetics, 43,* 291–297.

Henderson, N. D. (1979). Adaptive significance of animal behavior: The role of gene environment interaction. In J. R. Royce & L. P. Mos (Eds.), *Theoretical advances in behavior genetics* (pp. 243–299). Alphen aan den Rijn, The Netherlands: Sijthoof & Noordhoff.

Hess, E. (1959). Imprinting. *Science, 130,* 133–141.

Hess, E. (1973). *Imprinting: Early experience and the developmental psychobiology of attachment.* New York: Van Nostrand & Reinhold.

Hinde, R. A., & Stevenson-Hinde, J. (Eds.). (1972). *Constraints on learning.* London: Academic Press.

Hirsch, J. (1962). Individual differences in behavior and their genetic basis. In E. Bliss (Ed.), *Roots of behavior* (p. 3–23). New York: Paul B. Hoeber.

Hirsch, J. (1978). Evidence for equality: Genetic diversity and social organization. In W. Feinberg (Ed.), *Equality and social policy* (pp. 143–162). Chicago: University of Illinois Press.

Hirsch, J., & Boudreau, J. C. (1958). Studies in experimental behavior genetics: The heritability of phototaxis in a population of *Drosophila melanogaster. Journal of Comparative and Physiological Psychology, 51,* 647–651.

Hirsch, J., & Tryon, R. (1956). Mass screening and reliable individual measurement in the experimental behavior genetics of lower organisms. *Psychological Bulletin, 53*(5), 402–410.

Holt, E. B. (1931). *Animal drive and learning process: An essay toward radical empiricism.* New York: Holt.

Hotta, Y., & Benzer, S. (1970). Genetic dissection of the *Drosophila* nervous system by means of mosaics. *Proceedings of the National Academy of Sciences* (USA), *67,* 1156–1163.

Hoy, R. R., Hahn, J., & Paul, R. C. (1977). Hybrid cricket auditory behavior: Evidence for genetic coupling in animal communication. *Science, 195,* 82–83.

Immelmann, K. (1972). Sexual and other long-term aspects of imprinting in birds and other species. In D. S. Lehrman, R. A. Hinde, & E. Shaw (Eds.), *Advances in the study of behaviour* (Vol. 4, pp. 147–174). New York: Academic Press.

Iton, L. E. (1966). Inbreeding depression in the Japanese quail (*Coturnix coturnix japonica*) in three different environments. *Poultry Science, 46,* 1275.

Jacob, F., & Monod, J. (1961). Genetic regulatory mechanisms in the synthesis of proteins. *Journal of Molecular Biology, 3,* 318–356.

Jacobs, G. H. (1981). *Comparative color vision.* New York: Academic Press.

Janet, P. (1907). *The major symptoms of hysteria.* New York: Macmillan.

Jensen, A. R. (1970). Race and the genetics of intelligence: A reply to Lewontin. *Bulletin of the Atomic Scientists, 26,* 17–23.

Joffe, J. M. (1969). *Prenatal determinants of behavior.* New York: Pergamon Press.

Jones, J. M., Maloney, M. A., & Gilbreath, J. C. (1964). Size, shape, and color pattern as criteria for identifying *Coturnix* eggs. *Poultry Science, 43,* 1292–1294.

Jordan, H. E. (1914). Hereditary left-handedness with a note on twinning. *Journal of Genetics, 4,* 67–81.

Kety, S. S., Rosenthal, D., Wender, P. H., Schulsinger, F., & Jacobsen, B. (1978). The biologic and adoptive families of adopted individuals who became schizophrenic: Prevalence of mental illness and other characteristics. In L. C. Wynne, S. Cromwell, & S. Matthysse (Eds.), *The nature of schizophrenia.* New York: Wiley.

King-Smith, P. E. (1969). Absorption spectra and function of the colored oil drops in the pigeon retina. *Vision Research, 9,* 1391–1399.

Konishi, T. (1965). Developmental studies on the retinal oil globules in Japanese quail (*Coturnix coturnix japonica*). *Zoological Magazine (Dobutsugaku Zasshi), 74,* 119–131.

Kovach, J. K. (1974). Early color preferences in the coturnix quail. *Journal of Comparative and Physiological Psychology, 87*(6), 1049–1060.

Kovach, J. K. (1977). Binomial assessment of behavioral phenotypic variations: Constancy of choices, trial effects, and social interaction effects in mass-screened color preferences of quail chicks (*Coturnix coturnix japonica*). *Journal of Comparative and Physiological Psychology, 91*(4), 851–857.

Kovach, J. K. (1979). Genetic influences and genotype-environment interactions in perceptual imprinting. *Behaviour, 68,* 31–60.

Kovach, J. K. (1980). Mendelian units of inheritance control color preferences in quail chicks (*Coturnix coturnix japonica*). *Science, 207,* 549–551.

Kovach, J. K. (1983a). Constitutional biases in early perceptual learning. I. Preferences between colors, patterns, and composite stimuli of colors and patterns in genetically manipulated and imprinted quail chicks (*C. coturnix japonica*). *Journal of Comparative Psychology, 97*(3), 226–239.

Kovach, J. K. (1983b). Constitutional biases in early perceptual learning. II. Visual preferences in artificially selected, visually naive, and imprinted quail chicks (*C. coturnix japonica*). *Journal of Comparative Psychology, 97*(3), 240–248.

Kovach, J. K. (1983c). Perceptual imprinting: Genetic influences and genotype-environment interactions. *Developmental Psychobiology, 16*(5), 413–422.

Kovach, J. K. (1983d). Perceptual imprinting: Genetically variable response tendencies, selective learning, and the phenotypic expression of colour and pattern preferences in quail chicks (*C. coturnix japonica*). *Behaviour, 86,* 72–88.

Kovach, J. K. (1984a). The genetic context of early learning: Canalization by unconditional stimulus preferences. *Learning and Motivation, 15,* 394–416.

Kovach, J. K. (1984b). Do genes canalize behavior? And if so how? *Proceedings of Twenty-third International Congress of Psychology* (pp. 2–7). Acapulco, Mexico.

Kovach, J. K. (1985). Constitutional biases in early perceptual learning. III. Similarities and differences between artificially selected and imprinted color preferences in quail chicks (*C. coturnix japonica*). *Journal of Comparative Psychology, 99*(1), 35–46.

Kovach, J. K., & Wilson, G. C. (1981). Behaviour and pleiotropy: Generalization of gene effects in the colour preferences of Japanese quail chicks (*C. coturnix japonica*). *Animal Behaviour, 29*(3), 746–759.

Kovach, J. K., Wilson, G. C, & O'Connor, T. (1976). On the retinal mediation of genetic influences in color preferences of Japanese quail. *Journal of Comparative and Physiological Psychology, 90*(12), 1144–1151.

Kovach, J. K., Yeatman, F. R., & Wilson, G. C. (1981). Perception or preference? Mediation of gene effects in the colour choices of naive quail chicks (*C. coturnix japonica*). *Animal Behaviour, 29*(3), 760–770.

Kulenkamp. A. W. (1967). *The effects of intensive inbreeding on various traits in the Japanese quail (C. coturnix japonica).* Unpublished doctoral dissertation, Washington State University.

Lauber, J. K. (1964). Sex-linked albinism in the Japanese quail. *Science, 146,* 948–950.

Levy, J., & Nagylaki, T. (1972). A model for the genetics of handedness. *Genetics, 72,* 117–128.

Lewontin, R. C., Rose, S., & Kamin, L. (1984). *Not in our genes: Biology, ideology, and human nature.* New York: Pantheon.

Lorenz, K. (1935). Der Kumpan in der Umwelt des Vogels. *Journal für Ornithologie, 83,* 137–213.

Lucotte, G. (1975). Polychromatisme de la coquille de l'oeuf chez la Caille domestique *(Coturnix coturnix japonica)* III. Variabilite et mode de transmission des phenotypes a l'interieur de la forme dominante. *Comptes rendus des seances de la Societe de Biologie et des ses filiales, 169*(1), 30.

Maeda, Y., Hiromi, I., Tsutomu, H., & Manjiro, T. (1978). Protein polymorphism in the inbred strains of Japanese quail: Test of single gene heterosis. *Japanese Journal of Zootechnical Science, 49*(8), 607–613.

Manning, A. (1976). The place of genetics in the study of behaviour. In P. P. G. Bateson & R. A. Hinde (Eds.), *Growing points in ethology* (pp 327–343). Cambridge: Cambridge University Press.

Manosevitz, M., Lindzey, G., & Thiessen, D. D. (Eds.) (1969). *Behavioral genetics methods and research.* New York: Appleton-Century-Crofts.

Marler, P., & Peters, S. (1977). Selective vocal learning in the sparrow. *Science, 198,* 519–521.

Marr, D. (1982). *Vision.* San Francisco: W. H. Freeman.

Mather, K., & Jinks, J. L. (1971). *Biometrical genetics: The study of continuous variation* (2nd ed.). London: Chapman & Hall.

Mayr, E. (1965). *Animal species and evolution.* Cambridge, MA: Belknap Press.

McConnell, J. (1962). Memory transfer through cannibalism in planarians. *Neuropsychiatry, 3,* 42–48.

McGuire, T., & Hirsch, J. (1977). Behavior-genetic analysis of Phormia regina: Conditioning, reliable individual differences and selection. *Proceedings of the National Academy of Sciences* (USA), *75*(11), 5193–5197.

Meyer, D. G. (1971). The effect of dietary carotenoid deprivation on avian retinal oil droplets. *Opthalmology Research, 2,* 104–109.

Miller, C. A., & Benzer, S. (1984). Monoclonal antibody cross-reactions between *Drosophila* and human brain. *Proceedings of the National Academy of Sciences* (USA), *80,* 7641–7645.

Mullen, R. J., & Whitten, W. K. (1971). Relationship of genotype and degree of chimerism in coat color to sex ratios and gametogenesis in chimeric mice. *Journal of Experimental Zoology, 178,* 165–176.

Murphy, G. (1947). *Personality.* New York: Harper.

Murphy, G., & Kovach, J. K. (1972). *Historical introduction to modern psychology.* New York: Harcourt Brace Javonovich.

Nottebohm, F. (1970). Ontogeny of bird song. *Science, 167,* 950–956.

Oliverio, A., Eleftheriou, B. E., & Bailey, D. W. (1973). A gene influencing active avoidance performance in mice. *Physiology & Behavior, 11,* 497–501.

Pedler, C., & Boyle, M. (1969). Multiple oil droplets in the photoreceptors of the pigeon. *Vision Research, 9,* 525–528.

Poole, H. K. (1964). Eggshell pigmentation in Japanese quail. *Journal of Heredity, 55,* 136–138.

Poole, H. K. (1965). Spectrophometric identification of egg shell pigments and timing of superficial pigment deposition in Japanese quail. *Proceedings of the Society for Experimental Biology and Medicine, 119,* 547–551.

Poole, H. K. (1967). A microscopic study of uterine eggshell pigment in Japanese quail. *Journal of Heredity, 58,* 200–203.

Quinn, W. G., Harris, W. A., & Benzer, S. (1974). Conditioned behavior in *Drosophila melanogaster*. *Proceedings of the National Academy of Sciences* (USA), *71*(3), 708–712.

Rajecki, D. W., Lamb, M. E., & Obmascher, P. (1978). Toward a general theory of infantile attachment: A comparative review of aspects of the social bond. *Behavioural and Brain Sciences, 3*, 417–464.

Rife, D. C. (1940). Handedness with special reference to twins. *Genetics, 25*, 178–186.

Roaf, H. E. (1933). Colour vision. *Physiological Reviews, 13*, 43–79.

Rosenthal, D. (1970). Genetic research in the schizophrenic syndrome. In R. Cancro (Ed.), *The schizophrenic reactions: A critique of the concept, hospital treatment and current research: Proceedings of the Menninger Foundation conference on the schizophrenic syndrome* (pp. 245–258). New York: Brunner/Mazel.

Rothenbuhler, W. C. (1964). Behavior genetics of nest cleaning on honey-bees. IV. Responses of F_1 and backcross generations to disease-killed brood. *American Zoologist, 4*, 111–123.

Rubel, E. W. (1970). Effects of early experience on fear behaviour of *Coturnix coturnix*. *Animal Behaviour, 18*, 427–433.

Schaller, G. B., & Emlen, J. T. (1962). The ontogeny of avoidance behavior in some precocial birds. *Animal Behaviour, 10*, 370–381.

Searle, L. V. (1949). The organization of hereditary maze-brightness and maze dullness. *Genetic Psychology Monographs, 39*, 279–325.

Sefton, A. E., & Siegel, P. B. (1975). Selection for mating ability in Japanese quail. *Poultry Science, 54*, 788–794.

Shimakura, K. (1940). Notes on the genetics of the Japanese quail I. The simple, Mendelian autosomal recessive character, "brown-splashed white", of its plumage. *Japanese Journal of Genetics, 16*, 106–112. (in Japanese)

Sittman, K., & Abplanalp, H. (1965). White feathered Japanese quail. *Journal of Heredity, 56*, 220–223.

Sittman, K., Abplanalp, H., & Fraser, R. A. (1965). Inbreeding depression in Japanese quail. *Genetics, 54*, 371–379.

Skinner, B. F. (1971). *Beyond freedom and dignity*. New York: Knopf.

Sluckin, W. (1972). *Imprinting and early learning*. Chicago: Aldine.

Smith, F. V., & Templeton, W. B. (1966). Genetic aspects of the response of the domestic chick to visual stimuli. *Animal Behaviour, 14*, 291–295.

Strother, G. K. (1963). Absorption spectra of retinal oil globules in turkey, turtle and pigeon. *Experimental Cell Research, 29*, 349–355.

Thiessen, D. (1979). Biological trends in behavior genetics. In J. R. Royce & L. P. Mos (Eds.), *Theoretical advances in behavior genetics* (pp. 169–217). Alphen aan den Rijn, The Netherlands: Sijthoff & Noordhoff.

Tinbergen, N. (1951). *The study of instinct*. Oxford: Oxford University Press.

Tryon, R. C. (1940). Genetic differences in maze-learning ability in rats. *Thirty-Ninth Yearbook of the National Society for the Study of Education. Part I* (pp. 111–119). Bloomington, IL: Public School Publishing Co.

Ungar, G. (1973). Evidence for molecular coding of neural information. In H. P. Zippel (Ed.), *Memory and transfer of information* (pp. 317–341). New York: Plenum.

Uphouse, L., MacInnes, J. W., & Schlesinger, K. (1974). Role of RNA and protein in memory storage: A review. *Behavior Genetics, 4*, 29–81.

van Abeelen, J. H. F. (1975). Genetic analysis of behavioural responses to novelty in mice. *Nature, 254*, 239–241.

Vegni-Talluri, M. V., & Vegni, L. (1965). Fine resolution of the karyogram of the quail *Coturnix coturnix japonica*. *Chromosoma, 17*, 264–272.

Vom Saal, F. S., & Bronson, F. H. (1980). Sexual characteristics of adult female mice are correlated with their blood testosterone level during prenatal development. *Science, 208*, 597–599.

Waddington, C. H. (1940). The genetic control of wing development in *Drosophila*. *Journal of Genetics, 39*, 75–139.

Waddington, C. H. (1942). Canalization of development and the inheritance of acquired characters. *Nature, 150*, 563–564.

Waddington, C. H. (1957). *The strategy of the genes*. London: Allen & Unwin.

Wahlsten, D. (1979). A critique of the concepts of heritability and heredity in behavioral genetics. In J. R. Royce & L. P. Mos (Eds.), *Theoretical advances in behavior genetics* (pp. 425–481). Alphen aan den Rijn, The Netherlands: Sijthoff & Noordhoff.

Wald, G., & Zussman, H. (1938). Carotenoids of the chicken retina. *Journal of Biological Chemistry, 122*, 446–460.

Wallman, J. (1972). The role of the retinal oil droplets in the color vision of the Japanese quail. *American Zoologist, 10*, 506–507.

Ward, I. L. (1974). Sexual behavior differentiation: Prenatal hormonal and environmental control. In R. C. Friedman, R. M. Richart, & R. L. Vande Wille (Eds.), *Sex differences and behavior* (pp. 3–17). New York.

Watson, J. D. (1970). *Molecular biology of the gene* (2nd ed.). Menlo Park, CA: W. A. Benjamin.

Wilson, E. O. (1975). *Sociobiology: The new synthesis*. Cambridge, MA: Belknap Press.

Winokur, G. (1981). *Depression: The facts*. New York: Oxford University Press.

5 An Examination of Claims for Classical Conditioning as a Phenotype in the Genetic Analysis of Diptera[1]

Jeffry P. Ricker
Jerry Hirsch
Mark J. Holliday
University of Illinois at Urbana-Champaign

Mark A. Vargo
Brandeis University

An important aspect of research in behavior-genetic analysis is an examination of the nature of differences among individuals in the expression of behavior, especially of differences in learning ability. This chapter examines classical conditioning in two dipteran species, *Phormia regina* (the blow fly) and *Drosophila melanogaster* (the fruit fly). Because flies are more prolific, breed more rapidly, and are simpler biologically than those animals used traditionally to study learning, a successful analysis of learning in flies would provide a unique opportunity for the experimental study of heredity and experience. However, the study of classical conditioning in, and the genetic analysis of, these species has had difficulties associated with (a) the choice of control procedures, (b) the existence of a strong nonassociative effect (the central excitatory state) that may be confounded with conditioning, and (c) confusion about the implications of genetic analysis of populations for the ontogeny of conditioning ability in individuals. In our discussion, we consider the control procedures required to infer an association between events in a classical conditioning procedure. Then we analyze the evidence used to support claims of classical conditioning in the two

[1]Completed 9/20/84, revised 7/2/85, 11/19/85, 1/16/86, 7/17/86.

dipteran species in order to determine what evidence may be sufficient to infer conditioning so that it may then be used as a phenotype in the genetic analysis of populations.

OPERATIONAL AND THEORETICAL DEFINITIONS OF LEARNING

The scientific method involves associating differences in one event with those in another. In experimental science, one of these events is hypothesized to be a precursor (in a causal sense) of the second event and is manipulated by the experimenter to determine whether concomitant changes in the latter event occur. If changes in the two events are found to be associated, then one usually infers a causal relationship involving the changes. To detect effects of the causal event, however, measurement of the caused event must be reliable (Cronbach, 1951).

Similarly, the study of learning is concerned with changes in behavior resulting from changes in experience. Typically, its experimental study uses two types of conditioning procedure: classical and instrumental. Mackintosh (1983) distinguished theoretically between these two procedures:

> We can say that classical conditioning has occurred if the change in behaviour we record is, as a matter of fact, a consequence of the contingency between an external CS and a reinforcer; and we can say that instrumental conditioning has occurred if a change in behaviour is in fact a consequence of a contingency between that pattern of behaviour and a reinforcer. (p. 25)

Classical and instrumental conditioning usually are defined operationally, not theoretically. However, by using one or the other procedure, one takes a theoretical stand on the type of learning being studied. In this chapter, we consider only those studies using classical conditioning procedures to train flies.

The attempt to infer association from a classical conditioning experiment subsumes a number of questions, among the most important being (1) what changes in experience are to be induced; (2) what response systems are to be observed; (3) how relationships between experience and responses are to be detected. The first question considers the adequacy of control procedures for nonassociative effects. The control procedures that one chooses to study depend upon what aspects of experience are believed to be important in the development of association. The second question considers the distinction between learning and performance: a response system is needed that will reflect the learning that is to be detected. The third question overlaps partially the first two questions and, in addition, raises the problem of objective and reliable measurement of responses. That is, one must develop a procedure that is sensitive enough to detect the changes in which one is interested and to distinguish these from other changes that may occur.

Another question, important for behavior-genetic research, concerns the problem of genetic analysis with learning as the phenotype. This last question embraces all the others and raises other questions not considered in the study of learning. In this chapter, we are concerned more with an examination of claims for classical conditoning in *D. melanogaster* and *P. regina*. However, we do this with the intention that such a critique will have implications for the genetic analyses of these species. Therefore, we also discuss the types of genetic analysis appropriate for different types of evidence for learning, and the interpretations that may be made concerning the development of learning ability.

ANALYSIS OF CONTROL PROCEDURES

Learning in flies has been eagerly sought because it would provide many opportunities for fundamental biological research (Holden, 1985; McDonald, 1985). When learning has been claimed, it is usually the case that proper control procedures have not been performed. An adequate control procedure is one that keeps all extraneous variables constant, and allows only the variable of interest to change. In the present case, this would involve keeping all nonassociative effects constant and allowing to vary only the parameter that is thought to be important for association to occur. Therefore, a decision must be made about what may be this parameter. LoLordo (1979a) emphasized the subjectivity in this evaluation: "A researcher's approach to the assessment of non-associative factors will be theory-bound, i.e., it will depend upon his notions about the sorts of relationships between the [conditioned stimulus] CS and the [unconditioned stimulus] UCS which will promote the formation of an association between the two events (p. 34). In classical conditioning, it is usually believed that an important aspect of the associative process is the pairing (i.e., the temporal contiguity) between the CS and US. We agree with this and, further, consider that it is the effect of contiguity on the contingency between CS and US (i.e., on the probability that a US occurs given that the CS has or has not occurred) that is the basis of association (Rescorla, 1967; Rescorla & Wagner, 1972).

The distinction between contingency and contiguity may be thought of in terms of the "information" that the presentation of a CS (or of no CS) imparts about the probability of US presentation. For example, suppose that two different classical conditioning tests present a CS on 5 of 10 trials, with one presenting the US on every trial (i.e., 10 times), and the other presenting the US only after the presentation of CS (i.e., 5 times). In both tests, the number of pairings of CS and US (the contiguity) is identical; but, in the first test, the presentation of CS predicts the presentation of US only half of the time whereas, in the second test, the presentation of CS predicts the presentation of US perfectly. One would predict that the amount of conditioning would be greater in the latter test because, even though both tests have the same contiguity (i.e., the same number of CS-US pairings), they have different contingencies between CS and US (see LoLordo, 1979b, 1979c, for a lucid discussion of this model and its advantages and disadvantages for studying learning).

By using the notion of contingency as the cornerstone of our theoretical discussion of classical conditioning, we consider appropriate control procedures. In discussing the use of contingency to evaluate the appropriateness of control procedures, Rescorla (1967) argued, "We take as the logical criterion for an adequate control procedure that it retain as many features as possible of the experimental procedure while excluding the CS-US contingency" (p. 72). A summary of typical control procedures and the putative hypothesis tested by each one is presented in Table 5.1. Because avoidance experiments use a classical conditioning procedure to train subjects to avoid CS (i.e., avoidance is based upon the effects of a CS-US contingency; Overmier, 1979), these control procedures also apply to them.

Procedure (1), the unpaired control, changes the positive contingency in a classical conditioning procedure to a negative one by moving the US away from the CS (or vice versa). This decreases the contiguity between CS and US and, thereby, the strength of the "neural trace" of the CS when the US is presented. Eventually, the positive contingency between CS and US is eliminated when the presentation of CS is moved far enough away from the presentation of US so that the trace of CS has dissipated. Because it cannot be known at what point a positive contingency is eliminated (e.g., in conditioning a food aversion, a CS-US interval of several hours may result in an association), it would be wise to use a range of CS-US intervals and to determine whether the frequency of responses decreases with increasing intervals. That is, it is assumed that the

TABLE 5.1
Control Procedures and Putative Hypothesis Tested in Each Case

Control Procedure	*Hypothesis*
(1) Unpaired	(1) CS, US or CS-US interaction causes a nonassociative effect.
(2) CS-Alone	(2) CS causes a nonassociative effect.
(3) US-Alone	(3) US causes a nonassociative effect.
(4) (a) Replication (b) Blind Testing	(4) Situational variables cause nonassociative effects: (a) Chance or experimenter bias; (b) Experimenter bias.

strength of the trace is positively correlated with the strength of conditioning (see Mackintosh, 1983, pp. 202–210, for a discussion of the role of temporal contiguity in conditioning experiments). Rescorla (1967) argued that the unpaired procedure is not an appropriate control because it introduces a negative contingency; i.e., the animal may be learning that the presence of CS predicts that no US will be forthcoming. Rescorla advocates the use of a control procedure that has no contingency between CS and US (random or zero-contingency control procedure). If either the CS or the US is presented at any point within the intertrial interval (ITI) in a random manner, then the presence of the CS does not predict the occurrence of a US even though it is possible that pairings occur between them. This should allow a comparison of a group experiencing a positive contingency between CS and US with one experiencing no contingency. However, as Schneiderman (1973) noted, when the CS-US interval is short relative to the ITI, the unpaired and random control procedures are, for all practical purposes, identical because very few pairings (if any) will occur in the random procedure. Additionally, a random procedure will change CS-CS or US-US intervals so that, in the conditioning procedure, one must present a variety of ITIs to make it comparable to the control procedure. The unpaired procedure introduces no such complication because it changes only the CS-US interval: the parameter thought to be important for association to occur. Furthermore, Mackintosh (1973) argued that animals receiving a random procedure may be learning that the presentation of CS is independent of the presentation of US. This suggests that no control procedure containing both CS and US, even one with a contingency of zero, eliminates learning. However, these complications should not overwhelm us. For the immediate objectives of behavior-genetic analysis, it usually is necessary to show only that a given conditioning procedure has produced conditioning in a given species. At present, the goals of behavior-genetic analysis are different from those of learning theory.

Procedures (2) and (3), CS-alone and US-alone controls, respectively, remove more than the positive contingency between CS and US (Rescorla, 1967). The nonassociative effects of the two together are not held constant by presenting each separately. Therefore, we believe that control procedures (2) and (3) are inadequate for inferring association from a classical conditioning procedure (though they may be used to study responsiveness to the stimuli).

Grouped together under control procedure (4) are the last two controls, replication and blind testing. These controls are not limited, of course, to classical conditioning procedures but are, instead, requirements for an adequate experimental procedure of any kind. In conditioning experiments, they control for the nonassociative effects of specific situational variables. Situational variables are often ignored in research but may be very important in the results obtained (Rosenthal, 1976; Smith, 1970; Yeatman & Hirsch, 1971). Replication of a study (experimental and control procedures) by the same experimenter, and by different experimenters (ideally in different laboratories), controls for the subtle and ephemeral effects associated with a specific testing situation. Blind testing controls for the effects due to the desires and expectations that the experimenter may have concerning the outcome of a treatment. Automation of a procedure, both testing and scoring, is equivalent to blind testing. However, when the experimenter must be involved in either aspect of a procedure, blind testing should be performed using one of three methods: (a) procedure known by experimenter, subject unknown; (b) subject known by experimenter, procedure unknown; (c) both unknown by experimenter. The first method would involve, for example, using animals that differ genetically from each other, this genetic difference believed to affect conditioning in a particular manner (such as lines selected divergently for conditioning ability, inbred lines, or two stocks carrying different mutations). The "genetic identity" of the subjects is not known by the experimenter until after the experiment, although the procedure, whether conditioning or control, is known. Differences in the measure of conditioning (in the expected direction) between genetically different subjects would validate the conditioning procedure. Vargo and Hirsch (1985) used this method successfully to eliminate systematic bias (with a phenotype different from learning). The second method would involve withholding from the experimenter knowledge of the particular procedure (whether conditioning or control) being performed (an example of this is given below). The last method involves a combination of the first two.

Because they keep constant all possible nonassociative effects, control procedures (1) and (4) are sufficient to infer that association is occurring in a classical conditioning procedure. Of course, one may substitute a different control procedure for the unpaired control procedure as long as it is equivalent logically (i.e., it changes the contingency between CS and US from the one given to the experimental group). Until both sets of control procedures have been performed, one can conclude only that results are consistent with those to be expected when an association occurs between CS and US. Failure to reject the null hypothesis in either case allows only one conclusion: Association cannot be inferred. Because a discrimination procedure pairs the US with one CS (the CS+) but not with a second (the CS−), it is a combination of an unpaired control procedure and a conditioning procedure (Rescorla, 1967) and, therefore, meets our requirement for an adequate control procedure. However, as with the unpaired control, one cannot determine if learning involves a positive or a negative contingency between CS and US (i.e., whether conditioning is to the pairing of CS+ and US, or to the unpairing of CS− and US).

GENETIC ANALYSIS

The goals of behavior-genetic analysis include not only a behavioral analysis of individuals, but a genetic analysis of populations as well (Hirsch, 1967). Depending upon the interests of a researcher, a particular genetic analysis may take different forms. However, the ultimate (and, perhaps, unreachable) goal of genetic analysis with a behavioral phenotype is to isolate and identify in individuals primary products of genes that are associated with behavioral differences between individuals, and then to determine how these products interact with each other, with the organism, and with environmental variables in the development of behavioral differences. That is, the goals of behavior-genetic analysis involve the description of the development of an individual and its behavior, and the determination of the nature of differences in behavioral development among individuals. This is an extremely difficult enterprise even with relatively simple phenotypes and organisms (Lewin, 1984). It becomes even more difficult (if not impossible presently) with learned behaviors and with organisms as complex as flies. It was suggested above that, in a behavior-genetic analysis, it is usually enough to infer that conditioning has occurred in a given species by using a given method, and not to be concerned, at the moment, with a detailed behavioral analysis of conditioning. This is because presently the genetic analysis is of most importance in this area. However, for the genetic analysis to have meaning, beyond being an exercise in mathematics, a detailed behavioral analysis is necessary ultimately. The latter is, at this time, the domain of the learning theorist. The genetic analysis of populations is an early step in the developmental analysis of individuals (it may allow one to detect major gene effects in a population) and, therefore, should be of major concern for behavior-genetic analysis. We must emphasize that the conceptual analysis presented here is concerned only with the genetic analysis of a population using classical conditioning as a phenotype. The control procedures, and the goals of research, discussed here are applicable to this problem only and are not intended to imply anything about learning in general or research concerned with something other than genetic analysis.

Genetic analysis with conditioning as the phenotype is dependent upon the nature of the data collected; i.e., whether it is group or individual data. Learning is usually inferred from group data by observing changes in the average frequency of responses within one group, or differences among comparable groups, upon changing the contingency between CS and US. For different groups to be comparable, samples of individuals constituting different groups must be similar statistically with regard to the effects of changed contingencies (i.e., random samples from a population must be obtained). Although it may be assumed that at least some individuals have learned if learning can be inferred from group data, one cannot infer that any given individual has learned. Individual learning can be inferred only if the contingency between CS and US is changed for a given subject and concomitant changes in the number of responses to the CS are observed. However, this might prove difficult in practice because experiences in earlier tests may affect results of later tests in a given individual (i.e., the different measures of an individual may not be independent).

The form taken by a genetic analysis depends upon whether learning is inferred from group or individual data; and, with group data, whether genetic differences associated with differences in learning ability exist within the group. If such genetic differences are present, then one may perform quantitative genetic analyses with, for example, correlations between relatives (Falconer, 1981). If one can make the assumption that differences in learning ability are not associated with major gene differences within a group, but are associated with such differences between groups (such as between selected, inbred, or mutant lines), then one may perform crosses between groups and observe the distribution of learning ability in the hybrid generations. From such evidence, one may be able to infer the existence of major gene differences between groups. If learning can be inferred from the performance of an individual, then it may be worthwhile to perform mosaic analyses (Hotta & Benzer, 1970, 1972). Of course, in this last type of genetic analysis, it is the average effect of a mutation in a population that is described and not necessarily its effect in a given individual (see p. 189ff.). Because the type of genetic analysis used depends upon the distinction between inferring learning from group or individual data, it is important to be aware of the kinds of evidence appropriate for each.

However, the distinction between the two is not always appreciated. For example, Quinn, Harris, and Benzer (1974) used group data to measure avoidance by *D. melanogaster* of an odor paired with shock. Accepting, for now, their claim for learning from group data, one may infer that at least some individuals must have learned. When Quinn et al. divided this group into two subgroups on the basis of an arbitrary classification of learning (those avoiding on one test trial classified as learners, those not avoiding classified as nonlearners) and retrained the subgroups 24 hr later, a subsequent test showed no difference in avoidance behavior between them. Quinn et al. concluded that this "result suggests that the expression of learning is probabilistic in every fly" and that there is "no evidence for an 'intelligent' subset of the population" (p. 711). In other words, they claimed that all flies learn the contingency, but that there exists only a probability that this learning will be reflected in performance on any given trial. However, the results are inconclusive; therefore, the interpretation is questionable because individual learning cannot be inferred. That is, Quinn et al. did not show that changes in behavior of an individual were associated with changes in contingency between a response and reinforcer and, then, that this was associated with their arbitrary definition of individual learning. They cannot infer learning in any of those subjects classified as learners; nor can they infer the absence of learning in any of those subjects classified as nonlearners. The same criticism can be applied to Nelson (1971), Hirsch and McCauley (1977), and McGuire and Hirsch (1977). The last two studies, however, attempted to validate their classification of "good" and "poor" learners through selective breeding.

An associated problem in the attempt by Quinn et al. (1974) to show the nonexistence of individual differences in learning ability involves the reliability of their measure. As was noted above, measurement of the responses used to infer learning must be reliable (Mariath, 1985). Their instrument was designed to test groups, not individuals, for learning ability and does not have the reliability to detect individual differences with the number of trials used (Byers, 1980). Individuals cannot be reliably classified as learners and nonlearners on the basis of performance on one test trial, and the inability of those investigators to measure individual differences is understandable.

INSTINCT AND LEARNING

The following is a minor digression from the remainder of the paper, but it is important because it calls attention to a paradox that illustrates some general problems in the concepts and methodology used in behavioral research. Specifically, we find that the study of learned behavior has parallels in the field of ethology: The concept of association turns out to have no more analytic validity than does the concept of instinct used by some ethologists (e.g., Lorenz, 1950). Hebb (1953) criticized this concept of instinct: "Instinctive behaviour is what is not learned, or not determined by the environment, and so on. There must be great doubt about the unity of the factors that are identified only by exclusion" (p. 45). Hebb believed that instinct may not have validity as a singular concept because it is inferred from negative evidence. However, one may question this historical definition of instinct and argue that the concept of instinct presently is useful to describe certain aspects of behavior: movements or patterns of movement that are species-typical (Oyama, 1982). That is, we may remove the difficulty of using negative evidence to infer instinctive behavior by attributing to an observable event the term *instinct,* as has been done by Ricker and Hirsch (1985b) in a study of the "Evolution of an instinct under long-term divergent selection for geotaxis . . ." in *D. melanogaster*. By recognizing the divergent expressions of geotaxis in two selected populations to be instinctive (i.e., what was formerly species-typical has now evolved to become population-typical with opposite expression in two divergent populations [Hirsch & Erlenmeyer-Kimling, 1961]), questions were raised about correlations with reproductive fitness in laboratory habitat(s) and have resulted in further research into the adaptiveness of the behavior.

Similarly, negative evidence has been used to infer association. One does not observe directly the development of association, but only infers its existence when control procedures have excluded substantial effects of nonassociative factors. Therefore, one might also question its validity as a singular concept. For example, excitatory states are aroused in flies by the stimuli used in our conditioning procedures (see the following section). Therefore, it must be demonstrated that such states cannot account for all of the responses to CS. More generally, association is inferred from the part of the increment in responses to CS that remains after one has controlled for the parts due to sensitization, pseudoconditioning, habituation, excitation, and so forth (Médioni & Vaysse, 1975). Presently, it is not possible to remove this difficulty by redefining the term as we did with instinct, because there is no observable event that we might, by consensus, call association. If it were possible to demonstrate consistently a physical or physiological change that occurred whenever certain types of experience were presented to animals, then we might call that association. Apparently, many researchers believe this to be possible in their quest for the "engram" (Lashley, 1950; Thompson, 1983). Therefore, there is a historical parallel between the study of instinct and the study of learning that is paradoxical because these usually have been considered polar opposites: Instinctive is unlearned and learned is not instinctive.

EXCITATORY AND INHIBITORY STATES

One may be able to infer conditioning in Diptera without being concerned with measuring nonassociative effects as long as such effects are kept constant across experimental and control groups. However, there exist in Diptera nonassociative states—the central excitatory and inhibitory states (CES and CIS, respectively)—that may create difficulties in the development of an adequate conditioning procedure. Dethier, Solomon, and Turner (1965) first studied CES by observing responses to water stimulation in the blow fly, *Phormia regina*. They characterized CES in the following way: "The stimulation of a sucrose receptor either in a labellar hair or in a tarsal hair of a hungry blowfly will increase the subsequent responsiveness of the fly to water stimulation even though the fly is thoroughly water satiated, and even though it will not normally show a proboscis extension to water" (p. 311). The presentation of sucrose produces an excitatory change in the central nervous system that is greater with increasing sucrose concentrations and longer periods of food deprivation, and that decays over time (virtually gone after about 5 min). If an additional water or saline stimulation is presented between the sucrose and water stimulations, there are fewer responses to the later water stimulation indicating that the intercalated stimulation (incompletely) discharges CES.

In contrast, CIS is characterized by a reduced probability of proboscis extension to a stimulus after the animal has been presented with an "inhibiting" stimulus. Dethier, Solomon, and Turner (1968) used saline to set up a CIS. They inferred the existence of CIS from three effects of saline stimulation:

> (a) the failure of a fly to give normal maximal proboscis extension to an acceptable stimulus (e.g., sucrose) if this was closely preceded by an inhibitory stimulus; (b) the slower rate of retraction at the cessation of sucrose stimulation compared to the rate when inhibitory stimulation preceded sucrose removal; and (c) the failure of a fly to respond maximally to a behaviorally subliminal stimulus (e.g., water)

> superimposed on an induced CES when the induced CES was preceded by an inhibitory stimulus. (p. 148)

Conditioning procedures that use stimuli inducing CES or CIS may be causing nonassociative interactions among stimuli. We examine this possibility by first discussing the study of CES and its relationship with conditioning and then, in the following sections, analyzing critically some claims for classical conditioning.

In the CES procedure, distilled water (the pretest) is presented to the tarsi for 5 s. This measures the amount of water responsiveness. This is followed immediately by the presentation of sucrose for 1 s, which produces the excitatory state. An interstimulus interval of a given length (usually 45 s) follows, at the end of which a second water stimulus (the posttest) is presented for 5 s. The difference between the posttest and pretest responses is the measure of CES. In addition, the posttest discharges CES (Dethier et al., 1965) so that the sucrose stimulation from a previous trial should not affect responses on succeeding trials.

The first claim for classical conditioning of the proboscis extension in Diptera is that of Frings (1941), who presented to *Cynomia cadaverina* the odor of coumarin paired with sucrose stimulation of the tarsi. On the basis of unpublished observations on *Phormia* by Block, this study has been dismissed by Dethier (1966) who believed that the effect could be attributed to CES. Zawistowski (1984) suggested that this criticism is not valid because Frings washed the tarsi after each sucrose stimulation, which, as was shown by Dethier et al. (1965), discharges CES in *Phormia*. Because Block's study is unpublished, the nature of the criticism remains unknown. The study by Frings is unsatisfactory for other reasons. These reasons should become clear in following sections.

Nelson (1971) reported classical conditioning of the proboscis extension in *P. regina* and controlled for the effects of CES by incorporating two controls into her conditioning procedure. Because CES is virtually gone after about 5 min, Nelson used an ITI of 10 min. In addition, immediately before the presentation of the CS (CS_2), another stimulus (CS_1) was presented to discharge any CES still remaining. However, even though Nelson controlled for the effects of CES, she obtained evidence that it might be biologically correlated with conditioning. In her Experiment 7, CES was discharged by giving flies the opportunity to exercise between trials (mechanical stimulation has an effect similar to water stimulation in discharging CES), and it was found that the amount of conditioning decreased. Nelson believed that this could "mean that the presence of excitation increases the likelihood that flies will form an association between CSs and US" (p. 365) and concluded the following: "It is possible that CES and other preprogrammed mechanisms for controlling food intake . . . have overshadowed more long-term sorts of plasticity in *Phormia's* behavior, and that the classical conditioning described above is essentially an extension of the CES phenomenon" (p. 368). Studies were initiated in this laboratory to examine the possible relationship between CES and conditioning.

McGuire (1979; McGuire & Hirsch, 1977) selected divergently for conditionability and, in a different experiment, for CES with a modified version of Nelson's (1971) procedure (discussed later). He performed the CES and conditioning procedures with both the selected conditioning and CES lines and found CES scores to be positively correlated with conditioning scores in both instances. As an additional test of this relationship, McGuire (1983) performed hybrid crosses between lines selected divergently for conditionability and tested flies of the F_2 generation with both the CES and conditioning procedures. This experiment, called a hybrid-correlational analysis, is used to determine whether two traits are correlated genetically. A positive correlation indicates that either (1) differences in both traits are associated with differences in the same gene(s) or (2) both types of gene are closely linked. McGuire (1983) found a positive correlation, thereby providing additional evidence for a biological relationship between the two processes.

Tully (1982; Tully & Hirsch, 1982) selected divergently for CES using a procedure similar to McGuire's (1979). Tully, Zawistowski, and Hirsch (1982) performed a hybrid-correlational analysis with these lines and found a positive correlation between the two traits. This agreed with the finding by McGuire (1983) who used lines selected divergently for conditioning ability instead of for CES. However, there is a possible confound in the results of both Tully et al. and McGuire. Tully and Hirsch (1983) performed behavioral controls with the lines selected for CES (i.e., those used in the study of Tully et al.). They found that in the high line, but not in the low line, the water pretest increased the probability of proboscis extension to the water posttest (i.e., sets up a "water-induced CES"). Extending this finding to the classical conditioning procedure, it is possible that CS_1 sets up a "CS_1-induced CES." Therefore, responses to CS_2 may reflect this nonassociative effect and not only (or not at all) the development of an association based on the contingency between CS_2 and US. This possibility is discussed in more detail below.

So far, the discussion has focused on the role of excitatory states in conditioning *P. regina*. Vargo (1985; Vargo & Hirsch, 1982) has demonstrated the existence of a CES in *D. melanogaster* similar to that found in *P. regina* (also see Duerr & Quinn, 1982; Kemler, 1975). In both species, CES is initiated by sucrose, discharged by the water posttest, and increased by longer periods of food deprivation. However, differences between the two species exist in that in *P. regina*, CES is gone after about 5 min whereas, in *D. melanogaster*, CES persists for at least 10 min without dissipating completely. In addition, in a line of *D. melanogaster* selected for a high level of CES, no water-induced CES is present.

Holliday (1984; Holliday & Hirsch, 1984) has developed a conditioning procedure for *D. melanogaster* similar to that used by McGuire and Hirsch (1977) in conditioning *P. regina*. An unpaired control has shown that in an unselected line, the increase in responses to CS over trials cannot be due to

CES. A hybrid-correlational analysis, using the lines selected for high and low expressions of CES, showed a positive correlation between CES and conditioning in the F_2 generation. These results suggest that at least some of the genes associated with differences in CES and conditioning are the same, or are linked, in *D. melanogaster*.

A control experiment performed by Vargo (1985; Vargo & Hirsch, 1985) with the line selected for a high expression of CES raises some questions about the hybrid-correlational analysis. Because, in the high line but not in an unselected line, there is an increase over trials in responses to the pretest, and because the CES procedure may be seen as a conditioning procedure with a contingency of .5 between the water stimulations and sucrose, it was postulated that conditioning might be occurring in the high line. To validate the CES measure, the high line was tested with an unpaired control procedure (pretest was moved 45 s away from sucrose). Using the unpaired procedure, an increase in responses to the pretest was still present, apparently validating the CES measure by ruling out the possibility of conditioning to the pretest. However, this interpretation assumes that there is no neural trace of the pretest present when sucrose is given 45 s later. In an unselected line, this would be considered unlikely, but selection for a high expression of CES may have resulted in an increase in the length of the trace. Another possibility is that the posttest does not discharge CES completely in the high line, and hence CES increases over trials in that line. As a result, the correlation between CES and conditioning observed in the hybrid-correlational analysis may be due to residual CES and not a biological relationship between CES and conditioning. At this point, the question of whether CES is necessary for an association to occur in the classical conditioning of the proboscis extension is unresolved.

The central inhibitory state (CIS) may have effects on the classical conditioning of the proboscis extension similar to those of CES, but opposite in sign. Dethier et al. (1968) showed that saline decreases the probability of responses to water stimulation. Médioni and Vaysse (1975) used as aversive reinforcers both electric shock and quinine to train *D. melanogaster* in a counterconditioning procedure: sucrose was paired with one of the two reinforcers and a decrease in proboscis extensions to the sucrose was observed over trials. DeJianne, McGuire, and Pruzan-Hotchkiss (1985) replicated the experiment using quinine. In light of the results of Dethier et al., it is possible that the decrease in responses is the result of a CIS set up by the aversive reinforcers and that excitatory conditioning is not occurring. However, an unpaired control procedure performed by Médioni and Vaysse eliminates this possibility. They found that if the CS and US are unpaired, the probability of a response to the CS is greater than when they are paired.

Using the discussion of CES and CIS, and the theoretical analysis of concepts and procedures in the study of learning and of genetics, we now proceed to a critical analysis of claims for classical conditioning and avoidance learning in

P. regina and *D. melanogaster*. Emphasis is put on work performed in this laboratory because we know it best and because it illustrates many of the problems discussed in the previous sections.

CLASSICAL CONDITIONING IN *PHORMIA REGINA*

Working in Dethier's laboratory, and using the analysis of CES by Dethier et al. (1965) to develop a classical conditioning procedure, Nelson (1971) attempted to train food-deprived and water-satiated flies to extend their proboscises to a CS (either 1 M saline or distilled water) presented for 4 s to their tarsi (see Fig. 5.1). The US was 0.5 M sucrose and was presented to the labellum on the last second of CS stimulation. The CS-US presentation constituted 1 trial of the 15-trial conditioning procedure. To eliminate the effects of CES, Nelson used a 10-min ITI and presented an additional stimulation (termed CS_1) to the tarsi

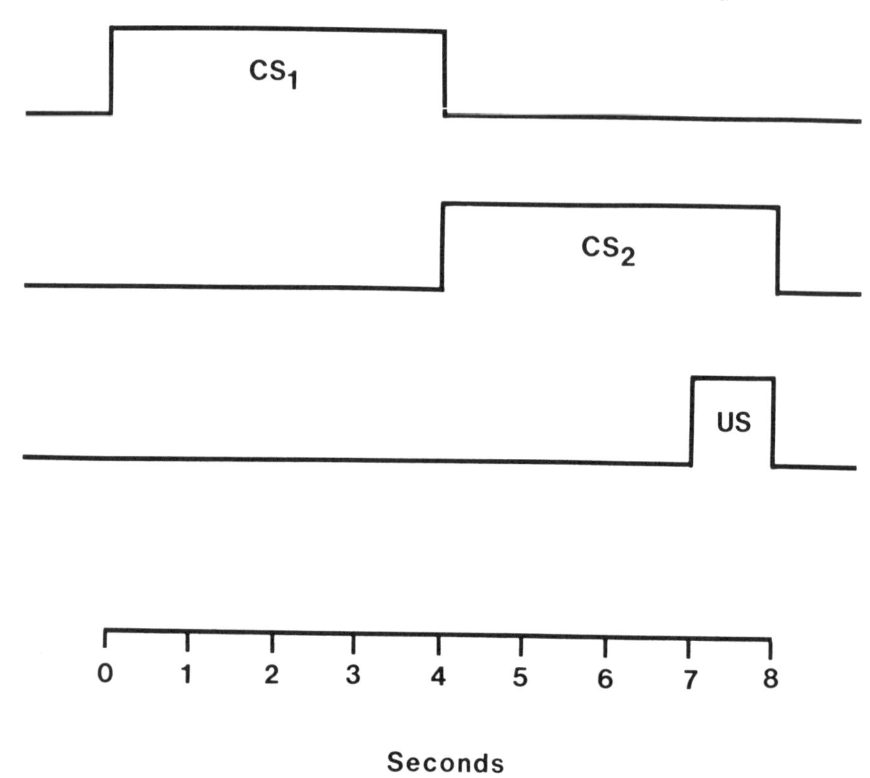

FIG. 5.1 Schematic representation of the timing of stimulus presentations in one trial of the 15-trial conditioning procedure used by Nelson (1971).

for 4 s prior to the CS (CS_2) stimulation (if SC_1 was water, CS_2 was saline, and vice versa). The preceding describes what usually is called a second-order, or combined trace and delay, conditioning procedure. However, it may also be seen as a discriminative conditioning procedure where the CS− and CS+ are contiguous temporally. This would be a correct description only if no first- or higher order conditioning occurred to the CS_1 (i.e., that it was, in fact, a CS−). However, to the extent that the amount of conditioning to CS_1 is less than that to CS_2, the procedure is adequate for inferring association (in the following, we shall call this procedure a discriminative conditioning procedure, with the realization that this may not be the label given to it typically). The CS solutions were contained in separate watchglasses and proboscis extensions were scored by sight. Flies were mounted to tackiwax by their wings and were moved manually to the CSs.

Only three of Nelson's (1971) experiments are of interest in showing the effect of changed contingencies using group data. Her Experiments 1, 2, and 4 presented results for the conditioning procedure just discussed, an unpaired control, and a more traditional discriminative conditioning procedure (i.e., CS_1 and CS_2 not contiguous), respectively. Figure 5.2a-b presents the percentage of flies responding over trials to the CS_1 and CS_2 in Experiment 1. Responses to both CS_1 and CS_2 increase over trials with a greater increase obtained for CS_2 regardless of stimulus order (i.e., whether CS_2 is saline or water). Because Nelson's procedure was designed to eliminate the effects of CES, the fact that both CS_1 and CS_2 increase suggests that conditioning may be occurring to both, though to a greater extent to CS_2 (owing, it is assumed, to its closer contiguity with US). And, because a discriminative conditioning procedure is the combination of an unpaired control and a classical conditioning procedure, Nelson's basic conditioning procedure tests the hypothesis of control procedure (1) (see Table 5.1).

However, it is possible that in a discriminative conditioning procedure, the effects of the presentation of, or the responses to, one CS may not be independent of the responses to the other CS. This interaction may be especially important when the two CSs are contiguous. The interaction may be associative (i.e., higher order conditioning) or nonassociative. Considering the latter type of interaction, it may be that CS_1 is "exciting" the fly so that responses to CS_2 increase over trials. As discussed above, Tully and Hirsch (1983) showed that in a line of *P. regina* selected for high levels of CES, a water stimulation will increase responses to a second water stimulation following shortly thereafter. In addition, Holliday (unpublished data) has found saline-induced CES in an unselected line: The number of responses to a water stimulus increased when preceded by the presentation of saline. This contradicts the work of Dethier et al. (1968), which shows that saline sets up a CIS. However, there is a difference in mounting techniques (discussed later) that Zawistowski and Hirsch (1982) found to affect

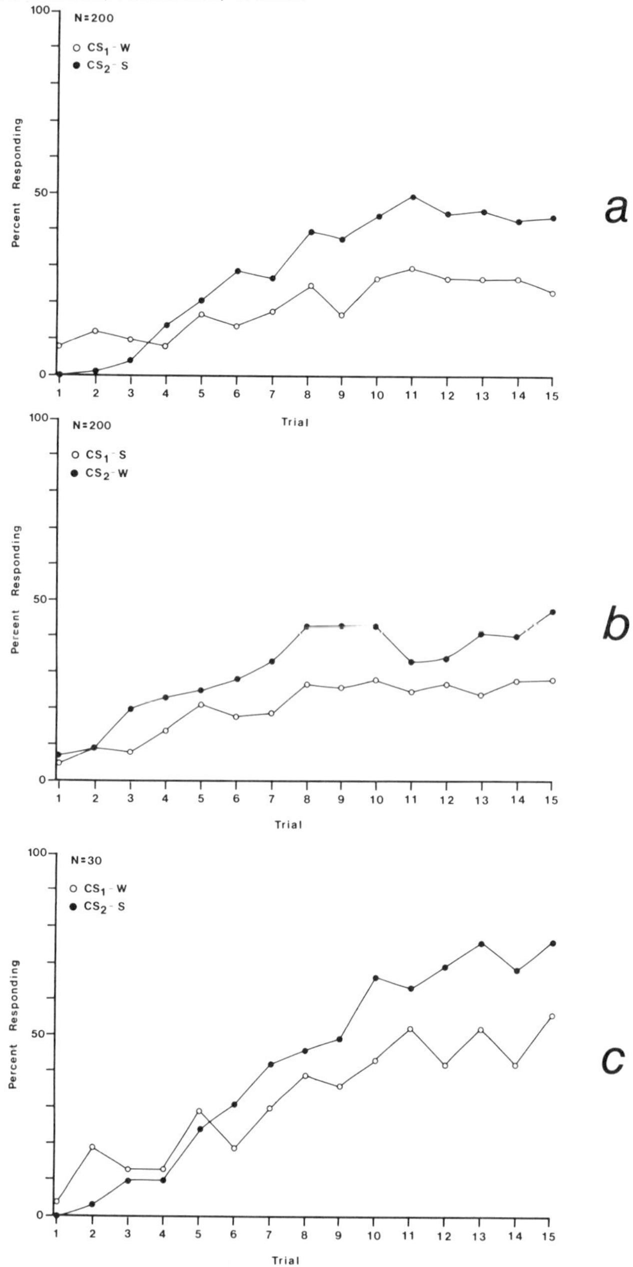

FIG. 5.2 Percentage of flies responding to CS_1 and CS_2 over trials in the conditioning procedure used by Nelson (1971): (a-b) Experiment 1; (c) Experiment 3; (d) Experiment 5; (e) Experiment 7. (*N* = sample size; *W* = distilled water; *S* = 1 M saline. Figures are redrawn from Nelson.)

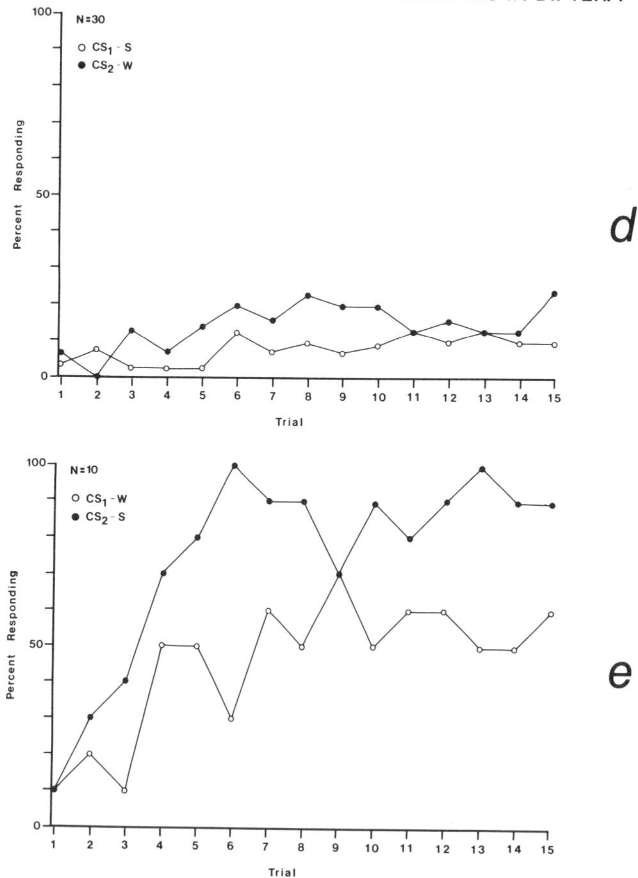

responsiveness to saline. Yet, using the same unselected line that was used by Holliday, Tully (unpublished data) found no evidence for either saline- or water-induced CES. Therefore, the evidence for a water- or saline-induced CES in the unselected line is equivocal. The effects of water-induced CES are evident only in the line of *P. regina* selected for a high expression of CES.

Testing for the presence of CS_1-induced CES in the conditioning procedure would involve one of two methods: removing either the contiguity between CS_1 and CS_2, or the contiguity between CS_2 and US. In performing a discriminative conditioning procedure that presented a CS+ (followed by US) or a CS− in a random order every 5 min for 15 trials, Nelson (1971) tested (unknowingly) the possibility of a CS_1-induced CES. (Because there was an interstimulus interval of only 5 min, CES may have had an effect on responses to CS+ or CS− when either was preceded 5 min previously by a CS+ presentation.) Using this procedure, Nelson observed a large increment in responses to CS+ suggesting that the increment in responses to CS_2 in her Experiment 1 was the result of conditioning and not CS_1-induced CES. The fact that no increment was seen in

CS − suggests that the increment in responses to CS_1 in the original procedure may have been the result of first- or higher-order conditioning.

Experiment 2 in Nelson (1971) presented results of an unpaired control procedure. Both water and saline were presented in each of 15 trials either 5 or 45 s before or after US presentation. The problem with this procedure is that responses to any stimulation following shortly after sucrose will show effects of CES. Therefore, only those stimuli presented before sucrose on a given trial show the nonassociative effects in which we are interested. Nelson observed no increase in responses to either saline or water. Again, results are consistent with an hypothesis of conditioning.

In four different experiments (1, 3, 5, and 7), Nelson (1971) tested conditioning groups that, because they represent replications of the conditioning procedure, allow us to test for the nonassociative effects of some situational variables. Figure 5.2 presents response curves for each of these four experiments. Although statistical tests of slopes are not presented by Nelson, there is an obvious increase in responses to CS_2 in three of the four experiments (four of five stimulus orders). Percentages of flies responding by Trial 15 (and, therefore, the slopes) appear to differ between the four experiments, reaching about 45% (both stimulus orders), 75%, 25%, and 90% in Experiments 1, 3, 5, and 7, respectively. The conclusions to be drawn from this comparison are uncertain. Do the differences in Nelson's study represent chance deviations or are they the result of other factors (including perhaps experimenter bias)? Table 5.2 shows that in measures of geotaxis for *D. melanogaster*, with the multiple-unit classification maze (which is an objective measure of behavior), there are significant differences upon repeated testings of an unselected line (from Erlenmeyer-Kimling, 1961). Because

TABLE 5.2
Mean Geotactic Scores and Standard Errors in Repeated Testings of an Unselected Line

	Sex	
Group	*Males*	*Females*
1	6.8 ± .25	7.2 ± .34
2	8.4 ± .35	4.2 ± .30
3	8.4 ± .36	5.9 ± .30
4	6.5 ± .23	6.2 ± .27
5	6.5 ± .31	5.3 ± .25
6	7.6 ± .28	
7	7.4 ± .34	
8	7.1 ± .31	
9	6.9 ± .39	
10	6.0 ± .26	
11	6.7 ± .29	

Note. From *A genetic analysis of geotaxis in Drosophila melanogaster* (pp. 69–70) by L. Erlenmeyer-Kimling, 1961.

these results cannot be attributed to the result of experimenter bias, they caution us against concluding anything from the large variance observed in Nelson's results.

Can learning be inferred from Nelson's (1971) group data? The nonassociative effects of the CS, US, and CS-US interaction were tested in Experiments 1, 2, and 4. The results indicate that these nonassociative effects either were not present or were negligible. Replications of the original conditioning procedure in Experiments 1, 3, 5, and 7 showed a large variation in the slopes, which makes one hesitant to conclude that situational variables had no effect. Because no blind tests were performed, there is no measure of the effects of expectancy. Blind testing could have been accomplished by performing the conditioning procedure, but not allowing the experimenter to know whether the CS_1-CS_2 presentation was water-saline, saline-water, water-water, or saline-saline. If flies discriminate between CS_1 and CS_2 in the first two groups, but not in the last two groups, then one may dismiss the possibility of experimenter bias. Therefore, we must conclude that although the results of group data presented by Nelson are consistent with an hypothesis of learning, they are not sufficient to infer learning.

In attempting to infer learning from individual data, Nelson (1971) performed two analyses. In the first, flies were arbitrarily classified as good, fair, and poor performers (those responding six to eight, three to five, and zero to two times on the last eight trials were classified as good, fair, and poor performers, respectively). However, no attempt was made to show that this classification of performers had any validity as a method of classifying different degrees of learning in individuals. To do this, one would have to test the same individual with different contingencies and then correlate these differences with the performance classification. In the second analysis, for each fly, the responses to CS_1 and to CS_2 were summed separately over the 15 trials, and then the number of responses to CS_1 was subtracted from that to CS_2. Because CS_2 had a closer contiguity to US than did CS_1, any positive difference score should indicate discrimination. However, the reliability of the difference score was not calculated so that one can not know the probability of obtaining a given difference score simply by chance. In addition, both methods of analysis suffer from the same problem that was discussed above for group data: the degree of experimenter bias and the consistency of replication (related to the problem of reliability) are not known. Therefore, as with group data, learning cannot be inferred from individual data in Nelson's study.

McCauley (1973; Hirsch & McCauley, 1977), in this laboratory, attempted to replicate Nelson's (1971) conditioning procedure using a different strain of *P. regina* and with changes in rearing conditions. Response curves over 15 trials for the conditioning procedure are presented in Fig. 5.3. Responses to CS_2 are very similar to those presented in Experiment 1 of Nelson's paper. Unlike Nelson's study, however, responses to CS_1 either do not change over trials (Fig. 5.3a) or they decrease (Fig. 5.3b). But, because different strains and rearing conditions were used, and because it is not the number of responses to CS_1 but the difference between it and the number of responses to CS_2 that is of importance here, the

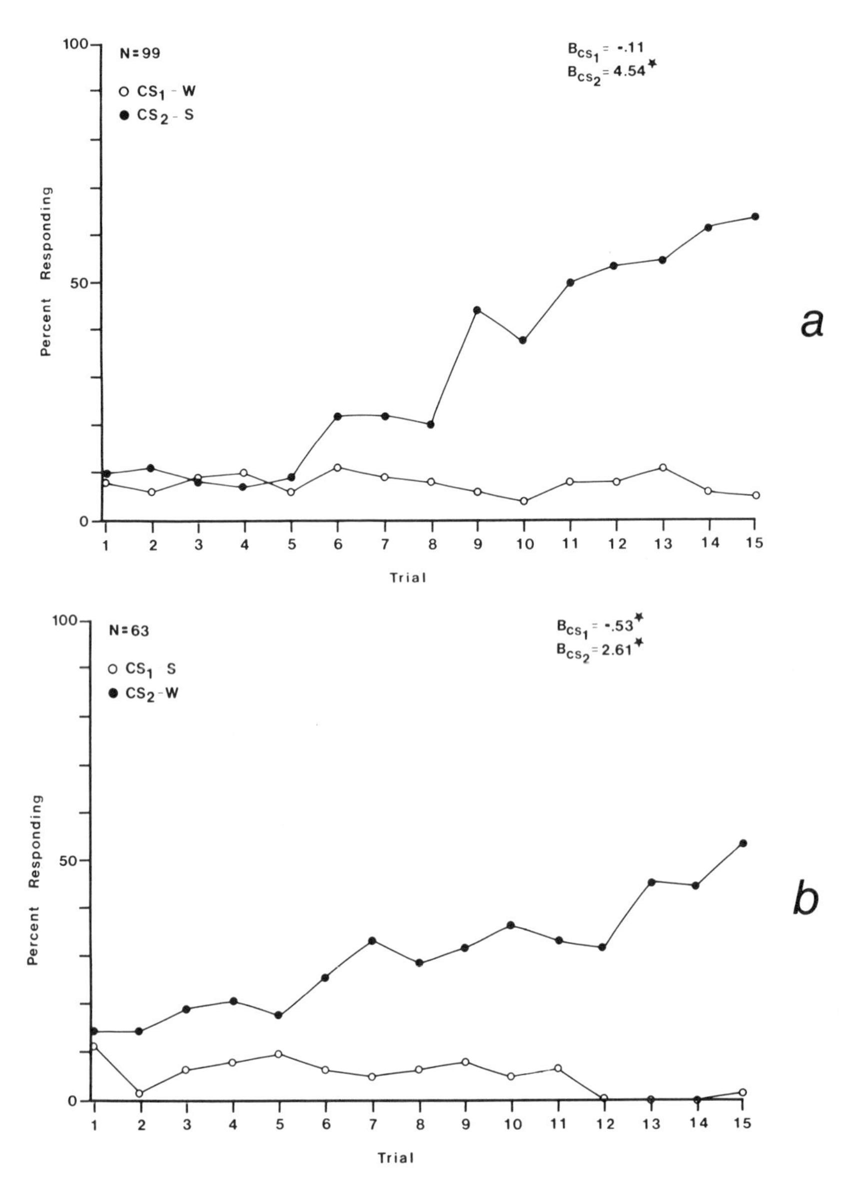

FIG. 5.3 Percentage of flies responding to CS_1 and CS_2 over trials in the study by McCauley (1973). (N = sample size; W = distilled water; S = 1 M saline; B = regression coefficient. Statistical significance is indicated by a star.)

differences between Nelson's and McCauley's studies are of little concern. In fact, McCauley's results present evidence consistent with conditioning that is stronger than that claimed by Nelson because there is a greater difference between CS_1 and CS_2. However, as in Nelson's study, the fact that there were no blind tests performed makes the results inconclusive. Therefore, we can conclude only that McCauley's results are consistent with an hypothesis of learning, but not sufficient to infer learning.

McGuire (1979) used the conditioning procedure of Nelson (1971) and the strain and rearing conditions of McCauley (1973). However, instead of mounting flies on tackiwax to restrain them, he enclosed them in plastic micropipet tips so that only the head and front legs projected from the tapered end (see Fig. 5.4). No assessment was made by McGuire of the possible effects of this change in procedure on responses to the CSs. However, Zawistowski (1979; Zawistowski & Hirsch, 1982) compared the different mounting techniques and found that the responses to saline were lower with Nelson's than with McGuire's technique. Hence, the change in methods had an effect on the responses of flies to stimulation. Whether there was an effect on the amount of conditioning is a question that cannot be answered yet. We must examine first whether conditioning has occurred at all. A study of this question must use appropriate control procedures.

From McGuire's data, response curves over 15 trials were calculated and are presented in Fig. 5.5. Comparing the results of the stimulus order water-saline with those of Nelson and McCauley presented in Figs. 5.2 and 5.3, respectively,

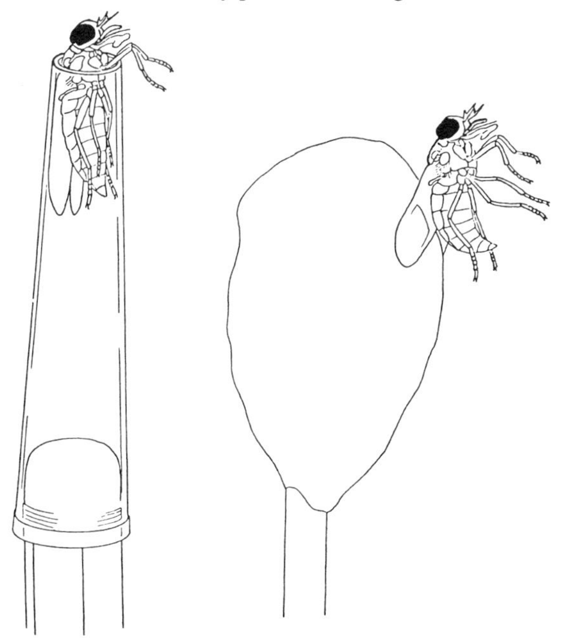

FIG. 5.4 Comparison of mounting techniques: pipette-tip vs. tackiwax (from McGuire, 1979).

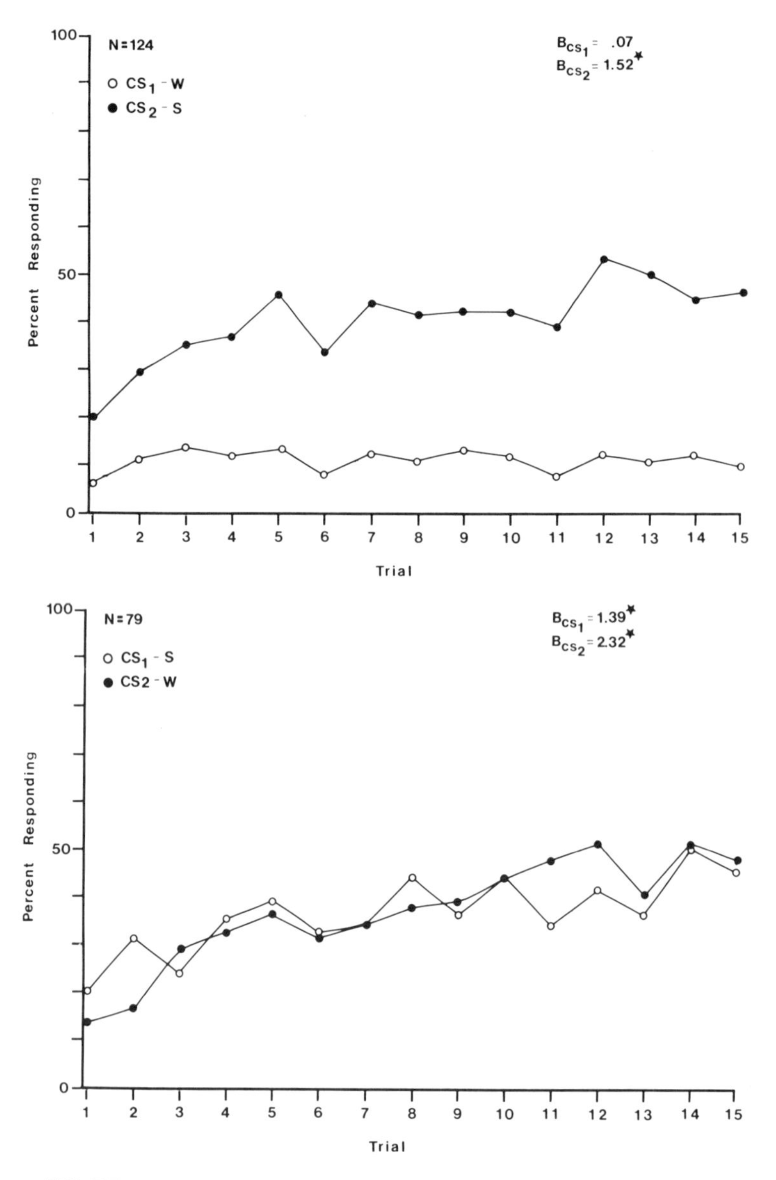

FIG. 5.5 Percentage of flies responding to CS_1 and CS_2 over trials in the study by McGuire (1979). (See caption to Fig. 5.3 for explanation of abbreviations and symbols.)

we can see that the increase in responses to CS_2 in McGuire's study is smaller than that obtained by McCauley and by Nelson, though there still appears to be a discrimination between CS_1 and CS_2. Figure 5.5 presents little evidence for discrimination between the CSs for the stimulus order saline-water, even though responses to both show an increase. The increase in responses to CS_2 in McGuire's study is equal to that of McCauley and similar to that of Nelson. Given the variability in the curves obtained by Nelson, it is possible that McGuire's results were only chance deviations toward the lower end of a distribution of possible response curves. To test this possibility, we review response curves from other researchers in this laboratory who used the same procedures and strain as McGuire.

Figure 5.6 presents a response curve from Jackson (1976), who used the stimulus order water-saline. Whereas a statistically significant decrease is observed in responses to CS_1, no change is observed in responses to CS_2. Therefore, no evidence is shown for conditioning to CS_2, although conditioned inhibition to CS_1 could be occurring. The results of Nelson (1971), McCauley (1973), and McGuire (1978/1979) were not replicated successfully by Jackson.

Figures 5.7 and 5.8 present response curves from Zawistowski (1979 and unpublished observations, respectively). The two figures represent data collected approximately 2 years apart. The curves in Fig. 5.7 are very similar to those of McGuire (1979) presented in Fig. 5.5: a small increase in responses to CS_2 and a small discrimination between it and CS_1 are observed with the stimulus order

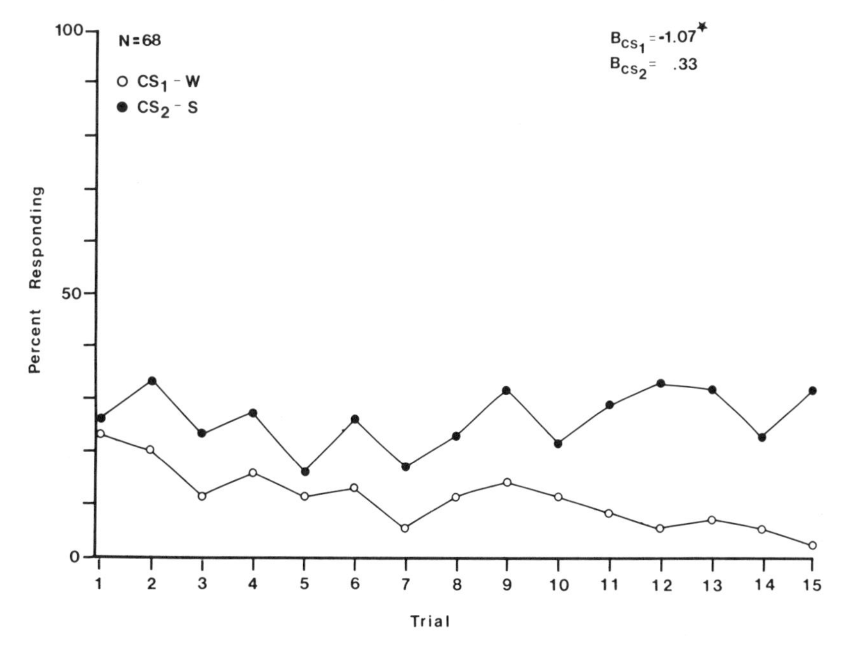

FIG. 5.6 Percentage of flies responding to CS_1 and CS_2 over trials in the study by Jackson (1976). (See caption to Fig. 5.3 for explanation of abbreviations and symbols.)

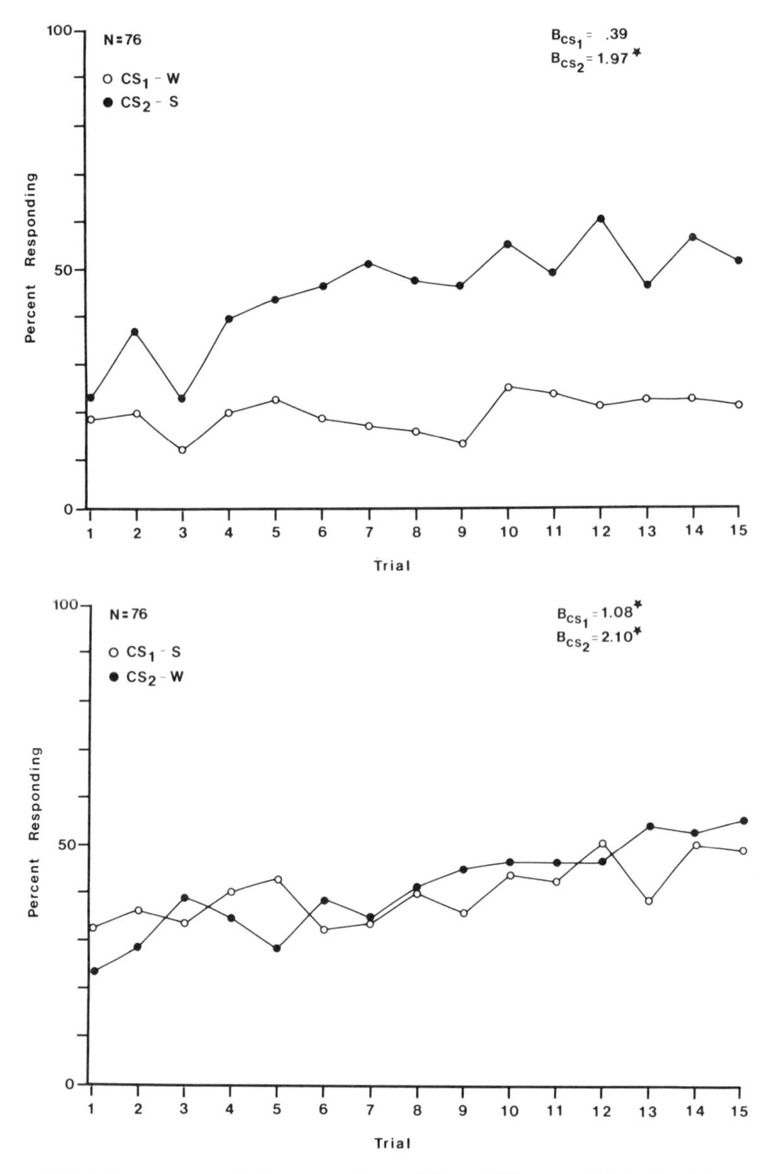

FIG. 5.7 Percentage of flies responding to CS_1 and CS_2 over trials in the study by Zawistowski (1979). (See caption to Fig. 5.3 for explanation of abbreviations and symbols.)

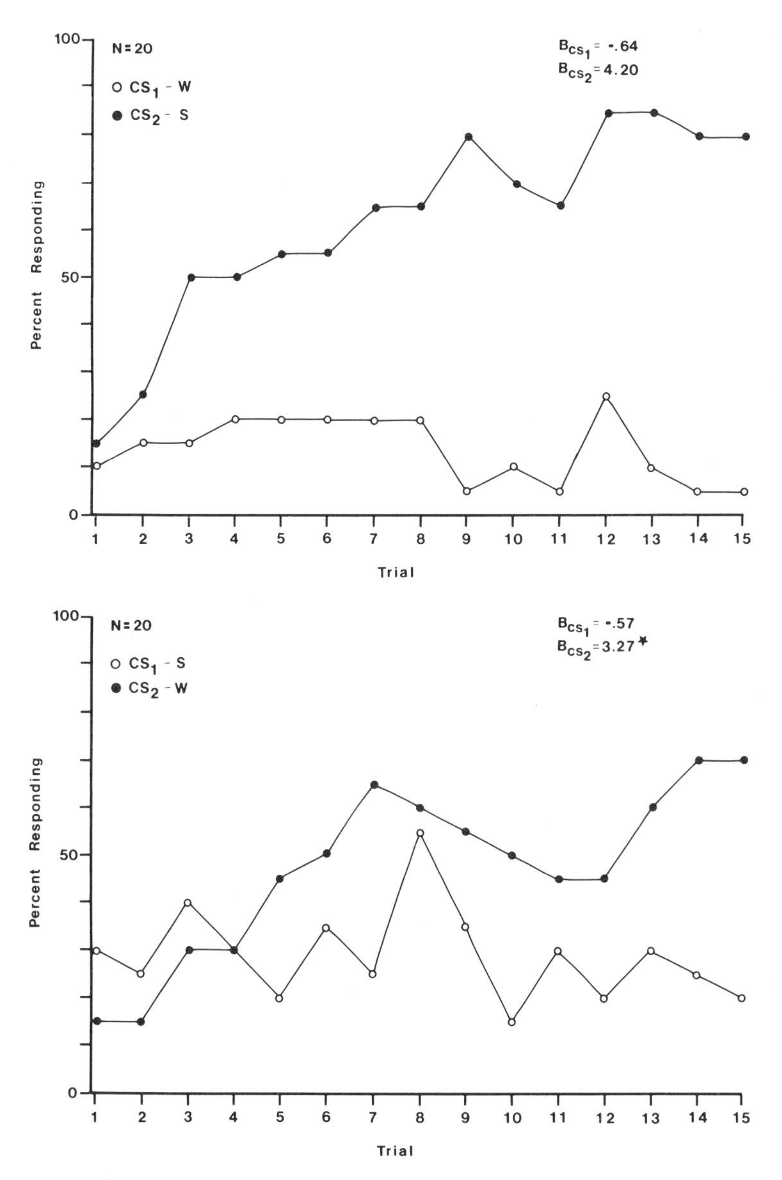

FIG. 5.8 Percentage of flies responding to CS_1 and CS_2 over trials in the study by Zawistowski (unpublished data). (See caption to Fig. 5.3 for explanation of abbreviations and symbols.)

water-saline; but no discrimination is apparent between CS_1 and CS_2 with the stimulus order saline-water. The curves in Fig. 5.8 are very different from those of Fig. 5.7. Both stimulus orders give evidence of strong conditioning, although the discrimination between CS_1 and CS_2 appears to be better with the stimulus order water-saline. Therefore, Fig. 5.7 appears to replicate McGuire closely, whereas Fig. 5.8 does not, though it is closer to some of the curves of McCauley (1973) and Nelson (1971). The curves in Figs. 5.7 and 5.8 do not replicate the curve of Jackson (1976) presented in Fig. 5.6. In fact, even though they represent data collected by the same experimenter, the curves in Figs. 5.7 and 5.8 do not appear to replicate each other.

Figure 5.9 presents response curves from Ricker (1980). No statistically significant changes are observed in responses to either CS_1 or CS_2. Therefore, the curves present no evidence for conditioning and do not replicate results of Nelson (1971), McCauley (1973), McGuire (1979), or Zawistowski (1979), although they are consistent with results of Jackson (1976).

Figure 5.10 presents results collected by many of the researchers in this laboratory over the years. Instead of presenting response curves, we present the data in terms of the slope (on the abscissa) versus the intercept (on the ordinate) of the CS_2 response curves. The 95% confidence intervals of the slopes are also graphed. We shall consider only the studies using the procedure of McGuire (1979) because they represent the majority of the work done in this laboratory. No increase in responses to CS_2 are observed in 7 of 19 studies, and in 8 of 13 studies, with the stimulus orders water-saline and saline-water, respectively. The results presented here indicate that different experimenters, and even the same experimenter, can not replicate consistently the increase in responses to CS_2 first shown by Nelson (1971). In fact, in almost half of the cases, no statistically significant increase is found. Therefore, there is great variability in the results obtained from this conditioning procedure, even within experimenters, that makes any interpretation very difficult.

We have videotaped four different experimenters in this laboratory to determine whether there were differences in techniques used to present the stimuli (Ricker, unpublished observations). Obvious differences between experimenters in the presentation of stimuli, but not in the scoring of responses, were found. However, the differences could not be related consistently to differences in the ability to obtain increases in responses to CS. These observations led to the development of an automated procedure for presenting stimuli (see below). Because of the inability to replicate consistently and the absence of blind tests, learning cannot be inferred in unselected lines of *P. regina* using this conditioning procedure.

It was stated earlier that, with a more traditional discriminative conditioning procedure, Nelson (1971) found an increase in responses to CS+ but not to CS−. Using an automated testing procedure (but still requiring an experimenter to judge a response), Zawistowski (1984; Zawistowski & Hirsch, 1984) has replicated this overall effect with a modification: Instead of requiring a discrimination between water and NaCl, as did Nelson, he required the flies to

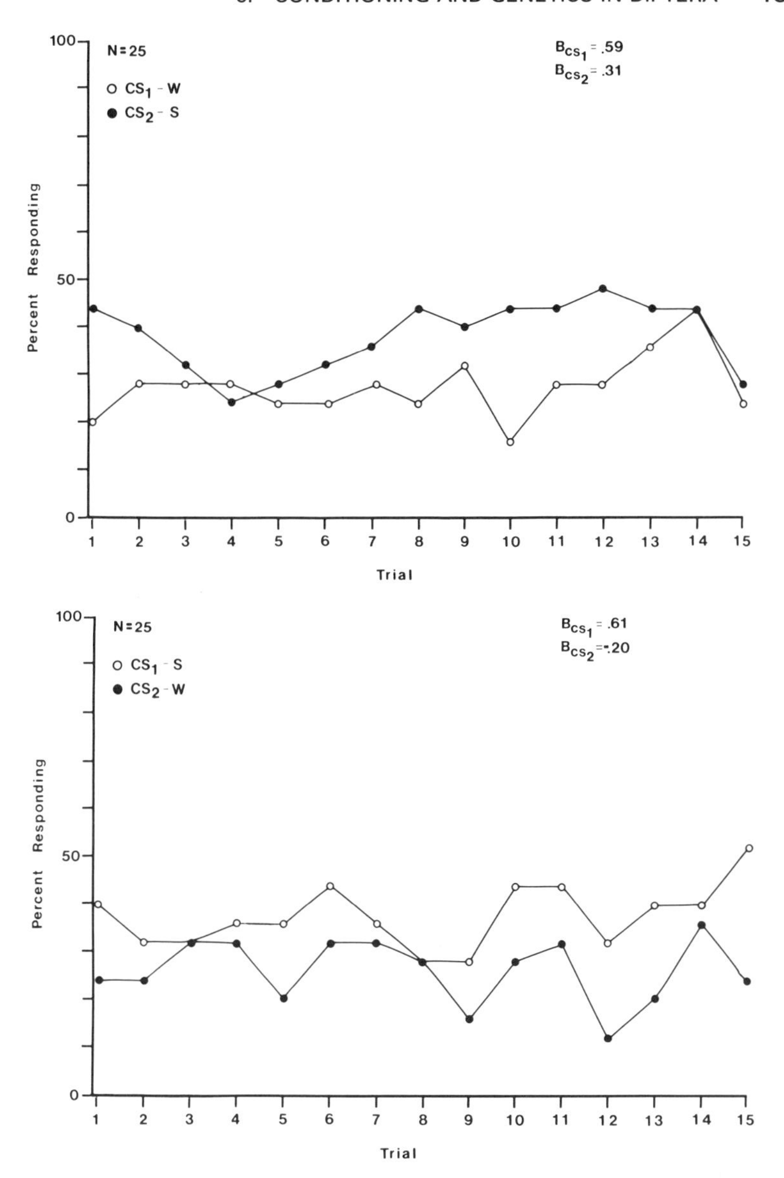

FIG. 5.9 Percentage of flies responding to CS_1 and CS_2 over trials in the study of Ricker (1980). (See caption to Fig. 5.3 for explanation of abbreviations and symbols.)

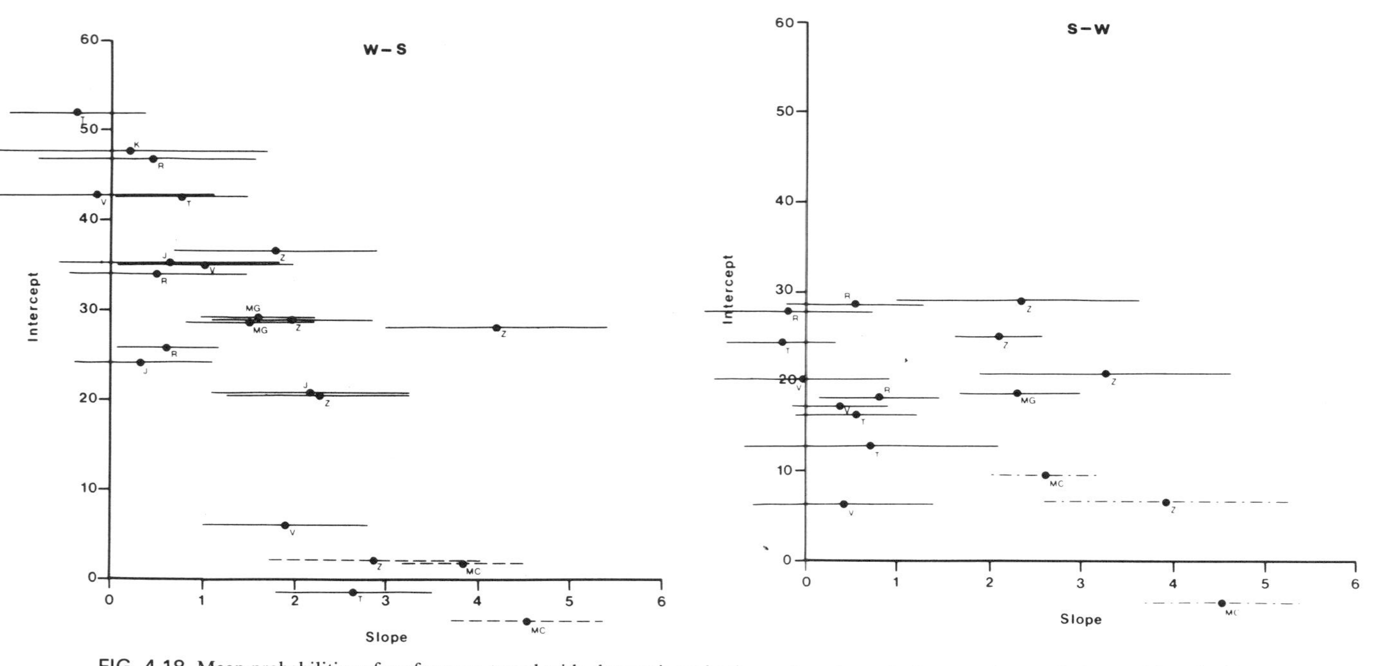

FIG. 4.18 Mean probabilities of preferences tested with chromatic and achromatic patterns in five genetic populations, with and without prior exposures to colors. Compare performances within and across populations, and note the linear to U-shaped relationship between performances elicited by colored and achromatic patterns. For more detail see Kovach, 1983d.

discriminate between two salts, KCl and NaCl. Zawistowski and Hirsch (1982) had noted that in the Nelson procedure, flies responded more to saline than to water when flies were mounted in micropipet tips. Therefore, in order to make the initial responsiveness to the two CSs more equal, two salts were used in the discrimination procedure. Thirty trials were presented with an ITI of 5 min. Flies received one of two possible stimulus configurations: KCl and NaCl being CS+ and CS−, respectively, or vice versa. Flies were retrained 24 hr later using the opposite stimulus configuration (reversal) to control for response bias. It was found that the probability of responses to the CS+ increased over that to the CS− in the first training session; and, during reversal, responses to the stimuli reflected the changed contingencies. In addition, because the differential probability of response is believed to result from the positive contingency between CS+ and US and/or the negative contingency between CS− and US, the difference between CSs in contingencies was removed in a control group (tested by hand instead of with the automated procedure) by arranging for each CS to be contiguous with the US on half the trials (i.e., both contingencies were equal to .5). Because presentation of either CS predicts presentation of the US equally well, there should be no discrimination between them. As predicted, no evidence for conditioning was obtained. However, no blind testing was performed. Hence, the results are not sufficient to infer learning.

Recently, the observations of Zawistowski (1984) have been replicated by Brzorad (Brzorad, 1985; Ricker, Brzorad & Hirsch, in press) with the automated testing procedure. In addition, we performed blind testing using the second method described in the second section of this chapter: the genetic identity of the subject was known by the experimenter although the procedure was not. Four stimulus configurations were tested, two of which were discriminative conditioning procedures (KCl and NaCl being CS+ and CS−, respectively, and vice versa), and the other two configurations presenting only one of the salt solutions (i.e., either KCl or NaCl occupying both stimulus positions), thereby presenting a .5 contingency. The discriminative conditioning procedure produced higher conditioning scores than did the procedure with a .5 contingency. This study is important because all control procedures that we consider to be sufficient to infer learning have been performed. The discrimination procedure is the combination of a paired and an unpaired procedure. The automated testing device, in conjunction with blind testing, controls for experimenter bias. Therefore, we conclude that *P. regina* may be conditioned with this procedure.

SELECTION FOR CONDITIONABILITY OF *PHORMIA REGINA*

McCauley (1973) and McGuire (1979) each attempted divergent selection of *P. regina* for conditionability. Using Nelson's (1971) classification of individual performance, they selected divergently by breeding good performers (six or more responses to CS_2 on the last eight trials) for a "Bright" line, and by breeding poor performers (two or fewer in McCauley's study, and two or one in McGuire's study) for a "Dull" line. Although the terms "Bright" and "Dull" are inaccurate in this context, we use them because they were used in the original studies. We would not use such terminology.

Figures 5.11 and 5.12 present response curves from the Bright line in Generation 7 of McCauley's study and Generation 8 of McGuire's study, respectively. They both show a large increment in responses to CS_2 over trials and no change in responses to CS_1. As stated in the previous section, we believe that this procedure is similar to a discriminative conditioning procedure and, therefore, may be used as a control for the nonassociative effects of the first control procedure in Table 5.1. Because it appears that there is little if any conditioning to CS_1 (see foregoing), this seems to be an acceptable way of labeling the procedure. However, this would not be the case if an interaction occurs between the two CSs because then this procedure would confound these nonassociative effects with those due to conditioning. As was stated earlier, there is no evidence for a CS_1-induced CES in an unselected population. However, selection may magnify effects not readily seen in an unselected population. In fact, Tully and Hirsch (1983) discovered water-induced CES in a line selected for high levels of CES, though this was not true for the unselected line. Selection for a Bright line may result in an increase in excitatory effects of the CS_1 that are negligible in an unselected line. Results consistent with this hypothesis were obtained by

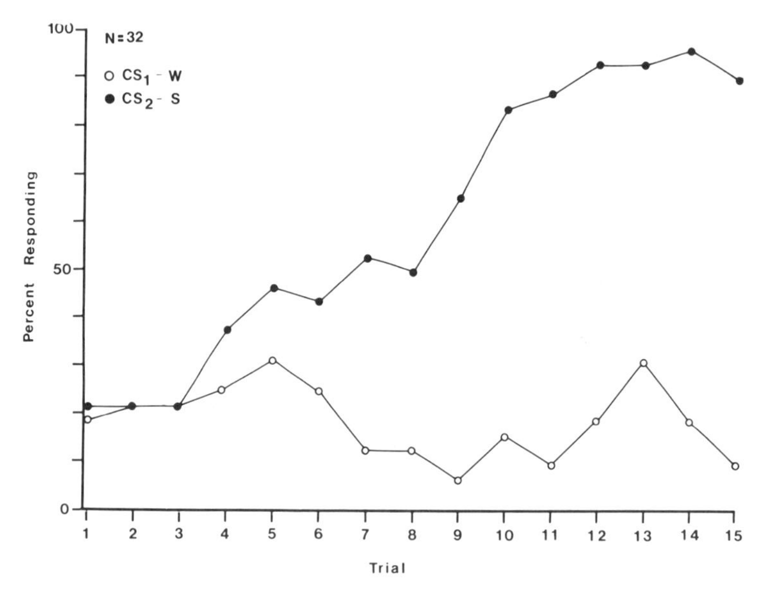

FIG. 5.11 Percentage of flies responding to CS_1 and CS_2 over trials in Generation 7 of the Bright line in the study by McCauley (1973). (See caption to Fig. 5.3 for explanation of abbreviations.)

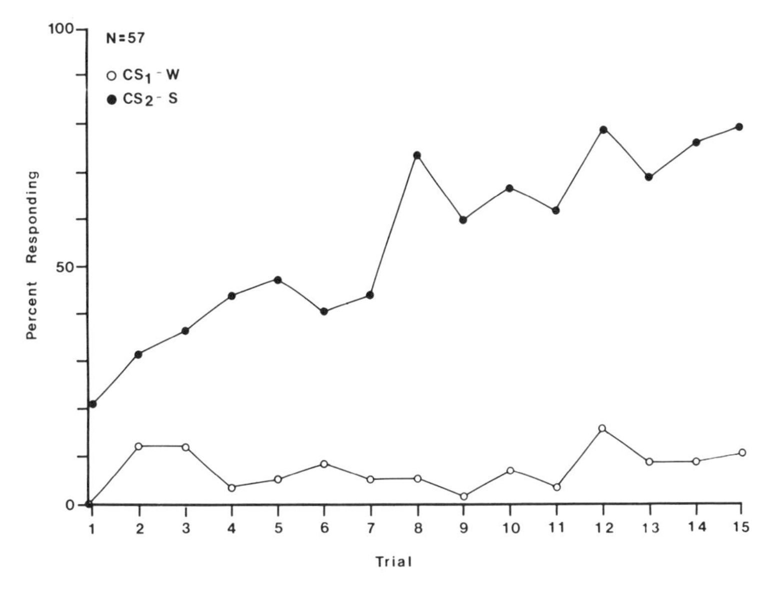

FIG. 5.12 Percentage of flies responding to CS_1 and CS_2 over trials in Generation 8 of the Bright line in the study by McGuire (1979). (See caption to Fig. 5.3 for explanation of abbreviations.)

Zawistowski (1984), who tested a Bright line selected by McGuire (1983), but different from the one discussed here. Zawistowski found that when the CSs were presented alone, increments in responses were observed that were not statistically different from those observed in a conditioning procedure. Although a control procedure that removed only the contingency would be more appropriate, the results suggest that CS_1 may be setting up an excitatory state that increases the probability of responses to CS_2.

The hypothesis that selection for a Bright line may increase the excitatory effects of the CS_1 was tested in a study by Ricker (1980). Using Nelson's (1971) classification of individual performance, he selected *P. regina* divergently with a procedure containing only one CS that was presented in the position of CS_2, that is, CS_1 was not present in the procedure. Therefore, because CS_1 was not present, it could not set up a CS_1-induced CES. Increases in response to the remaining CS should reflect only conditioning (assuming that all other nonassociative effects are negligible). In Generation 7, the selected lines were tested with the conditioning procedure that included both CS_1 and CS_2 so that the effects of selection with only one CS could be compared to the results of McCauley (1973) and McGuire (1979), who used two CSs in selection. Figure 5.13 presents response curves for Ricker's Bright line. On Trial 1, responses to CS_2 occur in about 90% of the flies and then decrease over trials. Responses to CS_1 show a similar effect but begin lower. Apparently, the absence of CS_1 in previous generations had changed the effects of selection from those observed in the

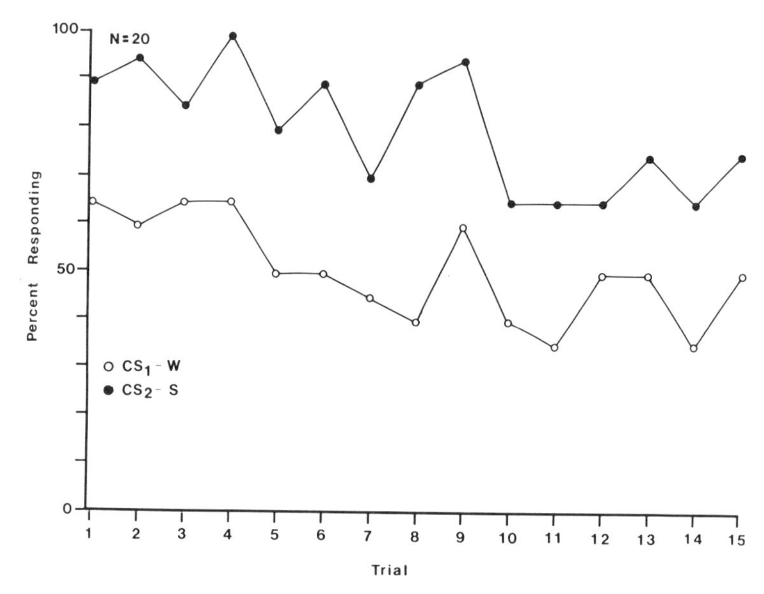

FIG. 5.13 Percentage of flies responding to CS_1 and CS_2 over trials in Generation 7 of the Bright line in the study by Ricker (1980). (See caption to Fig. 5.3 for explanation of abbreviations.)

studies of McCauley and McGuire. That is, when CS_1 is absent, selection results simply in an increase in responsiveness of flies to stimulation. Therefore, one might hypothesize that when CS_1 is present during the selection of a Bright line, instead of an increase in conditionability, selection results in an increase in the excitatory effects of CS_1. These excitatory effects would not necessarily be reflected in increased responses to CS_1. Because no control procedures using the selected lines were performed by either McCauley or McGuire, no conclusion can be made regarding the effects of selection. Therefore, we cannot conclude that either study has resulted in changes in the average conditionability of their populations.

CLASSICAL CONDITIONING IN *DROSOPHILA MELANOGASTER*

Thorpe (1939) attempted to condition *D. melanogaster* preimaginally by rearing larvae on a food medium containing 0.5% essence of peppermint. This may be seen as a classical conditioning procedure with the CS being the peppermint scent and the US being the medium (Thorpe stated that the US need not be the food but simply the "favorable environment" experienced by the fly in the presence of the odor, p. 430). Flies had a strong tendency to avoid the scent if they had never experienced it before. Given a choice between scented and unscented arms of an olfactometer, 65.1% of flies, on the average, chose the unscented arm. However, when reared in peppermint odor, only 33.3% chose the unscented arm if taken directly from the scented medium and tested. Yet, if isolated from the scented medium for a period of time before the test, flies tended to choose the unscented arm with an increasing frequency the longer the period of isolation. Thorpe concluded that the decreased avoidance of the scented arm by flies reared in scented medium reflects an associative process. Later, Thorpe (1943) reinterpreted the decreased avoidance as *habituation,* defining the term as an "activity of the central nervous system whereby innate respones to mild shock and warning stimuli wane as the stimuli continue for a long period without unfavourable results" (p. 221). Thorpe defined learning broadly as a process that "produces adaptive changes in individual behaviour as a result of experience" (p. 220), a definition that includes both associative and nonassociative processes. Hence, habituation was included as a form of learning, although it was not considered to be associative in nature (also see Thorpe, 1944).

Hershberger and Smith (1967), however, argued that the results of Thorpe (1939) are due to an associative process. They reared flies in either scented or unscented medium and, after emergence, kept flies of each group in an empty bottle that was either scented or unscented with peppermint. Regardless of whether they had been reared in scented or unscented medium, adult flies kept for 1 hr in an empty unscented bottle showed decreased avoidance of peppermint odor. Those flies reared in scented or unscented medium and kept in a scented bottle did not show decreased avoidance. Hershberger and Smith interpreted these findings in the following way:

> Presenting the conditioned olfactory stimulus in the absence of food results in extinction. . . . In [the group reared in scented medium and kept in an empty scented bottle] the conditioned acceptability of the peppermint scent in which the insects were reared is extinguished when the scent is presented in the absence of food. In [the group reared in unscented medium and kept in an empty unscented bottle] the conditioned acceptability of the neutral odour . . . in which the insects were reared is extinguished when [it] is presented in the absence of food. (p. 261)

Therefore, Hershberger and Smith claimed that evidence consistent with extinction of conditioning (to both the scented and unscented medium) proves that associative learning has occurred. In light of the conceptual discussion presented above, this evidence is not adequate to infer learning, because more than just the contingency between CS and US has been changed when the US is removed. An unpaired control is needed to determine whether learning has been observed or not. Yeatman and Hirsch (1971) emphasized a similar notion (although not using the same terminology) by noting missing controls; food should have been placed in the bottles in which flies were kept following emergence. This would have been equivalent to an unpaired extinction procedure.

Manning (1967) questioned the reliability of the measure used by Thorpe (1939) and Hershberger and Smith (1967). He argued that because flies were tested only once (thereby not obtaining a measure of the consistency of the effect), one cannot choose between the hypotheses of association and habituation:

> The most critical test of positive conditioning versus habituation is to run these flies through the olfactometer a second time. . . . On the habituation hypothesis one supposes that turning towards geraniol [the odor used by Manning in place of peppermint] on the first trial represents only random choice and that such flies will segregate at random . . . on the second run also. According to the conditioning hypothesis most flies choosing geraniol on the first run do so because they associate its smell with food and there should be an increased proportion of choices to geraniol on the second run. (p. 339)

Manning found that flies choosing geraniol on the first run were no more likely to choose it on the second run than another group that failed to choose it on the first run. Manning dismissed the study of Hershberger and Smith, concluding that Thorpe's (1943) interpretation (that habituation, and not an associative process, was occurring) was correct. The problem with this argument is the same as that described earlier in the discussion of Quinn et al. (1974): the reliability of the measure is unknown; thus, the number of trials needed to classify individuals reliably also is unknown. Therefore, we do not know whether Manning's results were due to unreliability of the measure or to the correctness of the habituation hypothesis.

Of more importance is Manning's (1967) inability to replicate the results of Hershberger and Smith (1967). He found that those flies kept in a scented bottle prior to testing showed decrease avoidance of the scent in an olfactometer regardless of whether they had been reared in scented or unscented medium. Thus, the extinction interpretation discussed by Hershberger and Smith becomes questionable.

Quinn et al. (1974) described a discriminative conditioning procedure for *D. melanogaster* that paired one odor (usually 3-octanol or 4-methylcyclohexanol) with an aversive reinforcement (usually 90 V ac, 60 Hz, electric shock) and presented another odor unpaired with reinforcement. Their conditioning measure was calculated by subtracting the proportion of the population avoiding the unpaired odor from that avoiding the paired odor. Thus, they tested for the hypothesis of control procedure (1) (see Table 5.1). Experimenter bias was ruled out through the use of a blind-testing procedure. The results of pooled experiments in different publications showed effects that were very similar, thereby meeting the requirement of replicability. These results seem to be conclusive evidence for conditioning (but see below).

The conditioning effect produced by the procedure described in Quinn et al. (1974) was not large. Of 44 lines tested (Dudai, 1977), in only 8 were there 30% or more of the population, on the average, avoiding the paired more than the unpaired odor. And no lines had more than 38% of the population avoiding the paired odor more than the unpaired odor.

Based on the experiment described earlier, Quinn et al. (1974) concluded that "all the flies have the same apparent capability" (p. 712); i.e., they claimed that all flies learned the contingency, but only a subset of the population showed the effect of learning in any one test trial. We have argued that the unreliability of the measure (Byers, 1980) does not allow one to test this hypothesis. This is similar to the problem in Manning's (1967) work previously discussed. Quinn et al. used a design similar to that used by Manning, obtained similar results, and concluded that all individuals learned. Quinn et al. did not suggest the same interpretation for Manning's experiment but, instead, accepted his interpretation: "Exposure of *Drosophila* larvae to odor altered their behavior as adults [they cite Thorpe, 1939]. This was interpreted as associative learning [Hershberger and Smith, 1967], but has since been shown to result from habituation [Manning, 1967] (p. 708). Obviously, Manning's results could be interpreted as having shown that all flies in his population had learned, thereby invalidating his criticism of Hershberger and Smith. (As mentioned before, however, there are other problems that make the claim for conditioning by Hershberger & Smith inconclusive.)

It seems likely that the major reason for the conclusion by Quinn et al. (1974) that their flies were identical behaviorally was that they had already assumed that flies of the Canton-S line were identical genetically. In the first paragraph of their paper, Quinn et al. stated that "Many flies of identical genotype are readily produced, so that behavioral measurements may be made on populations rather than individuals, yielding instant statistics" (p. 708). A similar assertion was made in Aceves-Pina et al. (1983) where the Canton-S line is described as an "inbred wild-type stock" (p. 831). These are claims that have not been subjected to empirical tests (see McGuire & Hirsch, 1977). Seymour Benzer (1976) has informed us that no inbreeding regimen was used to increase homozygosity in Canton-S. Apparently, Quinn et al. and Aceves-Pina et al. were basing their assertion on the argument from population genetic theory that a closed population of finite size will, after many generations, become homozygous at all loci (Falconer, 1981, p. 52). However, this argument assumes a number of simplifying conditions (Falconer, p. 48)—conditions unlikely to be met in reality. Ricker (1984; Ricker & Hirsch, 1985a, 1985b) has found genetic variation in small populations of *D. melanogaster* subjected to strong but intermittent selection pressures over 26 years (about 600 generations). Whether a particular population is isogenic or not is a hypothesis that needs to be tested in each case and is not something that may be assumed. The fact that Quinn et al. made this assumption with Canton-S resulted in their neglect of alternative hypotheses.

The assumption of behavioral identity made by Quinn et al. (1974) has had a strong influence on the interpretations of subsequent genetic analyses. These studies have used a strong mutagen, ethylmethanesulfonate, to induce mutations (Dudai, Jan, Byers, Quinn, & Benzer, 1976), thereby allowing the isolation of several "learning mutants" on the X chromosome. These mutations are believed to affect either learning ability or memory. We stated earlier that the ultimate goal of behavior-genetic analysis involves a description of behavioral development and an examination of the nature of behavioral differences among individuals. These problems have been studied with the mutants collected by Quinn and his colleagues. Two hypotheses may be made concerning the effects of these mutations on development of learning ability. (a) A mutation may have the same effect on learning ability (or memory) in each individual. Differences between individuals, then, are due to chance—this chance factor being, perhaps, a function of test reliability, but not important biologically. (b) A mutation may not have the same effect on learning ability (or memory) in each individual. Differences between individuals are due to differences in development of learning ability and, hence, are important biologically. Other factors (genetic, organismic, and environmental) may affect the behavioral expression of the gene (even though the primary gene product may remain the same). That is, the first hypothesis assumes that behavioral development is the result of an unfolding of a preformed entity (the gene), whereas the second hypothesis assumes that it is the result of a very complex interaction of many factors, with the final expression depending upon the specific values of these factors present during development.

These hypotheses affect the design of experiments. If Hypothesis (a) is accepted, then one may study the effect of a mutation on individual development by measuring the average performance of a group. That is, the average performance of the group is indicative of the performance of each individual. However, if Hypothesis (b) is accepted, then one must acknowledge that the effect of a mutation on individual development may not be the same in all individuals. That is, measuring the average performance of a group will give only the average effect of that mutation on individual development and not its effect on a given individual's development. It is apparent that accepting the first hypothesis would simplify the study of behavioral development tremendously.

Quinn and his colleagues appear to have accepted this simplification. Aceves-Pina et al. (1983) stated that most "mutations do not seriously alter learning behavior, but if a relevant one is present, it will affect all of the flies of a given population" (p. 831). However, they contradicted this assertion in other parts of the paper. For example, Booker and Quinn (1981) tested wild-type flies and "learning mutants" with a different conditioning procedure (avoidance of an electric shock by leg flexion) and found that the mutants performed poorly, on the average, whereas the wild-type flies performed well, on the average, which was the result expected. However, in discussing this study, Aceves-Pina et al. stated that "some individuals of all genotypes [i.e., even mutants] did learn. As in other tests, the learning disabilities in the mutants appear to be relative, not absolute" (p. 833). This passage seems to be reflecting a belief that different flies from a given line may differ in their learning ability; i.e., that they may not be identical behaviorally. In fact, in the paper by Booker and Quinn, it is stated that a "few individuals of each mutant type did learn [in the leg-flexion task]. In these instances, their behavior appeared similar to that of normal flies" (p. 3942).

Furthermore, considering the limitations of procedures such as the one used by Quinn et al. (1974), Booker and Quinn (1981) stated that "up till now mosaic mapping has not been applicable to learning—it requires that mutant behavior be reliably distinguishable from normal on the level of individuals (mosaics), and *only about a third of the flies learned in any previous paradigm*" (p. 3944; italics added). This last statement is not asserting that only a third of the flies performed what they had learned, but that only a third actually learned the contingency. We submit that these contradictions between published statements reflect a fundamental confusion in the thinking of Quinn and his colleagues about the nature of individual differences in their measure of behavior. We submit further that they have not shown behavioral identity of individuals in their populations and that, therefore, the interpretations drawn from their genetic analyses concerning the development of learning ability in particular individuals are questionable.

This confusion is made manifest in statements about the implications of the genetic analysis of populations containing "learning mutants" for the ontogeny of learning in individuals. For example, Aceves-Pina et al. (1983) assert the following: "Because genes often specify enzymes, one can, with mutants and luck, jump directly from a behavior to a molecule. . . . Work with [mutants], properly interpreted, can lead directly from an animal's behavior to its molecular heart of hearts" (p. 838; these two statements are separated by 33 lines of text). Let us analyze this passage. The first statement makes an argument on which the assertion in the second statement rests. The argument is:

If: genes produce enzymes
Then: enzymes produce behavior

However, this argument is illogical because the conclusion itself is an assumption (i.e., it does not follow from the premise). If one does assume the conclusion, then something else is implied:

If: genes produce enzymes
And: enzymes produce behavior
Then: genes produce behavior

It appears that Aceves-Pina et al. *assumed* the truth of the last statement and then deduced the middle statement. That is, the argument is illogical and, except for the first statement, without an empirical base. An isomorphism of genotype and phenotype has been assumed: given a specific gene, a specific expression of behavior will develop. This is simplistic because it rests on untested assumptions about the development of behavior. The study of development is still in its infancy despite a long history (for discussions see Hinde, 1968; Lehrman, 1953, 1970; Lewin, 1984; Oyama, 1982; Waddington, 1975). Little is known and even less may be assumed. Presently, the only conclusion that may be made about the results of the genetic analyses performed by Quinn and his colleagues is that differences in certain enzymes are *correlated with* differences in learning ability between groups with different genotypes at a given locus. Although this is itself an important discovery, one must be aware of what it does and does not imply about individual development. Otherwise, one might conclude more about the development of learning ability in specific individuals than the data can support.

Recently, we have become disturbed by differences between, on the one hand, claims for conditioning that have appeared in published accounts using the procedure described in Quinn et al. (1974) (e.g., Aceves-Pina et al., 1983; Byers, 1980; Dudai, 1977; Dudai et al., 1976; Quinn & Dudai, 1976; Quinn, Sziber, & Booker, 1979) and, on the other hand, results of studies using this same procedure that have not appeared in the literature. For example, John Ringo (1983), while working with J. Hall at Brandeis University, informed us that he was not able to replicate consistently the effect claimed for this conditioning procedure. He reported the following: "I am now trying to use the Quinn paradigm for 'conditioning'. . . . The experiment is simply very difficult to replicate. I get the expected results with Canton-S only some of the time." We have learned informally (Hirsch, 1983, 1986) that, in Quinn's laboratory at Princeton University, the learning effect in this apparatus could not be obtained with predictable consistency for about two years. Therefore, a new apparatus was developed (using dc instead of ac current), which has been reported to produce a much stronger and more consistent effect (Aceves-Pina et al., 1983; Tully & Quinn, 1985). Also, it has now been reported (Tully & Quinn, 1985, p. 268) that a humidity control at Princeton was belatedly discovered to have long been malfunctioning and might have affected earlier experiments. This is disturbing because the published accounts gave no indication of such problems. It is hoped that Quinn and his colleagues will discuss these problems in future publications because results of their genetic analyses, which have become increasingly complex and detailed, must rest on a solid behavioral foundation; otherwise, these genetic analyses become meaningless.

Using an instrumental conditioning procedure for testing Canton-S and several of the "learning mutants" isolated by Quinn and his colleagues, Mariath (1985) has reported reliable learning in individual *D. melanogaster*. Test-retest correlations of about 0.8 were obtained for all populations (i.e., Canton-S, *dunce, amnesiac, rutabaga*) and little difference was found in the learning indices among populations. Although little detail was presented concerning the procedure and, hence, little can be concluded about its adequacy, it does represent one of the few instances in which reliable individual learning has been reported in dipterans.

Holliday (1984; Holliday & Hirsch, 1984) has reported classical conditioning of *D. melanogaster* using an automated procedure (see Vargo, Holliday, & Hirsch, 1983, for details) similar to the hand-testing procedure developed by McGuire (1979). The automated procedure removes most of the problems concerning possible experimenter bias in the hand-testing procedure, although the experimenter must still judge whether or not a response has occurred. However, blind testing was also done. The conditioning procedure involved the presentation of a CS (.5 M NaC1) to the foretarsi for 5 s immediately followed by a US (.25 M sucrose). An interstimulus interval of 170 s ensued, and was followed by the presentation of an intertrial stimulus (ITS; distilled water) to discharge CES. Although this is similar to a discriminative conditioning procedure, the putative CS – (i.e., the ITS) is not known yet to be an appropriate discriminative stimulus because CES is still strong 170 s after sucrose presentation. This could confound responses due to inhibitory conditioning, CES, or a number of nonassociative factors other than CES.

Figure 5.14 presents results for the conditioning procedure. There is a statistically significant increase in responses to the CS and a statistically significant decrease in responses to the water stimulation. The latter suggests inhibitory conditioning, although, as stated above, this has yet to be tested. An unpaired control (CS preceding US by 90 s) was tested and results are presented in Fig. 5.15. Over trials, the unpaired group shows a much smaller increase in responses to CS than does a conditioning group tested at the same time.

The conditioning interpretation has been further validated by Holliday (unpublished), who has done blind testing with control method 4b. In a 15-trial experiment, he found significantly higher acquisition responding with a 1.0 (CS:NaC1, ITS:$H_2$0) than with a 0.5 (CS:NaC1, ITS:NaC1) contingency.

Furthermore, by varying the CS-US contingency between acquisition and extinction (i.e. paired during acquisition and unpaired during extinction), it has been possible to do excitatory conditioning with identified individuals (Holliday

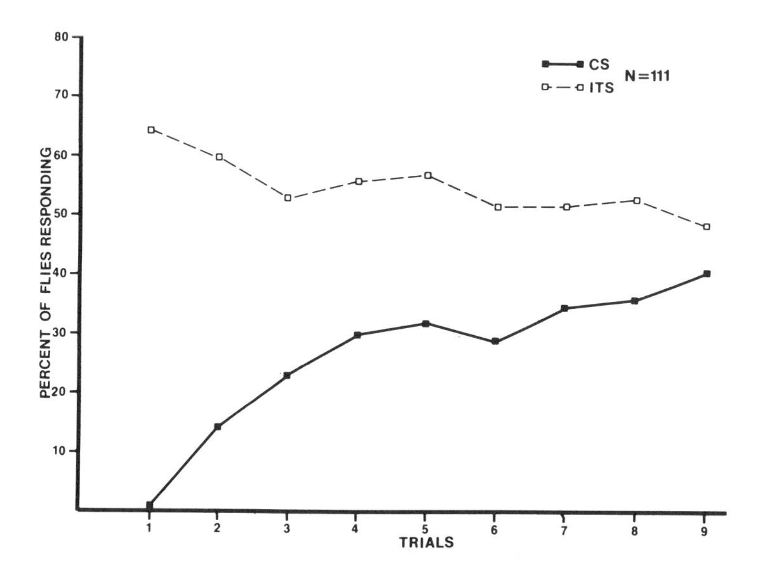

FIG. 5.14 Percentage of flies responding to CS and ITS over trials in the study by Holliday (1984). (*N* = sample size; *CS* = 0.5 M saline; *ITS* = distilled water.)

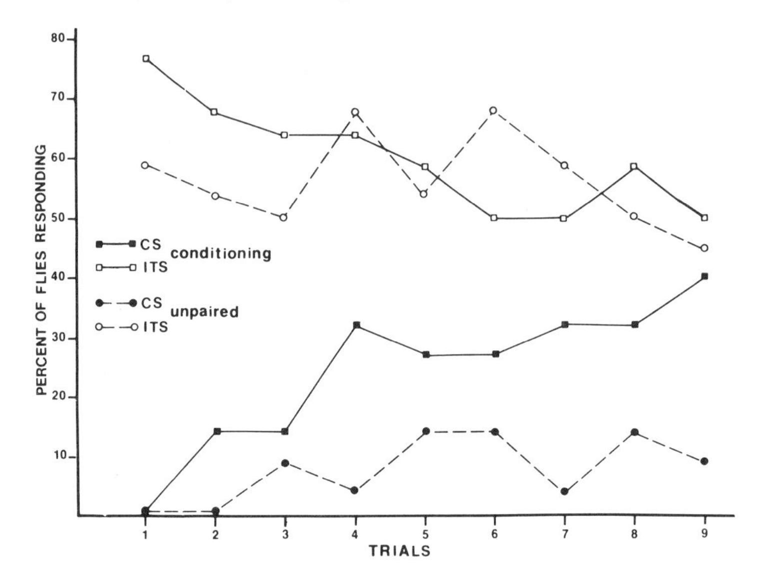

FIG. 5.15 Percentage of flies responding to CS and ITS over trials in the classical conditioning and unpaired control procedures in the study by Holliday (1984). (*N* = sample size; *CS* = 0.5 M saline; *ITS* = distilled water).

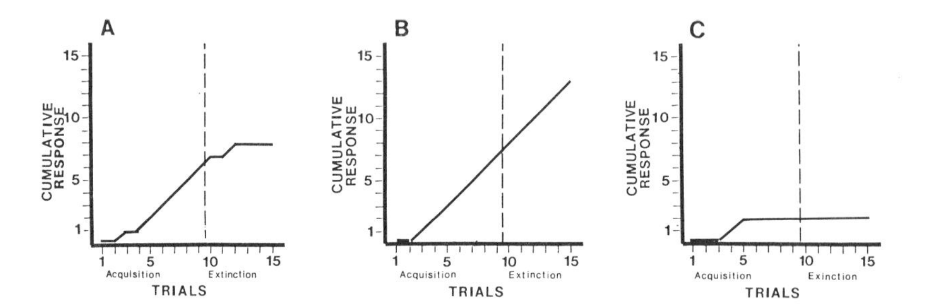

FIG. 5.16 Cumulative responses to CS over nine trials of conditioning and six trials of unpaired extinction for (A) an individual representing those flies (8 out of 34, 24%) showing both conditioning and extinction, (B) an individual representing those flies (5 out of 34, 15%) showing acquisition without extinction, and (C) an individual representing those flies (12 out of 34, 35%) responding, but showing neither conditioning nor extinction (Holliday & Hirsch, 1986a).

& Hirsch, 1986a). This procedure has permitted distinguishing four classes of individual differences (see Fig. 5.16): (a) 24% showing acquisition with extinction, (b) 15% acquisition without extinction, (c) 35% neither acquisition nor extinction, and (d) 29% not responding (not shown). Only class (a) qualified as showing associative conditioning and should be used in a selection study to breed a population with good conditioning performance.

CONCLUSION

This review of learning in Diptera shows that apparently both *P. regina* and *D. melanogaster* can be classically conditioned, but that the effect usually is not very strong. Additionally, conditioning has been impossible to infer consistently with different procedures and experimenters. There has been great variability in the results obtained. Is this because of experimenter bias, instability in the learning ability of flies, and/or an extreme sensitivity to minor changes in the experimental situation?

Intuitively, it seems reasonable that learning would be of little adaptive value for flies in nature (though proving a null hypothesis is not possible). Yet, this does not mean that, in another environment, one might not be able to construct conditions that favor learning. If we assume that such conditions have been constructed, what might we expect to see? It seems possible that association would tend to be a "hit-or-miss" affair: Sometimes, if conditions were just right, it would occur and could be inferred from the results. But a certain degree of inconsistency might be expected. Because learning ability may not have resulted in increased reproductive success in nature, its development in the animal may not be well "buffered" against changes in the experimental situation (Waddington, 1975). Therefore, changes in conditions that ordinarily might be imperceptible to the experimenter might cause large changes in the expression of learning ability. In fact, this may be the reason why there has been so much variability in the results of conditioning procedures discussed in preceding sections. However, the evidence overall appears to be conclusive: flies can learn.

This argument does not preclude the development of a conditioning procedure that allows one to infer association consistently; it only suggests that this might be very difficult. If such a procedure could be developed, the reasons for its success would raise questions of interest from an evolutionary perspective; i.e., what are the aspects of the procedure that allow such consistency and how do they relate to conditions in nature associated with reproductive success?

The intent of this chapter is to advocate a position of extreme caution in interpreting the results of conditioning experiments in Diptera. Dethier (1966) has expressed well the conclusion to which we, at an earlier stage, were led: "the entire question of learning in the fly is puzzling" (p. 125). It was puzzling because for every claim of successful conditioning, there often appeared to be a counterclaim (though usually unpublished). Claims for conditioning have been, perhaps, too hasty in many instances. We do not exclude ourselves from such criticism and have taken pains to show where we may have been misled. Yet, we have learned from these mistakes and hope that others perform a similar "soul-searching" analysis of their evidence. Recent attempts by McGuire (1984) and Tully (1984) to review the literature of fly learning have not been adequate because they have considered few of the problems examined here. Their discussions have now been corrected by Holliday and Hirsch (1986b), to which

Tully (in press) has since contributed another comment that merely repeats and further compounds their original confusion, once again misunderstanding the fundamental distinction in behavioral analysis between the type of evidence required to infer the presence of learning in a specified individual and the use of group data for demonstrating an effect in the species—a distinction Mariath (1985) appears to have made correctly. Learning in Diptera has problems not usually seen in studies of mammals. A more careful approach with respect to these problems may lead to interpretations that have more validity and generality. One factor that would be of benefit to those studying such problems is the availability of primary data. The analysis of the research presented here used the primary data available publicly in the theses or dissertations by McCauley (1973), Jackson (1976), McGuire (1978), Tully (1982), Zawistowski (1984), Holliday (1984), Vargo (1985), and Brzorad (1985). Therefore, the evidence is permanently on record for rechecking their analyses and interpretations using other assumptions or perspectives. In fact, we have disagreed with some of these interpretations and have formulated new ones. We believe that placing the primary data in readily accessible form should become a common practice.

In conclusion, the study of learning in Diptera is indeed puzzling. Yet, it is fraught with so many possibilities for biology and psychology that the enormous difficulties seem well worth the efforts to overcome them.

REFERENCES

Aceves-Pina, E. O., Booker, R., Duerr, J. S., Livingstone, M. S., Quinn, W. G., Smith R. F., Sziber, P. P., Tempel, B. L., & Tully, T. P. (1983). Learning and memory in *Drosophila*, studied with mutants. *Cold Spring Harbor Symposia on Quantitative Biology, 48,* 831–840.

Benzer, S. (1976). Letter to Jerry Hirsch, 26 October 1976. In J. Hirsch Papers, University of Illinois Archives, No. 15/19/22.

Booker, R. & Quinn, W. G. (1981). Conditioning of leg position in normal and mutant *Drosophila*. *Proceedings of the National Academy of Sciences, USA, 78,* 3940–3944.

Brzorad, J. N. (1985). *Conditioned discrimination in Phormia regina.* Unpublished master's thesis, University of Illinois, Urbana-Champaign.

Byers, D. (1980). Studies on learning and cyclic AMP Phosphodiesterase of the dunce mutant of *Drosophila melanogaster* (Doctoral dissertation, California Institute of Technology, 1980). *Dissertation Abstracts International, 40,* 2955B-2956B.

Cronbach, L. (1951). Coefficient alpha and the internal structure of tests. *Psychometrika, 16,* 297–334.

DeJianne, D., McGuire, T. R., & Pruzan-Hotchkiss, A. (1985). Pavlovian conditioning of proboscis extension in *Drosophila melanogaster. Journal of Comparative Psychology, 99,* 74–80.

Dethier, V. G. (1966). Insects and the concept of motivation. *Nebraska Symposium on Motivation, 14,* 105–136.

Dethier, V. G., Solomon, R. L., & Turner, L. H. (1965). Sensory input and central excitation in the blowfly. *Journal of Comparative and Physiological Psychology, 60,* 303–313.

Dethier, V. G., Solomon, R. L., & Turner, L. H. (1968). Central inhibition in the blowfly. *Journal of Comparative and Physiological Psychology, 66,* 144–150.

Dudai, Y. (1977). Properties of learning and memory in *Drosophila melanogaster. Journal of Comparative Physiology, 114,* 69–89.
Dudai, Y., Jan, Y-N., Byers, D., Quinn, W. G., & Benzer, S. (1976). Dunce, a mutant of Drosophila deficient in learning. *Proceedings of the National Academy of Sciences, USA, 73,* 1684–1688.
Duerr, J. S. & Quinn, W. G. (1982). Three *Drosophila* mutations that block associative learning also affect habituation and sensitization. *Proceedings of the National Academy of Sciences, USA, 79,* 3646–3650.
Erlenmeyer-Kimling, L. (1961). A genetic analysis of geotaxis in *Drosophila melanogaster* (Doctoral dissertation, Columbia University, 1961). *Dissertation Abstracts, 22,* 1262–1263.
Falconer, D. (1981). *Introduction to quantitative genetics* (2nd ed.). New York: Longman.
Frings, H. (1941). The loci of olfactory end-organs in the blowfly, *Cynomyia cadaverina* Desvoidy. *Journal of Experimental Zoology, 88,* 65–93.
Hebb, D. O. (1953). Heredity and environment in mammalian behaviour. *British Journal of Animal Behaviour, 1,* 43–47.
Hershberger, W. A., & Smith, M. P. (1967). Conditioning in *Drosophila melanogaster. Animal Behaviour, 15,* 259–262.
Hinde, R. A. (1968). Dichotomies in the study of development. In J. M. Thoday & A. S. Parkes (Eds.), *Genetic and environmental influences on behaviour* (pp. 3–14). Edinburgh: Oliver & Boyd.
Hirsch, J. (1967). Behavior-genetic analysis. In J. Hirsch (Ed.), *Behavior-genetic analysis* (pp. 416–435). New York: McGraw-Hill.
Hirsch, J. (1983, 1986). Two private communications about the Quinn laboratory 24 February 1983, 1 February 1986. In J. Hirsch Papers, University of Illinois Archives, No. 15/19/22.
Hirsch, J., & Erlenmeyer-Kimling, L. (1961). Sign of taxis as a property of the genotype. *Science, 134,* 835–836.
Hirsch, J., & McCauley, L. (1977). Successful replication of, and selective breeding for, classical conditioning in the blowfly *Phormia regina. Animal Behaviour, 25,* 784–785.
Holden, C. (1985). An omnifarious data bank for biology? Science, *228,* 1412–1413.
Holliday, M. J. (1984). *Classical conditioning of the proboscis extension reflex in Drosophila melanogaster.* Unpublished master's thesis, University of Illinois, Urbana-Champaign.
Holliday, M. J., & Hirsch, J. (1984). Excitatory (Pavlovian) conditioning. *Drosophila Information Service, 60,* 124–126.
Holliday, M., & Hirsch, J. (1986a). Excitatory conditioning of individual *Drosophila melanogaster. Journal of Experimental Psychology: Animal Behavior Processes, 12,* 131–142.
Holliday, M., & Hirsch, J. (1986b). A comment on the evidence for learning in Diptera. *Behavior Genetics, 16,* 439 to 447.
Hotta, Y., & Benzer, S. (1970). Genetic dissection of the *Drosophila* nervous system by means of mosaics. *Proceedings of the National Academy of Sciences, USA, 67,* 1156–1163.
Hotta, Y., & Benzer, S. (1972). Mapping of behaviour in *Drosophila* mosaics. *Nature, 240,* 527–535.
Jackson, D. (1976). *Individual learning (classical conditioning) stability in the blowfly* (Phormia regina). Unpublished master's thesis, University of Illinois, Urbana-Champaign.
Kemler, W. M. (1975). Non-associative modification of behavior in *Drosophila melanogaster* classical conditioning procedures (Doctoral dissertation, University of Nebraska, Lincoln). *Dissertation Abstracts International, 35,* 4220B.
Lashley, K. (1950). In search of the engram. *Symposia of the Society for Experimental Biology, 4,* 454–482.
Lehrman, D. S. (1953). A critique of Konrad Lorenz's theory of instinctive behavior. *The Quarterly Review of Biology, 28,* 337–361.
Lehrman, D. S. (1970). Semantic and conceptual issues in the nature-nurture problem. In L. R. Aronson, E. Tobach, D. S. Lehrman, J. S. Rosenblatt (Eds.), *Development and evolution of behavior* (pp. 17–52). San Francisco: W. H. Freeman.

Lewin, R. (1984). Why is development so illogical? *Science, 224,* 1327–1329.

LoLordo, V. (1979a). Classical conditioning: The Pavlovian perspective. In M. Bitterman, V. LoLordo, J. Overmier, & M. Rashotte (Eds.), *Animal learning: Survey and analysis* (pp. 25–59). New York: Plenum.

LoLordo, V. (1979b). Classical conditioning: Contingency and contiguity. In M. Bitterman, V. LoLordo, J. Overmier, & M. Rashotte (Eds.), *Animal learning: Survey and analysis* (pp. 61–97). New York: Plenum.

LoLordo, V. (1979c). Classical conditioning: Compound CSs and the Rescorla-Wagner model. In M. Bitterman, V. LoLordo, J. Overmier, & M. Rashotte (Eds.), *Animal learning: Survey and analysis* (pp. 99–126). New York: Plenum.

Lorenz, K. (1950). The comparative method in studying innate behaviour patterns. *Symposia of the Society for Experimental Biology, 4,* 221–268.

Mackintosh, N. (1973). Stimulus selection: Learning to ignore stimuli that predict no change in reinforcement. In R. A. Hinde & J. S. Hinde (Eds.), *Constraints on learning* (pp. 75–96). London: Academic Press.

Mackintosh, N. (1983). *Conditioning and associative learning.* New York: Oxford University Press.

Manning, A. (1967). "Pre-imaginal conditioning" in *Drosophila. Nature, 216,* 338–340. (Reprinted in J. Hirsch & T. R. McGuire [Eds.]. [1982]. *Behavior-Genetic Analysis.* Stroudsburg, PA: Hutchinson Ross).

Mariath, H. A. (1985). Operant conditioning in *Drosophila melanogaster* wild-type and learning mutants with defects in the cyclic AMP metabolism. *Journal of Insect Physiology, 31,* 779–787.

McCauley, L. (1973). *Classical conditioning in blowflies* Phormia regina, *replication and selective breeding for learning ability.* Unpublished master's thesis, University of Illinois, Urbana-Champaign.

McDonald, K. (1985, April 24). Panel urges use of 'lower' animals in experiments. *Chronicle of Higher Education,* pp. 1, 13.

McGuire, T. R. (1979). Behavior-genetic analysis of *Phormia regina*: Conditioning, central excitatory state, and selection (Doctoral dissertation, University of Illinois, 1978). *Dissertation Abstracts International, 39*(12), 5751B.

McGuire, T. R. (1983). Further evidence for a relationship between central excitatory state and classical conditioning in the blow fly *Phormia regina. Behavior Genetics, 13,* 509–515.

McGuire, T. R. (1984). Learning in three species of Diptera: The blow fly *Phormia regina,* the fruit fly *Drosophila melanogaster,* and the house fly *Musca domestica. Behavior Genetics, 14,* 479–526.

McGuire, T. R., & Hirsch, J. (1977). Behavior-genetic analysis of *Phormia regina:* Conditioning, reliable individual differences, and selection. *Proceedings of the National Academy of Sciences, USA, 74,* 5193–5197. (Reprinted in J. Hirsch & T. R. McGuire [Eds.]. [1982]. *Behavior Genetic Analysis.* Stroudsburg, PA: Hutchinson Ross).

Médioni, J., & Vaysse, G. (1975). Suppression conditionnelle d'un réflexe chez la Drosophile (*Drosophila melanogaster*): Acquisition et extinction. *Comptes Rendus des Séances de la Société de Biologie, 169,* 1386–1391.

Nelson, M. (1971). Classical conditioning in the blowfly (*Phormia regina*): Associative and excitatory factors. *Journal of Comparative and Physiological Psychology, 77,* 353–368.

Overmier, J. (1979). Avoidance learning. In M. Bitterman, V. LoLordo, J. Overmier, & M. Rashotte (Eds.), *Animal learning: Survey and analysis* (pp. 313–348). New York: Plenum.

Oyama, S. (1982). A reformulation of the idea of maturation. In P. P. G. Bateson & P. H. Klopfer (Eds.), *Perspectives in ethology* (Vol. 5, pp. 101–131). New York: Plenum.

Quinn, W. G., & Dudai, Y. (1976). Memory phases in *Drosophila. Nature, 262,* 576–577.

Quinn, W. G., Harris, W. A., & Benzer, S. (1974). Conditioned behavior in *Drosophila melanogaster. Proceedings of the National Academy of Sciences, USA, 71,* 708–712.

Quinn, W. G., Sziber, P. P., & Booker, R. (1979). The *Drosophila* memory mutant *amnesiac. Nature, 277,* 212–214.

Rescorla, R. (1967). Pavlovian conditioning and its proper control procedures. *Psychological Review, 74,* 71–80.

Rescorla, R., & Wagner, A. (1972). A theory of Pavlovian conditioning: Variations in the effectiveness of reinforcement and nonreinforcement. In A. Black & W. Prokasy (Eds.), *Classical conditioning II: Current research and theory* (pp. 64–99). New York: Appleton-Century-Crofts.

Ricker, J. (1980). Attempted replication of selection for classical conditioning in the blow fly, *Phormia regina.* Unpublished manuscript. In J. Hirsch papers, University of Illinois Archives, No. 15/19/22.

Ricker, J. P. (1984). *Selection and hybrid analysis of* Drosophila melanogaster *for differences in a measure of geotaxis.* Unpublished master's thesis, University of Illinois, Urban-Champaign.

Ricker, J., Brzorad, J. (1986). A demonstration of discriminative conditioning in the blow fly, *Phormia regina, Bulletin of the Psychonomic Society, 24,* 240–243.

Ricker, J. P. & Hirsch, J. (1985a). Evolutionary changes in laboratory populations selected for geotaxis. *Drosophila Information Service, 61,* 141–142.

Ricker, J. P. & Hirsch, J. (1985b). Evolution of an instinct under long-term divergent selection for geotaxis in domesticated populations of *Drosophila melanogaster. Journal of Comparative Psychology, 99,* 380–390.

Ringo, J. (1983). Letter to Jerry Hirsch, 12 February 1983. In J. Hirsch Papers, University of Illinois Archives, No. 15/19/22.

Rosenthal, R. (1976). *Experimenter effects in behavioral research* (Enlarged ed.). New York: Irvington Publishers, Inc.

Schneiderman, N. (1973). *Classical (Pavlovian) Conditioning.* Morristown, N. J.: General Learning Press.

Smith, N. (1970). Replication studies: A neglected aspect of psychological research. *American Psychologist, 25,* 970–975.

Thompson, R. F. (1983). Neuronal substrates of simple associative learning: Classical conditioning. *Trends in Neurosciences, 6,* 270–275.

Thorpe, W. H. (1939). Further studies on pre-imaginal olfactory conditioning in insects. *Proceedings of the Royal Society of London, 127,* 424–433.

Thorpe, W. H. (1943). Types of learning in insects and other arthropods: Part I. *The British Journal of Psychology, 33,* 220–234.

Thorpe, W. H. (1944). Some types of animal learning. *Proceedings of the Linnean Society of London, 156,* 70–83.

Tully, T. P. (1982). Behavior-genetic analysis of the black blow fly, Phormia regina, using the central excitatory state (CES). (Doctoral dissertation, University of Illinois, Urbana-Champaign, 1981). *Dissertation Abstracts International, 42,* 3568B.

Tully, T. (1984). *Drosophila* learning: Behavior and biochemistry. *Behavior Genetics, 14,* 527–557.

Tully, T. (in press). Measuring learning in individual flies is not necessary to study the effects of single-gene mutations in *Drosophila. Behavior Genetics.*

Tully, T. & Hirsch, J. (1982). Behaviour-genetic analysis of *Phormia regina* II. Detection of a single, major-gene effect from behavioural variation for central excitatory state (CES) using hybrid crosses. *Animal Behaviour, 30,* 1193–1202.

Tully, T. & Hirsch, J. (1983). Two nonassociative components of the proboscis extension reflex in the blow fly, *Phormia regina,* which may affect measures of conditioning and of the central excitatory state. *Behavioral Neuroscience, 97,* 146–153.

Tully, T., & Quinn, W. G. (1985). Classical conditioning and retention in normal and mutant *Drosophila melanogaster. Journal of Comparative Physiology, 157,* 263–277.

Tully, T., Zawistowski, S., & Hirsch, J. (1982). Behavior-genetic analysis of *Phormia regina:* III. A phenotypic correlation between the central excitatory state (CES) and conditioning remains in replicated F_2 generations of hybrid crosses. *Behavior Genetics, 12,* 181–191.

Vargo, M. A. (1985). The behavior-genetic analysis of *Drosophila melanogaster* using the central excitatory state. (Doctoral dissertation, University of Illinois, Urbana-Champaign, 1984). *Dissertation Abstracts International, 45,* 3655B.

Vargo, M., & Hirsch, J. (1982). Central excitation in the fruit fly (*Drosophila melanogaster*). *Journal of Comparative and Physiological Psychology, 96,* 452–459.

Vargo, M., & Hirsch, J. (1985). Behavioral assessment of lines of *Drosophila melanogaster* selected for central excitation. *Behavioral Neuroscience, 99,* 323–332.

Vargo, M., Holliday, M., & Hirsch, J. (1983). Automatic stimulus presentation for the proboscis extension reflex in Diptera. *Behavior Research Methods & Instrumentation, 15,* 1–4.

Waddington, C. H. (1975). *The evolution of an evolutionist.* Ithaca: Cornell University Press.

Yeatman, F., & Hirsch, J. (1971). Attempted replication of, and selective breeding for, instrumental conditioning of *Drosophila melanogaster. Animal Behaviour, 18,* 454–462.

Zawistowski, S. L. (1979). *Non-associative factors affecting response to saline in a classical conditioning paradigm for the blow fly,* Phormia regina. Unpublished master's thesis, University of Illinois, Urbana-Champaign.

Zawistowski, S. L. (1984). Conditioned discrimination in the behavior-genetic analysis of the blow fly, *Phormia regina.* (Doctoral dissertation, University of Illinois, Urbana-Champaign, 1983). *Dissertation Abstracts International, 45,* 389B.

Zawistowski, S., & Hirsch, J. (1982). Responsiveness of *Phormia regina* to saline. *Journal of Comparative and Physiological Psychology, 96,* 850–851.

Zawistowski, S., & Hirsch, J. (1984). Conditioned discrimination in the blow fly, *Phormia regina:* Controls and bidirectional selection. *Animal Learning & Behavior, 12,* 402–408.

6 Genetic Factors and Human Reactions to Alcohol

James R. Stabenau
University of Connecticut School of Medicine
Farmington

Encounter with alcohol spans most of man's social existence. Its use marks religious, cultural, and family events. For some individuals alcohol is used to alter states of consciousness in much the way other herbs and drugs have similarly been used over time. William James felt "drunkeness is a purely accidental susceptibility of a brain, evolved for entirely different uses, and its causes are to be sought in the molecular realm, rather than in any possible order of 'outer relations' " (Goodwin, 1976, p. 63). Observation of human behavior has led to categorization of the patterns of response to alcohol by the body and brain (Stabenau, in press). This chapter reviews data and their interpretation in an attempt to characterize the range of reactions in humans to alcohol and, where possible, to consider the way genetic factors influence the form and/or patterns of such reaction. The view covers the following areas with this goal in mind.

1. Epidemiology of use and misuse;
2. The definition and clinical methods used to measure or diagnose alcohol abuse, alcohol dependence, and alcoholism;
3. Genetic studies of alcohol intake, excretion, clearance, and metabolism;
4. Studies of transmission of genetic vulnerability to alcohol dependence and/or alcoholism among family members, twin populations, and adopted-away offspring;
5. The natural history of alcoholism including the antecedents and consequences of prolonged alcohol use/misuse;
6. The heterogeneous nature of genetic risk factors including assortative mating;
7. Studies of "genetic" markers in populations at high risk for the development of alcoholism;
8. Gene-Environment interaction and models of genetic transmission.

This sequence will take the reader through definitions, natural history, and genetically controlled studies of clinical forms of alcohol misuse. The order of the sections attempts to expand from a framework of observed alcohol use and misuse data to a conceptual network involving gene-environment interactional theory.

EPIDEMIOLOGY OF USE AND MISUSE OF ALCOHOL

The biological, psychological, social, cultural, and environmental reasons for use of alcohol have been catalogued. The primacy of any one of these factors or their interaction in explaining alcohol abuse is not clear. Alcoholism, defined by the dictionary as a chronic pathological condition, chiefly of the nervous and gastroenteric systems, caused by habitual excessive alcohol consumption, or as the end state of excessive or abnormal use of alcohol, is a protein disorder that transcends time and culture, having one common determinant: chronic repetitive excessive drinking. In the Western World throughout most of the 19th century alcoholism was seen as part of a spectrum of disorders termed "hereditary degeneracy" so that alcoholism, criminality, prostitution, insanity, epilepsy, and mental handicap were all viewed as features of one syndrome (Whalley, 1980). Modern psychiatric diagnostic measures applied in epidemiologic samples have characterized a variety of clinical patterns of appearance and etiology for the numerous members of this hereditary degeneracy syndrome.

Patterns of alcohol use have been observed to be different for cultures and appear to remain relatively stable over time. However, when alcohol consumption rates have been considered in addition to cultural differences, and possible genetic defects, the distribution of drinking behavior has been found to be not bimodal (i.e., "alcoholism" and "normal") but unimodal (Murray & Gurling, 1982a). In Western industrialized society a small percentage (5%–10%) of the population drinks a third or even up to a half of all the consumed alcohol (VonWartburg & Buhler, 1984). The majority of the people in alcohol-consuming cultures show a lifelong pattern of little or moderate drinking without developing alcohol-related problems.

The question is, why do individuals drink to an alcoholic state? The answers include the biochemical, psychologic, physiologic, psychodynamic behavioral, social, and cultural determinants of "drinking." The range and modal characteristics of the alcoholic state must be defined before advancement may be made as to different etiologies. Ludwig (1983) noted three major models: (1) *The disease model* in which certain individuals, genetically or biologically vulnerable, develop alcoholism when exposed to environments conducive to high consumption of alcohol. Once addicted, superimposed physical, neurological, and psychosocial problems appear as potential complications and a part of the disease. (2) *The motivational model* where excessive drinking was mainly attributed to intrapsychic conflict and use of alcohol for drive reduction. (3) *The learning (psychosocial) model*, although minimizing the importance of both biological and psychologically predisposing factors, accounted for drinking behavior mainly on the basis of learning, through parental or peer models through operant conditioning or through classical conditioning. These models are not mutually exclusive, but each in part is involved in accounting for the life course of alcoholism, and information from all three is needed to account for the total picture of alcoholic behavior. Most data in this chapter concern a review of hypotheses associated with the disease model.

To estimate prevalence of alcoholism and alcohol-related problems, Jellinek's method extrapolating from the incidence of hepatic cirrhosis is now generally regarded as invalid. Current viable methods include estimates based on direct assessment or reporting of alcohol mortality, personal and national alcohol consumption, data from general population surveys, and data from clients in treatment.

A major national survey of drinking practices and attitudes about drinking in America found "the highest proportions of heaviest drinking to be among those about age 40, of lower social status, living in larger cities and in the middle Atlantic and New England and Pacific areas and of Irish, British and/or Latin American extractions" (Cahalan, 1982, p. 99). The groups with the highest proportion of drinkers, however, did not have the highest proportion of heavy drinkers. For example, Jews and Episcopalians had the lowest proportion of abstainers and they also had extremely low rates of heavy or problem drinking. Abstainers were defined as those who do not drink alcohol beverages or who drink them less than once a year; heavy drinkers were those who sometimes drink two or more drinks of at least two beverage types per occasion and who report that they drink on at least 10 occasions per month (Cahalan, 1982). Per capita consumption by those 15 years and older, in the United States, has been reported to have increased by a third from slightly under 2 gallons of absolute alcohol in the early 1950s to nearly 2.7 gallons by the 1970s (Straus, 1983). In a second national survey in 1979, Cahalan reported no change in frequencies of heavy drinking. Rates for men were 13% in 1967 and 12% in 1979; for women they were 2% in 1967 and 3% in 1979 (Cahalan, 1982). Haglund and Schuckit (1977), using data from Cahalan and Cisin (1968) and Effron, Keller, and Gurioli (1974), reviewed the characteristics of American drinkers by drinking group. Heavy drinking of alcohol appears to be a necessary precursor for the development of alcohol dependence and alcoholism, and thus attention should be directed especially to the group of heavy drinkers. Heavy drinking is more frequent for men as compared to women. When sexes are combined, heavy drinking appears to be evenly distributed for ages 21 to 60 and for white and black races (Table 6.1). In a survey based on routine health examinations in 59,760 persons, reported use of three or more drinks daily was similar in whites, Latins, and blacks but was lower in Asian groups. Men of all races reported more drinking than women. Compared to data collected 15 years earlier, a substantive decline was reported in proportions of both abstainers and heavy

TABLE 6.1
Characteristics of American Adults by Drinking Group

	% Abstainers	*% Drinkers*			
		Infrequent	*Light and moderate combined*[a]	*Heavy*	*Total*
Total	32	15	41	12(18)	68
Sex					
Men	23	10	46	21(28)	77
Women	40	18	37	5(8)	60
Age					
21–29	24	15	47	14(18)	76
30–39	22	17	46	15(19)	78
40–49	29	12	44	15(21)	71
50–59	40	14	36	10(25)	60
60+	47	15	32	6(11)	53
Race					
White	31	15	42	12(17)	69
Black	38	12	36	14(23)	62

Source: Haglund and Schuckit, 1977, p. 32. Reprinted by permission.
Note. Based on data from Cahalan and Cisin, 1968; and Effron, Keller, and Guriolo, 1974. Percentage of heavy drinkers among all drinkers is given in parentheses.
[a](28 light, 13 moderate)

(3+) drinkers, as well as an apparent narrowing of race and sex differences (Klatsky, Siegelaub, Landy, & Friedman, 1983).

Lifetime prevalence rates for alcoholism have been made from epidemiologic samples of urban populations through home interviews using a structured diagnostic format (SADS, Schedule for Affective Disorders and Schizophrenia). Rates derived from a New Haven, Connecticut, sample studied in 1975–76 were 6.7% total with 10.1% for males and 4.1% for females (Weissman, Myers, & Harding, 1980).

In summary, for most world populations, groups with the highest proportion of drinkers did not have the highest proportion of heavy drinkers; rates of heavy drinking have been four to five times higher for men than women; and alcoholism rates are double for males compared to females.

THE PROBLEM OF WHAT TO DIAGNOSE

The study of human behavior and alcohol use and misuse is a powerful vehicle for exploring unified concepts and interactional models of behavior. Strauss (1983), has noted that:

> In the biological sphere we must conceptualize both genetic and experiential factors that affect the characteristics of the structure, function, and chemistry of the human body, all interacting to determine the potentialities and limitations of a particular individual's responses to alcohol. We must recognize the interaction between these varied somatic factors and the psychological meaning of alcohol, its reinforcing potential, and the reactions that an individual may seek for in alcohol: modifying mood, altering self-perception or projecting oneself differently to others. We must account for the input of social roles and cultural norms in defining the meaning of alcohol and in prescribing its use. We must also recognize that the activation of an individual's physical and psychological potentialities for responding to alcohol depends on exposure; including such factors as quantity, frequency and duration of drinking; the time, the place and the associated events and activities; and the form in which alcohol is ingested. All these factors are influenced by the physical environment; the prevailing level of technology; the social/cultural customs, values, attitudes and laws; and the occurrence of behaviors that complement drinking or that compete or conflict with it (p. 6).

The confusion in interpreting studies on human alcohol use parallels the confusion about the definition of alcoholism. The model set forth by Kaij and McNeil (1979) exemplifies the problem faced in clinical research of the alcoholism syndrome. These authors pragmatically distinguish alcohol addiction as defined by the occurrence of physical dependence on alcohol, loss of control over drinking, and blackouts during drinking bouts; and chronic alcoholism as a state in which medical and psychiatric complications have been added to the picture of alcohol addiction. A more inclusive formulation involving a progression may include three diagnostic stages or categories of alcohol misuse: *abuse*, which includes only heavy and problematic drinking; *alcohol dependence,* which includes physical and psychological dependence on, or addiction to, alcohol in addition to heavy or problematic drinking; and *alcoholism,* which involves dependence on alcohol but also the range of medical and psychosocial problems that accompanies prolonged alcohol misuse. This review focuses on the genetics of alcohol use, dependence and/or addiction, and the genetics of the syndrome of alcoholism with its associated range of psychosocial problems.

Jacobson (1983) has urged that continuing evolution of methods of detection, assessment, and diagnosis of alcohol abuse and dependence disorders is necessary if we are to alter effects of several static and outdated constructs and practices that may be impediments to progressive movement in prevention and treatment. It is equally clear that any measurement of the impact of genetic control over alcohol disorders must be based on diagnostic schema that will reliably and validly identify behaviors that can be shown to have genetic correlates.

Three different overlapping concepts are necessary in attempting to identify genetically controlled behavior. (1) *Detection*, or the process of identifying persons who are alcoholic or who have major significant life problems involving the use of alcohol. This is usually a binary decision: Is the person alcoholic or not? Although useful in screening, this level of behavioral description is empirical, is often not etiological, and may assume a unitary disease concept. An example of progress in detection is that of discriminate function analysis of screening tests, questions, and laboratory methods. One such formulation has achieved 100% sensitivity for excessive drinking and alcoholism, without any decline in the specificity of predictive value of a positive test result. To achieve this, Bernadt et al. used the three-question Reich-interview and blood levels of glutamate dehydrogenase activity (Bernadt, Mumford, & Murray, 1984). (2) *Assessment* may or may not involve a unitary concept but it does assume a continuum of severity on one or more dimensions along which the degree of seriousness may vary. There has been an underlying assumption in asserting the progressive nature of alcoholism. Caddy, Goldman, and Huebner (1976) have clustered eight prevailing models of alcoholism assessment into three: the disease model, the symptomatic model, and the behavioral model. Caddy (1977) supports the position of a behaviorally oriented multivariant account of alcoholism. This empirical modeling is different from, but may allow testing of Ludwig's (1983) Heuristic Modeling of Alcoholism. (3) *Diagnosis* is a labeling process that denotes the name of the disease a person has or is believed to have. The value of establishing a diagnosis is to provide a logical basis for treatment and a prognosis based on scientific and skillful methods that attempt to establish the course and nature of that disease, by evaluating the signs and symptoms present that distinguish it, and the natural history of the process (Jacobson, 1983).

Boyd, Weissman, Thompson, and Myers (1983) have applied seven different diagnostic criteria systems to a community sample of 510 subjects and found a lifetime prevalence of alcoholism in a range that varied from 3.5% using DSM III (American Psychiatric Association, 1980) criteria for alcohol dependence to 6.3% using SADS-RDC criteria for alcoholism and 5.3% for alcohol abuse according to DSM III criteria (Table 6.2). It is important to determine if alcohol abuse, alcohol dependence, and alcoholism are discontinuous and have separate

TABLE 6.2
Current and Lifetime Prevalence of Alcoholism in 510 Subjects
According to Seven Diagnostic Schemes

Diagnostic Scheme	*Current Prevalence (%)*	*Lifetime Prevalence (%)*
SADS-RDC	2.4	6.3
National Council on Alcoholism	2.4	6.3
DSM-III alcohol abuse	2.2	5.3
Feigner criteria	1.8	4.9
ICD-9 alcohol dependence syndrome	1.8	4.7
DSM-III alcohol dependence	1.6	3.5
Jellinek's gamma alcoholism	1.6	3.1

Source: Boyd et al., 1983, p. 1311. Reprinted by permission.

mechanisms (i.e., genetic/biochemical) that determine why an individual becomes alcoholic or whether they represent a progressive continuum of alcohol use over time by which anyone who drinks sufficient quantities of alcohol for long enough becomes alcoholic. Use of the SADS-RDC definition of alcoholism obscures such an issue, as that system does not discriminate between abuse and dependence. Boyd et al. (1983) note the hierarchical nature of the schema in the sense that some definitions identify, through restrictive criteria, smaller groups of subjects than do other definitions.

A summary of the biologic signs and symptoms that are required for the major schema used in the diagnosis of alcohol dependence and/or alcoholism appears in Table 6.3. All five of the systems require the detection of the dependence state for the diagnosis of "alcoholism." In addition, the detection and assessment of some, but not all, of the complications of alcoholic dependence (i.e., the physical, social, and medical complications) are a component of the diagnosis of the *alcoholism syndrome* by the Jellinek (1960), Feighner, Robins, Guze, Woodruff, and Winokur (1972), Spitzer, Endicott and Rubins (1978), and DSM III (1980) criteria (Table 6.3). The separation of the psychosocial consequences of dependence from the dependence syndrome is artificial. However, separate analysis of possible genetic control over the appearance of these "consequences" or "complications" in the course of alcohol use in the human lifetime is necessary to ascertain if subsets of alcoholics are at differential genetic risk for either psychosocial or medical consequences of prolonged alcohol use. For example, Mandell (1983) found that published data on retrospective recall support the position that there is a characteristic developmental sequence of clinical signs and symptoms in alcohol dependence illness. They cluster in three sequential, temporal phases: psychological dependence; physiological dependence; and neurological disorganization.

The application of operationally defined "states" of alcohol misuse may allow for effective means of assessing whether there are "traits" of alcohol misuse. A typological approach in the scientific investigation of alcohol misuse seems a necessary step. Jacobson (1983) has summarized the problem as follows. "Rather than continue the chimerical pursuit of a 'typical alcoholic' or a unitary 'alcoholism,' it would seem more reasonable and prudent to entertain the idea that there may be several alcoholisms in which once detected, assessed and diagnosed may be amenable to different treatments" (p. 379). Meyer, Babor, and Mirkin (1983) have provided a review of published typologies (Table 6.4).

In summary, alcohol misuse may involve three diagnostic stages: abuse, including heavy and problematic drinking; alcohol addiction, including physical and psychological dependence; and alcoholism, including medical and psychosocial consequences of prolonged alcohol misuse in addition to the alcohol dependence syndrome. Rates of abuse are almost twice those for dependence.

In the following sections, data are provided supporting the view that the alcoholism syndrome is moving from a unitary concept of disease toward a

TABLE 6.3

Comparison of Biologic Signs and Symptoms of Alcohol Dependence and Psychosocial/Medical Complications of Chronic Alcohol Misuse in Five Schemata for Diagnosis of Alcohol Dependence and/or "Alcoholism"

DSM III Criteria for Alcohol Dependence	*SADS, Research Diagnostic Criteria for Alcoholism*	*Feighner Criteria for Alcoholism*	*Jellinek Criteria for Alcoholism*	*W.H.O. Criteria for Alcohol Dependence*
I Pattern of Pathologic Alcohol Use				
Need for daily use for adequate functioning			2) Surreptitious drinking	1) Narrowing of repertoire of drinking behavior
Inability to cut down or stop	(3) Admits often can't stop (1) Says he drinks too much	Group 2. (1) Not able to stop Group 4. (1) Thinks he drinks too much	4) Avid drinking 8) Loss of control 13) Persistent remorse 14) Periods of total abstinence	2) Salience of drink-seeking behavior 6) Compulsion to drink (subjective awareness)
Repeated efforts to control by going on wagon or restriction to certain times of day		Group 2. (2) Allowing himself to drink . . . only after 5 p.m., only on weekends, only with others	15) Changing the pattern of drinking 18) Behavior becomes alcohol centered 22) Geographic escape	
Traffic accidents while intoxicated	11) Traffic difficulties due to drinking, reckless driving, accidents, speeding	(2) Traffic difficulties		
Arguments or difficulties with family or friends	(2) Others complain (7) Frequent difficulties with family members, friends, or associates (8) Divorce or separation where drinking is primary reason	Group 4. (4) Others object (2) Family objects (3) Lost friends	23) Change in family habits 24) Unreasonable resentments 16) Drop friends 29) Alcoholic jealousy	

	Binges (at least 2 days) Drinks more than one fifth per day	(9) Three occasions of 3 days drinking more than one fifth per day	Group 1. (4) Binges. 2 days with default of obligations	25) Protect supply 3) Preoccupation with alcohol 41) Obsessive drinking 5) Guilt feelings about drinking 6) Avoid reference to alcohol	
	Blackouts	(13) Frequent blackouts	Group 1. (3) Blackouts	1) Blackouts	
	Drinking exacerbates serious physical disorder				
	Drinks nonbeverage alcohol		Group 2. (4) Nonbeverage alcohol	36) Recourse to "technical products"	
II	Tolerance to alcohol				
	Need for increased amounts to achieve desired effect or diminished effect with same amount			0) Increased tolerance, response to alcohol as a needed drug	3) Increased tolerance to alcohol
III	Withdrawal symptoms				
	Morning shakes and malaise after cessation or reduction relieved by drinking	(14) Tremors	Group 1. (1) Tremulousness		4) Repeated withdrawal symptoms
		(4) Drinking before breakfast	Group 2. (3) Drinking before breakfast	30) Regular morning drinking	5) Relief drinking
IV	Impairment in social or occupational functioning				
	Violence while intoxicated	(10) Physical violence	Group 3. (4) Fighting	11) Grandiose behavior	
	Absence from work	(5) Missed work, impaired job performance, unable to take care of household responsibilities	(3) Trouble at work	12) Marked aggressive behavior 19) Loss of outside interests 21) Marked self-pity	

TABLE 6.3
(*continued*)

DSM III Criteria for Alcohol Dependence	*SADS, Research Diagnostic Criteria for Alcoholism*	*Feighner Criteria for Alcoholism*	*Jellinek Criteria for Alcoholism*	*W.H.O. Criteria for Alcohol Dependence*
Loss of job	(6) Job loss		17) Quit job	
Arrest for intoxicated behavior	(12) Picked up by police due to behavior	(1) Arrests for drinking		
V Medical Complications				
	(18) Cirrhosis	Group 1 (2) Cirrhosis		
	(19) Polyneuropathy	(2) Polyneuropathy		
		(2) Gastritis		
	(15) Delerium Tremens	(1) Delirium		
	(16) Hallucinations	(1) Hallucinations		
	(2) Korsakoff's psychosis	(2) Korsakoff's psychosis	34) Alcoholic psychosis	
	(17) Withdrawal seizures	(1) Convulsions		
		(2) Myopathy		
		(2) Pancreatitis		

Source: Adapted from Robins, 1982, pp. 44–46 and Mandell, 1983, pp. 417, 418. Reprinted by permission.

TABLE 6.4
Operational Definitions and Research Hypotheses Associated with Some Typological Formulations of Alcoholism

Primary Source	*Subtypes Postulated*	*Operational Criteria*	*Research Hypotheses*
Jellinek (1960) Negrete (1973)	Gamma–Delta	American alcoholics scoring high on "loss of control" scale, low on "inability to abstain" scale items French alcoholics scoring low on loss of control items and high on inability to abstain items	Gamma alcoholics will report more frequent intoxication, more severe withdrawal symptoms, more psychological "escape" reasons for drinking. Delta alcoholics will manifest greater social and psychological adjustment, and drink with greater frequency, less variability, and more for social reasons. Gamma alcoholics are expected to have a better prognosis.
Levine & Ziegler (1973)	Essential–Reactive	STEN scores on the 16 Personality Factor Inventory measuring high (essential) and low (reactive) emotional stability	Reactive alcoholics will manifest more variable drinking patterns, drink more for psychological escape reasons, manifest more current and past psychopathology, respond better to treatment.
Tomosovic (1974)	Binge–Steady	Self-report measure of quantity, frequency, and variability of consumption; previous month, previous 6 months, lifetime	Binge drinkers will indicate more social consequences drink more for escape reasons, manifest more current and past psychopathology, relapse more rapidly after resuming drinking.

Winokur et al. (1971)	Primary–Secondary	Age of onset of alcoholism and other psychopathology	Primary alcoholics will drink more for social reasons, manifest fewer social consequences, begin drinking earlier, come into treatment at a later age.
Schuckit et al. (1969)	Depressed–Antisocial Personality	Psychiatric diagnosis	Secondary and depressed-antisocial personality alcoholics will report more loss of control, more drinking.
Penick et al. (1978)	Positive–Negative Family History of Alcoholism	Family History Interview	Patients with positive family history will tend to have an earlier age of onset, indicate more psychopathology; female alcoholics, especially binge drinkers, should have more family history.
Shelly & Goldstein (1976)	Positive–Negative Organic Brain Dysfunction	Halstead-Reitan Neuropsychological Evaluation	Positives will report a longer history of heavy alcohol consumption, and indicate more numerous physical dysfunctions.
Smart (1979)	Male-Female	Gender identification	Males will report higher average daily intake, a longer history of daily drinking, greater tolerance and dependence, more social consequences. Females will drink more for psychological reasons, and indicate a greater prevalence of depression and a lower prevalence of antisocial personality.

Source: Meyer et al., 1983, pp. 238, 239. Reprinted by permission.

multifactorial model of illness that recognizes various subtypes of alcoholism. Such an approach may define researchable components of the heterogenous nature of alcoholism and may permit the determination of the etiology of these alcoholism subtypes.

GENETIC STUDIES IN HUMAN ALCOHOL USE

If a small portion of the population consumes the major quantity of alcohol, are there genetic factors that contribute to this alcohol use pattern and subsequent development of alcoholism in some of those individuals?

Our understanding of genetic control over the behavioral consequences of drug intoxication is surprisingly limited. Clinical studies during intoxication have shown a range of drug effects that includes joy, pain, euphoria, and dependency. No way as yet exists to predict which of the range of responses will occur following a particular drug dose at a particular time. Nongenetic issues such as expectancy, experience with the drug, tolerance, and social setting may be factors that modulate the behavioral effects of intoxication (Mello, 1983).

Although studies in humans and in experimental animals have shown a wide degree of individual variation in the biological responses to alcohol, Erwin and McClearn (1982) have concluded that a major share of this variation is of genetic origin. Animal study of response to alcohol factors in attaining dependence and tolerance employs inbreeding and selective breeding to heighten genetic contributions. Human studies must rely on family pedigree models, monozygotic (MZ) and dizygotic (DZ) twin comparison models, and adoption or cross-fostering models. Family and twin studies are confounded by familial and cultural effects even though twin studies control for genetic differences between siblings. Adoption cross-fostering models provide a measure of genetic differences and may also control for cultural and environmental rearing factors.

The National Institute of Alcoholism and Alcohol Abuse (NIAAA) (Fifth Special Report, 1983) has listed a number of biological responses to alcohol that could define their pathophysiological significance in the induction and maintenance of the chronic state of alcohol abuse in humans: alcohol craving or alcohol-seeking behavior; sensitivity of the central nervous system (CNS) to alcohol; sensitivity of other body systems to alcohol; rate of alcohol metabolism or elimination; rate of acquisition of tolerance to alcohol; rate of development and severity of physical dependence on alcohol; and sensitivity to the medical complications of clinical alcohol consumption.

Alcohol-Seeking Behavior

It is not known how much alcohol consumed over what given time periods is necessary to lead to alcohol dependence in humans. Estimates of alcohol use based on frequency of different patterns of alcohol beverage drinking have been compared in samples of differing genetic composition (Table 6.5). There is no significant difference in percentage of heavy drinking among those at genetic risk for alcoholism (i.e., having an alcoholic first- or second-degree family relative) as compared to control groups. It is not clear what the natural history of drinking styles is for individuals at risk for alcoholism as compared to those not at risk for alcoholism. One question is whether those who are destined to become dependent upon alcohol move from social drinking into heavy drinking, and then to problem and alcoholic drinking if they drink at "heavy" rates long enough. Further, do those with a genetic vulnerability move more rapidly through these stages into alcoholic drinking?

Filmore (1975) described a prospective series of college students followed into adulthood. Thirty-five percent of the variance in drinking problems at follow-up (time 2) were explained by drinking problems at initial assessment (time 1). Women had two pathways into drinking problems: (1) early psychological dependence and early frequent intoxication in college that predicted frequent intoxication and symptomatic drinking with few social complications in middle age; (2) early binge drinking associated with early frequent intoxication that predicted social complications, frequent intoxication, and symptomatic drinking. The prospective study also found for men that 19% of the variance of problem scores at time 2 was explained by time 1 problems. Filmore described a combination of early symptomatic drinking and either binge drinking or frequent intoxication that was a significant predictor for later time 2 problems. She notes that this group of young men already displayed drinking problems that are more frequently reported by advanced alcoholics and that are usually seen in middle age. The author also describes a typology in a subgroup of men similar to that described for women, in which social complications added to significant drinking and frequent intoxication are highly predictive across time. She hypothesizes that "frequent intoxication from early manhood to middle age has existed as a fairly steady drinking lifestyle" (p. 901). Data from other sources are not totally clear on the role of heavy drinking and alcoholic outcome. Goodwin concluded from his adoption study data (Table 6.5) that men with genetic vulnerability for alcoholism were more often "alcoholic" drinkers rather than "heavy drinkers" or "problem drinkers" (Goodwin, Schulsinger, Hermansen, Guze, & Winokur, 1973).

Two twin studies have provided within-pair variance estimates for monozygotic and dizygotic twins as to several drinking variables. Partanen, Bruun, and Markkanen (1966) studied 902 pairs of males from Finland aged 28–37 years and produced a factor analysis of drinking habits. Density, or the pattern of intake of alcoholic beverages, showed a genetic effect in that there was a smaller within-pair variance (W^2) for MZ twins compared to DZ twins. The estimated H, or heritability, as the proportion of total variance that may be attributed to genetic sources was .38 for this drinking variable. Amount of alcohol consumed also showed a genetic effect as W^2_{MZ} was smaller than W^2_{DZ}, with an H of .34.

TABLE 6.5
Frequency of Drinking Style for Males and Females in Family, Twin, and Adoption Studies (in percent)

Study, (N), Authors and Year	*Abstainer*		*No Excess, moderate or social drinker*		*Heavy drinker*		*Problem drinker, probable alcoholic drinker*		*Alcoholic drinker*	
	M	*F*	*M*	*F*	*M*	*F*	*M*	*F*	*M*	*F*
Siblings of Alcoholics (767)	16	33	44	52	19[a]	7[a]			22	8
Siblings of Spouse of Alcoholic (541) Hall et al., 1983a	22	38	46	51	19[a]	6[a]			12	5
Adults with childhood behavior problems (508)			41	54	16	10	14	4	7*	4
Adults without childhood behavior problems (99) Robins et al., 1962			66	79	19	11	3	0	0	0
Adoptees: Biologic parent alcoholic (55)			51		22		9		18**	
Adoptees: Biologic parent non-alcoholic (78) Goodwin et al., 1973			45		36		14		5	
Twin pairs (729)										
MZ concordant					58		43			
DZ concordant					41		48			
Partanen et al., 1966										
Twin pairs (111)										
MZ concordant	0		47		54				83	
DZ concordant	0		28		30				71	
Kaij, 1960										

[a]includes probable alcoholic drinkers.
$*p < .001$, $**p < .02$.

Lack of control, however, had similar within-pair variance for MZ and DZ twins. The British Twin Register Study by Clifford, Fulker, Gurling, and Murray (1981) demonstrated that for male and female MZ and DZ twin pairs the weekly consumption of alcohol was highest for males, 24.2 versus 8.5 cL of absolute alcohol, as compared to females. For weekday alcohol consumption (but not weekend) intraclass correlation, *r* (an estimate of genetic effect) for males was .80 for MZ pairs and .41 for DZ pairs (*p* .05), whereas for females *r* MZ was .63 and *r* DZ was .02 (NS, as within-pair variance was equal to between-pair variance). The authors concluded that genetic influences seem to be important in alcohol consumption in both males and females. However, psychological effects of alcohol appeared to be mainly environmentally determined.

In summary, males are consistently more frequently described as heavy drinkers; early frequent alcohol use is predictive of late frequent alcohol use for both males and females; drinking style is established early and, if heavy and enduring, is predictive of later life drinking problems. Family studies show no difference in heavy drinking rates in biologic family members versus marital family members. Twin studies show that the amount of alcohol consumption in males is under a degree of genetic control. Finally, although heavy drinking appears to be required for the development of alcoholism, most of the data suggest that individuals with the highest biologic genetic risk for alcoholism who are heavy drinkers move into alcoholic drinking, whereas others appear to be able to continue in heavy and even problem drinking.

Sensitivity of CNS to Alcohol

The quantity of alcohol consumed by an individual may be influenced by a series of positive and negative reinforcers for the continued use of alcohol. Response pattern may be based upon differences in the systemic effects of alcohol on the CNS (i.e., the state of pleasurable or unpleasurable sensation that accompanies initial alcohol intake) and the preservation and maintenance of the pattern of use or avoidance by classical conditioning. Most work in this area has been done in animals. Erwin, McClearn, and Kuse (1980), using HS/Ibg stock mice, demonstrated that voluntary ethanol consumption and acquisition of acute tolerance to ethanol were positively associated, whereas those measures were not significantly related to CNS sensitivity to alcohol. Selective breeding has produced mice with genetically controlled differences in CNS response to alcohol (long and short sleep) (McClearn, 1983) and to ethanol dependence (Wilson, Erwin, DeFries, Pedersen, & Cole-Harding, 1984). By infusion of chemical substances (tetrahydro-isoquinolines or beta-carboline) in the brains of rats and monkeys, Myers and Ewing (1980) showed that aversion to oral ingestion of alcohol was overcome. In contrast, infusions of high doses of tetrahydro-papaveroline inhibited the animals' involuntary intake of alcohol, even in weak concentrations. Although not replicated by others, the authors suggest that the biology of alcohol tolerance and dependence follows the CNS opiate receptor model and that changes in neuroreceptor sites may regulate sensitivity to alcohol aversion and regulate its consumption.

Studies of sensitivity to alcohol in humans with control for genetic differences are few. Schuckit (1984b) has reported a study of 23 male drinkers aged 21–25 with family history positive for alcoholism (FHP) who were matched to 23 controls with negative family history (FHN). All subjects were evaluated for subjective feelings of intoxication during a placebo phase, and when drinking 0.75 mL/kg and 1.1 mL/kg of ethanol. The FHP subjects reported less intense feelings of subjective intoxication after drinking, especially during the 2 hours following the peak blood alcohol concentration (BAC), and most marked at the 0.75 mL/kg dose. BAC and pretest expectation differences were not significant for the two groups. These data support the hypothesis that a decreased sensitivity in genetically vulnerable men may lead to excess use and then to dependence.

Alcoholics, when compared to non-alcoholics, have been observed to be less able to estimate BAC accurately even after discrimination training focused on changes in internal sensations during intoxication (Lanskey, Nathan, & Lawson, 1978). The genetic effects were evaluated in a study of 24 males aged 18–30 whose alcohol use patterns were more than extremely light and less than very heavy or problem drinking. Half the subjects had a family history of alcoholism in one parent or two second-generation family members. Family history or drinking pattern did not correlate with BAC estimation accuracy; however, "low-tolerant" subjects were substantially more accurate than "high-tolerant" subjects in post-training BAC estimation. The authors suggest that the development of tolerance may relate to the inability to discriminate BAC, but find no support for a family history positive genetic effect (Lipscomb & Nathan, 1980).

Wilson, Erwin, McClearn, et al. (1984) used 24 brother pairs to measure the genetic basis of individual differences in CNS sensitivity and acute behavioral tolerance to ethanol. Fifteen behavioral tests were made at time 1 immediately following a dose of ethanol that produced a BAC of 100mg/100mL, and at time 2 when BAC had fallen to one-half the time 1 peak. For 10 baseline tests there were significant intraclass correlations across brother pairs. No correlations were found for measures of sensitivity, and only one for acute tolerance; i.e., hand steadiness ($r = .45$, p .05). However, since baseline r for hand steadiness was already .60 (p .001) the data are not a strong demonstration of genetic control of either sensitivity or tolerance to alcohol in humans.

Rate of Alcohol Metabolism and Elimination

Alcohol enters the body and is absorbed through the stomach and small intestine. A number of factors, some of which may be genetically controlled, affect the absorption of ethanol: the concentration gradient of ethanol; blood flow at the site of absorption; the irritant properties of ethanol; the rate of ingestion; the type of beverage; presence or absence of food; emptying time of the stomach; protein deficiency; bodily temperature; physical activity; menstrual cycle; age and ethnicity (VonWartburg, 1971).

Only small amounts (2%) of ethanol are excreted unchanged in urine, expired air, and sweat. The liver is the principal site of ethanol metabolism. Clearance studies show approximately 75% of a dose of ethanol is eliminated by hepatic metabolism through degradation via a liver alcohol dehydrogenase (ADH) (Fig. 6.1). Human ADH, a dimer, is thought to exhibit polymorphism through combinations of monomers coded by three separate genes with polymorphism at one gene locus (Thacker, Veech, Vernon, & Rutstein, 1984). ADH polymorphism differences among Mongoloids with a high percentage of "atypical" ADH and the Asian alcohol flush have been described (Table 6.6). However, "atypical ADH" now appears to be an artifact of incomplete electrophoretic separation of isozymes. A direct genetic role of ADH in the etiology of alcoholism does not appear to be likely in view of the similarity of the rates of alcohol disappearance in alcoholics and non-alcoholics (Thacker et al., 1984).

Three separate twin studies have shown that variability in the degradation and elimination of ethanol has a genetic component (Vogel & Motulsky, 1979) (Table 6.7). The intrapair correlations for alcohol elimination from the blood for MZ twins were higher than for DZ twins. Similarly, the intrapair correlations for rate of alcohol degradation were higher for MZ as compared to DZ twins. Alcohol elimination rates were studied in 10 adoptees with an alcoholic biologic parent and 10 matched control adoptees. No significant differences were found between the two groups (Utne, Hanse, Winkler, & Schulsinger, 1977). Thus, although degradation and elimination may be in part genetically controlled, these metabolic rates may or may not be a vulnerability factor in alcoholism.

The second reaction of metabolism of ethanol in the liver is catalyzed by ALDH (acetaldehyde dehydrogenase) (Fig. 6.1). There are at least two kinetically distinct forms of hepatic ALDH: ALDH I, a low K_m mitochondrial form, and ALDH II, a high K_m cytoplasmic form. The steady state level of acetaldehyde

ALCOHOL METABOLISM

1) $CH_3CH_2OH + NAD^+ \xrightarrow[\text{(cytoplasm)}]{\text{ADH}} CH_3C(H){=}O + NADH + H^+$

Ethanol → Acetaldehyde

2) $CH_3C(H){=}O + NAD^+ + H_2O \xrightarrow[\text{(mitochondria)}]{\text{ALDH}} CH_3C(O^-){=}O + NADH + 2H^+$

Acetate

FIG. 6.1

TABLE 6.6
Atypical Alcohol Dehydrogenase (ADH) and Acetaldehyde Dehydrogenase (ALDH) Isoenzymes in Different Populations

Population	*% with Atypical ADH Isozyme*[a]	*% with ALDH Isozyme Deficiency*[b]	*% with "Somatic" Symptoms to Alcohol*[d]		*% with Facial Flushing to Alcohol*[c]
			I	*II*	
Caucasoid	5–20		0–6	0–9	
USA	3–5				
Europe		0			4–10
Great Britain and Greece	5				
Denmark and Sweden	10				
Egypt, Sudan, Liberia		0			
Chile	20				
Ecuador		39			
Mongoloid			26–53	18–37	
Japan	80–90	48			58–85
China		35			57–83
Indochina and Korea		40			

[a]Von Wartburg, 1980, p. 140
[b]Goedde, Agarwal and Harada, 1983, p. 183
[c]Goedde et al., 1983, p. 184
[d]Goedde, 1978, p. 181
I. Not in stomach, palpitation, tachycardia, $p < .001$
II. Muscle weakness, dizzy, sleepy, falls asleep, $p < .01$

is determined by the kinetics of ALDH. Facial flushing and somatic symptoms in alcohol-sensitive individuals are determined by differences in ALDH activity; i.e., in about 50% of Orientals ALDH I isozyme is missing (Thacker et al., 1984). Genetic difference between Caucasoids and Mongoloids appears to be important in differences of sensitivity to alcohol and its degradation products (Table 6.6). Suwaki and O'Hara (1985) found that 50.9% of 1,646 Japanese men had facial flushing upon alcohol ingestion. They concluded that there was a relationship between flushing and other indices of sensitivity to alcohol and low rates of alcohol-related problems in the subjects, and that alcohol-induced flushing acts as an inhibitory factor against excessive alcohol use. Harada, Agarwal, Goedde, Takagi, and Ishikawa (1982) reported that flushers were deficient in ALDH isozyme I, and that the frequency of this deficiency was 41% in Japanese non-alcoholics but only 2.3% in Japanese alcoholics.

Investigators of blood acetaldehyde levels have found values higher for alcoholics than non-alcoholics (Eriksson, 1980; Korsten, Matsuzaki, Feinman, & Lieber, 1975; Lindros, Stowell, Pikkarainen, & Salaspuro, 1980). In healthy subjects with alcohol first-degree relatives, Schuckit and Rayses (1979) found blood acetaldehyde values following ingestion of alcohol higher than those of family history negative controls. Other studies have found no differences for

TABLE 6.7
Studies of Alcohol Elimination and Degradation in Twins

Author (year) (alcohol dose)	No. of twin pairs	Alcohol elimination from blood (mg/ml h)				Alcohol degradation rate (mg/kg h)			
		Range	r_{MZ}	r_{DZ}	h^2	Range	r_{MZ}	r_{DZ}	h^2
Luth (1939) (0.5 g/kg)	10MZ, 10DZ	0.051–0.141	0.64	0.16	0.63	50.0–109.6	0.77	0.45	0.67
Vesell et al. (1971)(1 mL/kg)	7MZ, 7DZ	0.11–0.24	0.96	−0.38	0.98				
Kopun and Propping (1977) (1.2 mL/kg)	19MZ, 21DZ	0.073–0.255	0.71	0.33	0.46	57.6–147.6	0.76	0.28	0.41

Source: Vogel and Motulsky, 1979, p. 515. Reprinted by permission.
r_{MZ} = Intraclass correlation coefficient in MZ twins.
r_{DZ} = Intraclass correlation coefficient in DZ twins.
h^2 = Heritability estimate from comparison between MZ and DZ twins.

blood acetaldehyde level between alcoholics and controls (Eriksson & Peachy, 1980) or between healthy children with alcoholic parents compared to control children (Behar et al., 1983).

Thacker et al. (1984) have concluded that most studies of acetaldehyde are fraught with analytic and technical errors in acetaldehyde metabolism detection and "it would appear that the maximum total acetaldehyde content in blood of normal persons metabolizing alcohol is 2–3 uM and is unchanged in alcoholics hospitalized for detoxification, these reduced levels of acetaldehyde are consistent with basic kinetic and thermodynamic principles" (p. 379).

Rates in Development of Tolerance and Physical Dependence

Most work in the area of genetic factors associated with tolerance of and dependence upon alcohol is found in reports of animal study. Erwin and McClearn (1982) have reviewed animal data of their laboratory and those of others and conclude that studies suggest "a genetic influence on the acquisition of both acute and chronic tolerance to ethanol and upon the display of withdrawal symptoms following the discontinuance of chronic alcoholic exposure" (p. 414). However, the critical role of environmental differences in the expression of postulated genetically controlled tolerance in animals has only recently been examined. An observation not unique to alcohol is that the same regimen of daily alcohol (by intubation) produces tolerance much more rapidly in a rat that receives the alcohol just before a daily test performance than in one receiving the drug just after the daily test. Both develop tolerance on a much lower dosage than is needed by a rat receiving ethanol in its home cage without any performance required (Chen, 1968; LeBlanc, Kalant, & Gibbins, 1973). Tabikoff, Melchoir, and Hoffman (1984) using an inhalation methodology of administration, confirmed environment-dependent and environment-independent forms of alcohol dependence in rats.

In summary, some form of genetic influence has been attributed in humans to amount of alcohol consumed, sensitivity of the CNS and body function, and alcohol degradation and elimination. Genetic control over acquisition of tolerance and development of dependence has been most explicitly obtained from studies of animals selectively bred toward alcohol use.

GENETIC STUDIES OF THE TRANSMISSION OF THE VULNERABILITY TO ALCOHOLISM

The evidence suggesting that genetic factors are important in the transmission of a vulnerability toward the development of alcoholism lie in three observations: (1) frequencies of diagnosed alcoholism in male and female first- and second-degree relatives of known alcoholics exceed the rates for males and females in the general population; (2) rates of concordance for alcoholism are higher among monozygotic (MZ) than dizygotic (DZ) twins; and (3) children born to alcoholic parents but raised by foster parents with or without alcoholism have higher rates for developing alcoholism in adulthood than adoptees with biologic parents who are not alcoholic. The following examines such studies and their findings.

Rates of alcoholism in the Danish population as defined by chronic alcohol abuse and hospitalization for the alcohol dependent state are 3%–5% males and less than 1% for females (Goodwin et al., 1973). Recent estimates of American urban rates of probable and definite alcoholism defined by SADS-RDC criteria ascertained from epidemiologically controlled and selected samples by structured interview provide rates of 10.1% for males and 4.1% for females (Weissman et al., 1980). Similar studies in three U.S. urban sites of lifetime prevalence of alcoholism using the NIMH-DIS and DSM III criteria found rates of alcohol abuse *and* dependence as, 19.1 ± 1.1%, 24.9 ± 1.4%, and 28.9 ± 1.8% for males; and 4.8 ± 0.5%, 4.2 ± 0.4% and 4.3 ± 0.6% for females in New Haven, Baltimore, and St. Louis, respectively (Robins et al., 1984). The higher figures in U.S. rates are due mainly to more inclusive diagnostic criteria. The criteria for alcohol abuse principally include excessive use and thus are most inclusive, whereas criteria for dependence require either tolerance or dependence and thus include only a portion of abusers. The broad SADS-RDC criteria for alcoholism select a group of drinkers different in extent from those designated as abusive or dependent drinkers. Cotton (1979) reviewed 39 published studies of alcoholism in family members of alcoholics, psychiatric patients, and medical and surgical patients. A range of criteria was used for diagnosis of alcoholism in both the probands and the relatives. When mean frequencies of alcoholism in parents (known to have passed the period of risk for alcoholism of 15–55 years) were compared, 27.0% of fathers and 4.9% of mothers of alcoholics were alcoholic. Mean rates for parents of psychiatric patients were slightly higher than rates for non-psychiatric patients, but mean rates for both were in the range of rates of the general population for males and females, respectively (Murray & Stabenau, 1982) (Table 6.8).

Although family pedigree studies suggest a genetic pattern in alcoholism distribution, they do not control for effects of familial rearing by alcoholics upon children or for the distribution of genetic risk among siblings in a family. These two contributions to the variance of alcoholism are not equal or easily identifiable by family study methods. The study of alcoholism in pairs of MZ and DZ twins has provided some control for genotypic variation. The results of three major twin studies using three different methods of characterizing alcoholism are shown in Table 6.8.

Kaij (1960) characterized the responses to chronic alcohol use among pairs of male Finnish twins from a hospitalized alcohol treatment population and a temperance board population of registered alcohol offenders. He compared twins through clinical appraisals of drinking from social use to severe chronic abuse

TABLE 6.8
Comparison of Diagnosis of Alcoholism, Alcohol Abuse, and Alcohol Dependence in Samples with Control for Genetic Factors

Sample, Senior Author, and Year	*Sample Subsets +*		*Diagnosis and Criteria Used*
General Population, (*N* = 510) Weissman, 1980	Male 10.1	Female 4.1	Alcoholism (Probable and Definite), SADS-RDC
Family Pedigree (39 studies male and female)			Alcoholism,
Cotton, 1979	Fathers	Mothers	
Non-Psychiatric Patients	5.2	1.2	Mixed Clinical Methods
Psychiatric Patients	9.9	1.8	
Alcoholics	27.0	4.9	
Twin Studies (% Concordance)	Mz Twins	Dz Twins	
Gurling, 1981			
Male	33 (5/15)	30 (6/20)	Alcohol Dependence,
Male	8 (1/13)	13 (1/8)	W.H.O. Criteria
Hrubec, 1981 Male	27 (7/271)	12 (53/444)	Alcoholism, ICD-9
Kaij, 1960 Male	71 (10/14)	32 (10/31)	Chronic Alcoholism, Physical Dependence, Blackouts, and Pathologic Desire
Half-Sibling Study (male)	Biol. parent	Biol. parent	Primary Alcoholism,
Schuckit, 1972	Alcoholic	Non-alcoholic	Hospitalization and
Rearing Parents Alcoholic	46 (11/24)	14 (2/14)	Psychosocial Problems
Rearing Parents Non-Alcoholic	50 (11/22)	9 (9/104)	
Adoption Studies (male)			
Rearing Parents Nonalcoholic			
Goodwin, 1973	18 (10/55)*	5 (4/78)	Alcoholism, Feighner
Cadoret, 1978	40 (4/10)***	7 (5/72)	Alcoholism, Modified Feighner
Bohman, 1978	39 (35/89)**	14 (98/723)	Alcohol Abuse, Temperance Board Hospital Records

[a]percent, number of subjects in parentheses.
$*p < .02$. $**p < .01$. $***p < .001$.

(Grade 4) or symptomatic addiction to alcohol. Both samples were comparable, and the overall concordance for chronic alcoholism (71%) was twice that for DZ twins (32%). Two more recent twin control studies of alcohol use represent a departure from Kaij's work. Hrubec and Ommen (1981) utilized the large U.S. Veterans Twin Register, which included individuals healthy enough to be inducted into the service, but who also were identified as alcohol abusers through either outpatient or inpatient treatment. Rates of alcoholism defined by ICD-9 criteria were twice as high (26%) in MZ pairs as in DZ pairs (12%). A United Kingdom consecutive series of hospitalized alcohol abusers with identifiable twins were interviewed and compared using the WHO Alcohol Dependence Criteria. Although the authors did not report findings for chronic alcoholism, there was no significant difference in their report for alcohol dependence for either males or females when these samples of MZ and DZ twins were compared (Gurling, Clifford, & Murray, 1981) (Table 6.8).

These three twin studies suggest that a genetic vulnerability hypothesis is supported when chronic alcoholism is measured but not when dependence alone is measured. Chronic alcoholism as defined by Kaij and the ICD-9 criteria is based in part upon psychosocial complications that are not a part of the definition of the alcohol dependence syndrome as defined by the WHO Criteria (see Table 6.3).

Twin studies like family pedigree studies are subject to similar familial non-genetic factors that derive from a shared common familial milieu. Adoption and half-sibling studies, the human form of the cross-fostering model so common to animal genetic studies, allow not only estimation of genetic similarity or differences but control for the effect of differences in rearing.

Schuckit, Goodwin, and Winokur (1972), using clinical estimates of alcoholism based on St. Louis diagnostic methods that preceded the Feighner criteria, studied a series of siblings with and without an alcoholic biological parent who were raised in a family setting where one of the parents was or was not alcoholic (Table 6.8). Those children with an alcoholic biological parent had similar rates for alcoholism in adulthood whether raised by an alcoholic parent or stepparent (46%) or by parents or stepparents who had no diagnosable alcoholism (50%). Control siblings with non-alcoholic biologic parents had rates of alcoholism of 14% and 9%, respectively, when raised by alcoholic and non-alcoholic parents or stepparents. These data suggest that genetic vulnerability toward alcoholism was transmitted and occurred independently of alcoholism in the family of rearing.

Three adoption studies, one in the United States and two in Scandinavian countries, produced data that support the hypothesis of genetic transmission of vulnerability toward alcoholism as found in the half-sibling study. Goodwin et al. (1973), using Feighner criteria in a Danish birth registry sample, found that adopted-out males with a biologic parent hospitalized for alcoholism had significantly more alcoholism (18%) than male adoptees with non-alcoholic biologic parents (5%). Problem drinking and heavy drinking of alcohol were as frequent for adoptees with or without biologic parents who were alcoholic (Table 6.5). Adopted-away daughters from the same birth register study, however, had only 2% alcoholism ($N = 49$) if the biologic parent was a hospitalized alcoholic compared to 4% ($N = 47$) in adoptees with a biologic parent without alcoholism (Goodwin, Schulsinger, Knop, Mednick, & Guze, 1977). The authors suggest that because only 3% of Danish women abuse alcohol and only 1%–2% of daughters in this study were heavy drinkers, whereas 22%–36% of sons in this sample were heavy drinkers, vulnerability was not expressed, due to insufficient exposure to alcohol misuse by the vulnerable females in this study. The Goodwin study suggests that psychosocial components of chronic alcohol abuse as required in the Feighner definition of alcoholic drinking may be an integral part of the expression of genetic vulnerability toward chronic alcoholism.

Cadoret and Gath (1978) assessed a sample from American adoption agencies, using Feighner or modified Feighner criteria for diagnosis of alcoholism in interviewed adoptive parents and in records of biologic parents. Adoptees were interviewed and ICD-9 criteria for alcoholism were applied. The authors' data indicate that 40% (4 of 10) adopted-away children (3 males and 1 female) with one or more biologic first-degree or second-degree alcoholic relatives became alcoholic as adults. In contrast, only 7% of the 72 adoptees without known alcoholism among their first- or second-degree relatives were alcoholics. These findings support those of the Danish adoption study. Genetic vulnerability from first-degree relatives as demonstrated for children with hospitalized alcoholic parents in the Danish study and from second-degree relatives for children with alcoholic grandparents in the American study.

Other Psychopathologies and Alcoholism

Alcoholism associated with antisocial personality or behavior has appeared in differing fashions in the three adoption studies listed in Table 6.8. In a study from Sweden of alcoholics ascertained by state hospital and criminal records and temperance board offense registration, Bohman (1978) found a significantly greater frequency of alcohol abuse, characterized by repeated temperance board registration and/or hospitalization for alcoholic treatment or criminal alcohol-related convictions, among adoptees with a biologic parent designated as an alcohol abuser (Table 6.8). Cloninger, Bohman, and Sigvardsson (1981) characterized their adoptive male alcoholics as Type I (13%), those with a significant but weak genetic vulnerability for alcoholism; and Type II (4%), those with a criminal alcoholic father and a nine-fold increase in risk for alcoholism. The Type II alcoholic males of the Swedish adoption study (24% of the alcoholic abusers) may be what others have described as the male antisocial personality (ASP) alcoholic (Cloninger & Reich, 1983; Stabenau, 1984). Two large-scale U.S. studies of psychopathology in alcoholics found a rate for ASP in male alcoholics of 20% (Powell, Penick, Bingham, & Rice, 1982; Winokur, Reich, Rimmer, & Pitts, 1970). Similarly, the Danish adoptive study (Goodwin, Schulsinger, Hermansen, Guze, & Winokur, 1975) demonstrated an excess of antisocial behavior in the childhood of one sample of adopted-out male alcoholics, and the American adoptee study of alcoholism found that two of three fathers with antisocial diagnosis or behavior, but no alcoholism diagnosis by Feighner criteria, had adopted-out sons who were diagnosed as alcoholic (one with antisocial personality in addition to alcoholism) (Cadoret & Gath, 1978).

Both the Feighner and DSM III (American Psychiatric Association, 1980) classification schema place considerable weight on psychosocial behavior in the diagnosis of alcoholism and the alcohol dependent state. Current studies attempt to ascertain the relative contributions of, and possible interactions between, a "familial alcoholism genotype" and "alcoholism associated with antisocial personality" (Cloninger & Reich, 1983; Hesselbrock, Hesselbrock, & Stabenau, 1985; Hesselbrock, Stabenau, Hesselbrock, Meyer & Babor, 1982; Lewis, Rice, & Helzer, 1983; Stabenau, 1983; Stabenau, 1984).

The diagnosis of antisocial personality as determined by DSM III (American Psychiatric Association, 1980) criteria is made from a proscribed series of four sets of childhood and adult antisocial behaviors and conditions: a) the individual is at least 18 years of age; b) the onset and history include *3 or more* of the following before age 15: 1. truancy; 2. suspension from school; 3. arrests; 4. running away; 5. persistent lying; 6. repeated casual sex; 7. repeated drunkeness or substance abuse; 8. theft; 9. vandalism; 10. grade IQ discrepancy; 11. violations of rules; 12. initiation of fights; c) there have been *at least four* manifestations of the following since age 18; 1. inability to sustain consistent work; 2. lack of ability to function as responsible parents; 3. failure to accept social norms with respect to law; 4. inability to maintain attachment to a sexual partner; 5. irritability and aggressiveness; 6. failure to honor financial obligations; 7. failure to plan ahead; 8. impulsivity; 9. recklessness; d) there is a pattern of continuous of antisocial behavior with no intervening period of at least five years (DSM III, 1980). Some investigators believe a broader definition, less descriptive and more subjective, is needed to characterize the "sociopath," i.e., to include an inability to learn from experience and an inability to feel guilt or show remorse for deviant behaviors (Schuckit, 1984b).

The prevalence of antisocial personality in the general population has been estimated at 3.3% for white males and 1.0% for white females (Cloninger, Christiansen, Reich, & Gottesman, 1978). There have been few studies designed to measure the genetic contributions of underlying components to the antisocial syndrome. Evidence from twin studies that impulsivity is a temperament is not substantial (Buss & Plomin, 1978). However, a recent study of 12,898 unselected twin pairs in Sweden demonstrated a heritability index for psychosocial instability of .50 for men and .58 for women and for psychosocial extraversion of .54 and .66, respectively. The authors attributed about one half of the phenotypic variance to genetic factors (Floderus-Myrhed, Pedersen, & Rasmuson, 1980). It is not clear whether these characteristics are associated with the development of antisocial personality disorder.

Family pedigree studies of individuals with ASP have demonstrated genetic correlation between first-degree family members with the same diagnosis (Table 6.9). The tetrachoric correlation for diagnostic concordance of ASP among first-degree family members was $r = .49$ for singleton siblings and $r = .29$ for

TABLE 6.9
Frequency of Antisocial Personality Diagnosis in Family, Twin and Adoption Studies (%)

Sample, Senior Author, Year	*Male*	*Female*
General Population	3.3	0.9
Cloninger, 1978		
Family Study		
Cloninger, 1983		
First Degree Relatives of ASP Probands		
Male	17	4
Female	28	10
Twin Studies[a]	MZ Proband Concordance	DZ Proband Concordance
Christiansen, 1974	51	26
Dalgard, 1976	41	26
Adopted-Away Offspring[b]	ASP in Biologic Parent	Non ASP Biologic Parent
Crowe, 1974	13 (6/46)	0 (0/46)
Cadoret, 1978	22 (4/18)	0 (0/25)

[a]Criminality.
[b]Sample numbers in parentheses.

parent and offspring. For MZ twins with criminality, within-pair correlations were higher ($r = .70$) than between DZ twins ($r = .41$) (Cloninger & Reich, 1983). Adoption studies that control for effects of being reared with biologic parents and/or siblings have demonstrated a significant correlation between development of ASP in adoptees and the presence of ASP in their biologic parents. Control adoptees and their biologic parents did not show this relationship (Cadoret, 1978; Crowe, 1974) (Table 6.9). Considerable support for a hypothesis of genetic vulnerability for antisocial personality disorder has emerged from these family pedigree, twin, and adoption studies.

Until recently there has been confusion about the genetic separateness in the role of ASP and familial alcoholism in the transmission of alcoholism. Interview studies have now shown that both ASP and alcoholism have different and separate segregation patterns in families (Cloninger & Reich, 1983; Cloninger, Reich, & Wetzel, 1979). Specifically, first-degree relatives of alcoholics have significantly elevated rates of alcoholism (33% male, 4% female) and rates of ASP not too different from those in the general population (3% for males and 1% for females). Although rates for ASP in the general population are low, ASP morbidity rates are significantly higher for first-degree family members of ASP subjects (18% male, 8% female) whereas frequencies of alcoholism are in the expected range (6% male, 0–1% female) for the general population (Cloninger & Reich, 1983).

Winokur et al. (1970) found an excess of depression in female alcoholics and an excess of antisocial personality in male alcoholics. He and his associates included alcoholism as a component in a depression spectrum disorder (Winokur,

1974). Schuckit, Rimmer, Reich, and Winokur (1970), on the other hand, suggested that antisocial personality when found with alcoholism determined an early onset type of alcoholism. Cloninger and associates, however, using data based on Feighner diagnostic criteria from interviewed probands with primary alcoholism, manic depressive disorder, and antisocial personality disorder, concluded that each condition was a unitary, separately transmitted, and genetically determined trait (Cloninger & Reich, 1983; Cloninger, 1979). The distribution of these three diagnostic categories in probands and their first-degree relatives is shown in Table 6.10.

Although family, twin, half-sibling and adoptive studies with different degrees of control for genetic and rearing differences support a genetic vulnerability hypothesis for the transmission of alcoholism, undetermined biological and unspecified cultural differences appear to be operative in the phenotypic expression of genotypic vulnerability. Such differences may explain the reduced alcohol use and lower rate of chronic alcoholism for females as compared to males. Also, the Swedish adoption study has demonstrated that adverse early natal experience is critical and interactive with genetic factors in doubling the expression of vulnerability for some male and female adoptees who become alcoholic (Cloninger, Bohman, & Sigvardsson, 1981).

In summary, a genetic vulnerability hypothesis for chronic alcoholism has been suggested by twin-pair concordance comparison and controlled adoption cross-fostering studies.

THE NATURAL HISTORY OF ALCOHOLISM

We do not know what the natural history of alcohol dependence of chronic alcoholism is. Most studies of the progressive stages of the "disease" were of older chronic alcoholics (Table 6.11) (Glatt, 1961; Jellinek, 1946; Park, 1973;

TABLE 6.10
Psychiatric Disorders in First-Degree Relatives of Alcoholic Probands by Primary Diagnosis

Primary diagnosis of alcoholic proband	*Primary diagnosis of first-degree relatives*					
	Alcoholism		*Depression*		*ASP*	
	f/n	*%*	*f/n*	*%*	*f/n*	*%*
Alcoholism	184/807	23*	78/535	15	19/516	4
Depression	19/130	15	24/89	27[a]	3/70	4
ASP	18/106	17	14/72	19	15/68	22[a]

Source: Cloninger & Reich, 1983, p. 148. Reprinted by permission.
Note. Data from Sinokur et al. (1971). Diagnoses are based on all available data about subjects over 17 and age-adjusted by Stromgren method. Prevalence of ASP is given for male relatives only; other prevalences include both sexes.
[a]Relatives with same primary diagnosis as the proband are significantly increased compared to the others (contingency X^2, $p < .05$).

TABLE 6.11
Stages in Symptom Emergence of Alcohol Abuse and Dependence
(Age Mean Years)

Sr. Author, Year	*Sex*	*N*	*1st Drink*	*1st Drunk*	*1st Dependence Symptoms AM Drinking*	*Tremors*	*1st Treatment Intervention*[a]
Jellinek (1946)	M	98	—	18.9	29.9	32.7	36.8
Amark (1951)	M	203	17.8	—	30.8	—	36.5
Trice (1958)	M	262	17.6	18.3	35.6	38.6	—
Glatt (1961)	M	192	17.6	20.1	35.3	37.2	40.2
Park (1973)	M	806	16.4	—	25.1	29.9	34.0
Schuckit (1979)							
Primary Alc.	F	154	16.9	21.2	34.1	—	—
Aff. Dis. Alc.	F	40	16.9	21.9	34.6	—	—
ASP Alc.	F	40	13.4*	14.8*	21.5*	—	—
Stabenau (1984)							
DSM III Alc. Dependence	M	156	13.9	17.1**	30.9	—	33.4
DSM III Alc. Dependence	F	54	15.0	19.0	31.4	—	33.7
ASP Alc.	M & F	91	12.7***	15.8***	28.7***	—	32.7
Non-ASP Alc.	M & F	119	15.3	18.9	32.1	—	34.2

[a]Hospital except for Temperance Board in Amark's study.
*$p < .05$ ASP (Antisocial Personality) vs. Aff. Dis. Alc. (Affective Disorder) or Primary Alcoholism.
**$p < .01$ M vs. F. ***$p < .001$ ASP vs. Non-ASP.

Taylor & Helzer, 1983; Trice & Wahl, 1958). Recent cross-cultural demographic studies that have included a wide range of age and explicit diagnostic methods of assessing alcohol misuse and other psychopathology have found stage omissions and reversals that have suggested alcoholism may alternatively be viewed as a heterogenous behaviorally oriented multivariant experience rather than a unitary concept embodying a progressive disease (Caddy et al., 1976; Cahalan, 1982; Clark & Cahalan, 1976; Cloninger & Reich, 1983).

Thus, to identify genetic factors and their mechanisms of influence on drinking behavior over time, focus has been directed toward assessment of known genetic variables and alcohol abuse patterns. Three such factors of genetic importance are male-female gender differences in alcoholism, differences between family history of alcoholism, and frequency of antisocial personality in alcoholics.

The genetic pathogenesis of alcoholism has been linked to two different overlapping bodies of clinical observations: One is that alcoholism is frequently associated with a history of alcoholism among biologic family members, and the other is that alcoholism in adulthood is often an outcome of childhood conduct disorder (Robins, 1966) and a concomitant of adult antisocial personality (ASP) disorder (Schuckit et al., 1970).

Goodwin (1979) cites Jellinek's work as the source for "familial alcoholism" as a form of alcoholism characterized by a family history for alcoholism, early onset, severe symptoms, and absence of other conspicuous psychopathology. Studies of this hypothesis have shown that alcoholics with a family history have early onset, more social problems (Penick, Reed, Crowley & Powell, 1978; Schuckit, 1984a), more severe alcohol-related symptoms, and more antisocial behavior (Frances, Timm, & Bucky, 1980); and that they are younger at the age of first intoxication. When both parents are alcoholic they proceed to treatment at an earlier age (McKenna & Pickens, 1981). However, the relationship between psychopathology, family of origin, and the natural history of alcoholism in the alcoholic is not clear.

The early onset of alcoholism has been linked to the presence of ASP in both male and female probands (Hesselbrock et al., 1984; Schuckit & Morrissey, 1979; Schuckit, Pitts, Reich, King, & Winokur, 1969; Stabenau, 1984). Robins (1966) demonstrated that the childhood of the alcoholic bore more resemblance to the childhood of the sociopath than to the neurotic. Lewis, Helzer, Cloninger, Croughan, and Whitman (1982) found ASP but not primary depressive illness associated with an increased risk for alcoholism in women and men. The mechanisms, however, by which ASP may influence the development of alcoholism have not been identified. An implicit hypothesis is that factors associated with ASP influence onset of alcoholism, whereas factors associated with family history of alcoholism more often determine the consequences of chronic alcohol abuse and dependence.

In a study of 350 male and female alcoholic inpatients, clinical issues such as the age at the various stages of dependence, the quantity and frequency of alcohol use, and the occurrence of psychological, social and/or physical impairment subsequent to chronic alcohol use have been evaluated by a detailed alcohol drinking and drug use history, a standardized psychiatric interview using the NIMH Diagnostic Interview Schedule (DIS) and the Family History Research Diagnostic Criteria (FHRDC) (Hesselbrock, Babor, Hesselbrock, Meyer, & Workman, 1983; Hesselbrock, Hesselbrock, & Stabenau, 1985; Hesselbrock et al., 1984; Stabenau, 1984; Stabenau & Hesselbrock, 1980).

Stabenau and Hesselbrock have used family pedigrees to assess family histories of alcoholism as follows: family history negative (FHN), no alcoholism on either side; family history positive unilineal (FHPU), alcoholism (a parent or sibling of a parent) on one side only; and family history positive bilineal (FHPB), alcoholism (a parent or sibling of a parent) on both sides of the family (Hesselbrock, Stabenau, Hesselbrock, Meyer, & Babor, 1982; Stabenau & Hesselbrock, 1980). This method of assessing family history of alcoholism includes methods that have been described by Penick et al. (1978); Francis et al. (1980); McKenna and Pickens (1981), and Schuckit (1980). To characterize a proband as alcoholic or an individual at risk, these investigators have used both parents, one first-degree relative, a first- or second-degree relative, or any biologic family member as alcoholic. The alcohol-dependence and antisocial personality disorder diagnoses in the 350 alcoholic probands were made according to DSM III criteria from NIMH-DIS individual interviews (Robins, Helzer, Croughan, Williams, & Spitzer, 1979) of the probands and a Family History Research Diagnostic Category (FHRDC) (Andreasen, Endicott, Spitzer, & Winokur, 1977) diagnosis of probands' relatives by FHRDC methods from proband information (Hesselbrock, Stabenau, Hesselbrock, Mirkin, & Meyers, 1982).

A comparative evaluation of stages of symptom emergence from a number of studies is given in Table 6.11. When alcoholics are compared by sex and type of additional psychopathology, both being male and having antisocial personality disorder are significantly associated with early onset of drinking and drunkenness (Schuckit & Morressy, 1979; Stabenau, 1984). However, there were no significant differences in the stages of alcohol misuse or dependence for family history positive alcoholics as compared to family history negative alcoholics (Hesselbrock, Hesselbrock, & Stabenau, 1985; Stabenau, 1984). Although ASP diagnosis in both males and females is associated with early onset of drinking and drunkenness, affective disorder in females is not (Lewis et al., 1983; Schuckit & Morrissey, 1979).

Understanding the nature of dependence including impaired control and psychological symptoms and social problems is tantamount to understanding the syndrome of chronic alcoholism. Studies of genetic factors in the natural history of the consequences of alcoholism should include evaluation of sex and psychopathology in the proband and history of alcoholism in the family. Table 6.12 lists the comparison of frequencies of four types of complications of alcohol dependence or chronic alcohol misuse in an analysis of covariance (Stabenau, 1984). Scale items used to define these four factors have been described (Hesselbrock, Hesselbrock, & Stabenau, 1985). When age and years of problem drinking were controlled, significant differences in impaired control and psychological problems were found in those alcoholics who had alcoholism on both sides of their biologic family (FHPB). There was also a significant interaction between family history and antisocial personality. Significant variance in social problems was found associated with ASP diagnosis for both males and females. Those alcoholics in the highest quartile of alcohol use (4 in Table 6.12) had significantly elevated mean scores on all four factor variables. However, only 20%–35% of total variance is accounted for by the variables of sex, family history, ASP, and amount of alcohol consumed in the previous 6 months.

These data suggest that gender, bilineal family history for alcoholism, and antisocial personality are all or in part genetically controlled and contribute to the risk of alcohol abuse and the nature of symptoms of chronic alcohol dependence.

CNS and Somatic Complications

The pathogenesis of alcohol-related degenerative medical complications in the various body systems is difficult to establish but is proceeding through technical advances in biochemical identification of enzymatic alteration of liver, blood,

TABLE 6.12
Consequences of Chronic Alcohol Use in Alcohol Dependent Subjects by Family History for Alcoholism, ASP Diagnosis, Sex, and Alcohol Use (Oz. Abs. Alc. Last Mo.) with Control for Age and Years of Problem Drinking ($N = 225$)

	Family History for Alc.[2]			*ASP Diagnosis*[3]		*Sex*		*Oz. Abs. Alc. Last Mo.*[4]				
Variable[1]	*FHN*	*FHPU*	*FHPB*	*No ASP*	*ASP*	*Female*	*Male*	*1*	*2*	*3*	*4*	r^2
Impaired Control	35.4	34.5	37.7[a]	34.8	35.3	35.3	34.9	27.0	33.8	39.0	38.5[d]	.350
Social Problems	13.6	14.0	15.0	13.2	15.4[b]	14.0	14.2	12.5	14.1	15.3	14.3[e]	.334
Psychological Problems	18.1	16.8	19.6[c]	17.1	18.0	19.5[f]	16.9	14.9	16.2	19.3	19.0[g]	.245
Dependence	10.9	10.9	12.1	11.3	11.0	11.5	11.0	8.8	12.0	12.1	12.8[h]	.209

Source: Stabenau and Hesselbrock, 1984a. Reprinted by permission.
[1]Mean frequencies of items comprising each factor variable. [2]FHN = Family History Negative, FHPU = Family History Positive, one side of family, FHPB = Family History Positive, both sides family. [3]Mean Age ASP females 29, males 34, non ASP females 39, males 44. [4]Quartiles of alcohol use, 1 lowest, 4 highest.

[a]$F = 3.6$, $df = 2$, $p = .03$. (2 way Family Hx × ASP, $F = 3.1$, df 2, $p = .05$).
[b]$F = 15.0$, $df = 1$, $p = .000$. [c]$F = 4.3$, $df = 2$, $p = .02$.
[d]$F = 24.9$, $df = 1$, $p = .000$. [e]$F = 5.5$, $df = 3$, $p = .001$.
[f]$F = 8.0$, $df = 1$, $p = .005$. [g]$F = 7.43$, $df = 3$, $p = .000$.
[h]$F = 11.5$, $df = 1$, $p = .000$. (Analysis of Covariance)
Note: Diagnosis determined by DSM III criteria.

muscle, and brain function, and with advances in computerized EEG spectral analysis and evoked potential measurement. Few studies of alcholics utilizing these methods control for family history of alcholism or antisocial personality psychopathology. When consumption for males and females is corrected for body weight, ASP alcoholics, both male and female, drink significantly more alcohol than do non-ASP alcoholics (Stabenau, Dolinsky, & Fischer, 1986). Ashley et al. (1981) have reported the lifetime incidence of selected diseases and complications from 1,001 alcoholics who were volunteer admissions to the Medical Unit of the Addiction Research Foundation of Ontario, Canada. Men had significantly more fatty liver, peptic ulcer, gastritis, ischemic heart disease than women, and women had significantly more anemia. Frequency of chronic brain damage in the sexes was not different (5.1%, 5.0% females). Hrubec and Omenn (1981) have recently analyzed alcohol use and misuse and health of 15,924 male Veteran twin pairs. Significantly different case-wise twin concordance percentages occurred for alcoholism, 26.3 (MZ), 11.9 (DZ); alcoholic psychosis, 21.1 (MZ), 6.0 (DZ) and liver cirrhosis, 14.6 (MZ) and 5.4 (DZ). No twin pairs were found concordant for pancreatitis. The authors believe these data provide evidence in favor of a genetic predisposition to organ-specific complications of alcoholism.

It has been estimated that Wernicke-Korsakoff psychosis or syndrome (WKS), a distinct part of the broad clinical syndrome of alcoholic psychosis, accounts for about 10% of alcohol-related organic brain syndromes (Horvath, 1975). Blass and Gibson (1977) have demonstrated that some patients with WKS have a demonstrable inborn enzymatic abnormality in transketolase. These genetically predisposed individuals develop thiamine insufficiency due to diets marginal in the vitamin, and while drinking heavily develop the psychosis of Wernicke-Korsakoff Syndrome.

The observation of increased frequency of concordance of alcoholic psychosis among MZ twin pairs and an identifiable biochemical enzymatic deficiency in one of the alcohol-related organic psychoses supports efforts to conceptualize alcoholism as a pharmacogenetic disorder (Omenn, 1975).

Deficits in liver and brain integrity have been clearly demonstrated in young alcoholics by computerized tomography (Lee, Moller, Hardt, Haubek, & Jensen, 1979). Indirect assessment of central nervous system integrity has included measurement of body sway with and without ingestion of alcohol. One study found that family history positive (FH+) non-alcoholic subjects had significantly greater body sway compared to family history negative (FH−) subjects (Lipscomb & Nathan, 1980), but another found the opposite (Schuckit, 1984b, 1984c). Abstract, task and problem solving, perceptual motor, and learning and memory tasks have been compared in middle-aged alcoholic and non-alcoholic males with histories of alcoholism in first-degree family members. Schaeffer, Parsons, and Yohman (1984) concluded that (1) performance deficit in abstract, task and problem solving, and possibly learning and memory tasks may antedate the alcoholic stage in FH+ individuals; (2) alcoholism and positive family history of alcoholism have independent, additive deleterious effects on cognitive-perceptual functioning; and (3) that, in the future, neuropsychological studies of alcoholism should consider the frequency of FH+ and FH− individuals in both alcoholic and control groups.

Compared to primary alcoholism there are only isolated reports of alcoholism complications associated with sociopathy, psychopathy, or antisocial personality disorder. Glatt (1961) reported a significantly greater occurrence of attempted suicide of young female psychopathic alcoholics among the 66 suicidal alcoholics in his sample of 268. All five of the males aged 21–30 in his sample, who tried suicide, were psychopathic (Glatt, 1961). Medical complications of cirrhosis, gastritis, pancreatitis in 50 ASP alcoholics from an outpatient criminal cohort were significantly less frequent than for 42 non-ASP alcoholics with other personality disorders (diagnosis of alcoholism and ASP were made using Feighner criteria) (Virkkunen, 1979). On the other hand, Penick et al. (1984) found 182 primary alcoholics had fewer alcohol-related medical complaints than ASP alcoholics.

It has become clear that the evaluation of the human response to alcohol in respect to the contribution of genetic factors requires appropriate measurement of family pedigree and psychopathology variables.

In summary, genetic factors influencing the course of alcoholism are male gender associated with highest risk for alcoholism; presence of ASP diagnosis associated with early onset of alcohol dependence; and presence of alcoholism on both sides of the pedigree associated with greater consequences of prolonged drinking.

HETEROGENEITY OF CLINICAL ALCOHOLISM

Family diagnostic interview studies have begun to examine the rule of the psychopathologies in the pathogenesis of alcohol dependence. Questions have arisen: Are there several alcoholisms with different genetic diatheses? What is the relationship between depression, drug abuse, antisocial personality disorder, and alcoholism? Do these conditions just appear together? Are these conditions interchangeable expressions of a single underlying type of vulnerability? Are these disorders genetically independent of each other? What, if anything, is their relationship to the risk for developing alcoholism?

Winokur et al. (1970) interviewed 259 alcoholic subjects and 172 female and 335 male relatives of probands, using the diagnostic criteria of Cassidy, Flanagan, Spellman, and Cohen, Wolfgram, McKinney, and Cantwell (1957) and Guze (1967). They found alcoholism most frequent among male relatives whereas affective disorder appeared more frequently in female relatives of alcoholic probands. They also noted an increase in sociopathy in the male relatives compared to the female relatives of alcoholic probands. These findings led to Winokur's hypothesis of a division of depressive illness into pure depressive disease and a depressive spectrum disorder. He speculated that alcoholism may represent a sex-linked variant of "depressive spectrum disease" (Winokur, 1974). Schuckit focused on the antisocial traits of alcoholics and suggested that alcoholism associated with antisocial personality was an early onset variant of alcoholism (Schuckit & Morrissey, 1979; Schuckit et al., 1970).

Other clinical studies noted the essential characteristics of primary alcoholics and affective disorder alcoholics (Schuckit et al., 1969), and the characteristics of ASP-diagnosed female alcoholics (Schuckit & Morrissey, 1979). Penick et al. (1984), using the psychiatric diagnostic interview (PDI) and family history data in a study of 594 male patients from five VA programs, compared psychiatric disorder in first-degree family members in a sample of 257 primary alcoholics, 98 depressive alcoholics and 104 ASP alcoholics. The percentage of patients with one or more relatives with an alcoholism diagnosis was 58% for primary alcoholic probands, 74% for depressed alcoholic probands, and 74% for ASP alcoholic probands. The frequency of depression in relatives of proband alcoholics, was 24%, 30%, and 43%, respectively, and for ASP 23%, 32% and 45%, respectively. Data such as these do not advance our understanding of the influence of psychopathology in the family and alcoholism. It is only by comparison of the expected and observed frequencies of different psychopathologies in entire pedigrees, preferably by direct interview but at least by family history methods, that clarification emerges. The above data only confirm the clinical observation of a diagnostic overlap between depression, ASP, and alcoholism. The diagnostic family interview study of Cloninger and Reich (1983) shown on Table 6.10 demonstrates the independence of transmission of alcoholism, ASP, and depression and has been replicated by Merikangas, Leckman, Prusoff, Pauls, and Weissman (1985) for depression, alcoholism, and ASP, diagnosed by SADS-RCD criteria in a sample of 124 male and 176 female probands and 1,331 first-degree relatives.

These same diagnostic categories of psychopathology have been evaluated in substance abusers (Mirin, Weiss, Sollogub, & Michael, 1984). When the expectancy rate for affective disorders in the relatives of alcoholic and non-alcoholic substance abusers was compared, there were no significant differences. Conversely, there was no significant difference in the expectancy rate for alcoholism in the relatives of patients with or without affective disorder. Thus, in the population of substance abusers, being alcoholic did not increase the probability of having a relative with affective disorder. The prevalence of alcoholism and/or affective disorder was highly correlated only with the same clinical entity in the proband. Mirin et al. (1984) reported 57% of their sample of 160 consecutively admitted hospitalized substance abusers had at least two diagnoses, substance abuse and alcoholism, or substance abuse and affective disorder, whereas

14% had all three disorders concurrently. The authors conclude that their data fit a model of inheritance in which alcoholism and affective disorder correspond to two or more different genotypes that are transmitted separately.

Lewis et al. (1982), using Feighner criteria, have compared risks for alcoholism and diagnosis of depression and ASP in three populations of females. Eighty-four were from a general hospital sample, 78 were felons, and 42 black and 42 white women were from a narcotic addiction hospital. The authors found that hospitalized women with primary depression show no increase in the rate of alcoholism over that of the general population. Female felons have very high alcoholism rates. Those with an ASP diagnosis have the highest rate for alcoholism, and the rate for alcoholism in women narcotic addicts is higher than in the general population. Rates are significantly higher for narcotic addicts with, than for those without, antisocial personality diagnosis. Addicts without ASP have a rate comparable to Cahalan's (1982) female population figure for alcoholism (Lewis et al., 1982).

No studies have clearly shown depression to significantly influence the risk of alcoholism in women. Because of the relatively low rate of primary depression in men, comparisons have not been reported between depression and risk for alcoholism in men (Hesselbrock, Hesselbrock, & Workman-Daniels, in press; Lewis et al., 1982).

The relative risk for alcoholism controlling for ASP, sex, diagnosis of the proband, and family history for alcoholism in the proband is seen in Table 6.13. Lewis and colleagues (1983) studied 131 men and 281 women who were referred

TABLE 6.13
The Risk of Alcoholism Controlling for Sex, ASP, and Family History of Alcoholism

	Number Alcoholic	*Number Nonalcoholic*	*Observed Percent Alcoholic*	*Predicted Percent Alcoholic*
Male				
ASP(−) FH ALC(−)	15	59	20	21
ASP(−) FH ALC(+)	9	13	41	40
ASP(+) FH ALC(−)	8	10	44	54
ASP(+) FH ALC(+)	13	2	87	75
Women				
ASP(−) FH ALC(−)	7	161	4	4
ASP(−) FH ALC(+)	7	59	11	10
ASP(+) FH ALC(−)	6	16	27	17
ASP(+) FH ALC(+)	6	18	25	34

Source: Lewis et al., 1983, p. 109. Reprinted by permission.
Note. ASP, Antisocial Personality Disorder in Probands. FH ALC, History of Alcoholism in the Family of the Proband.

for psychiatric evaluation by the medical and surgical services of a general hospital. Feighner et al.'s (1972) criteria for alcoholism and Guze's (1976) criteria for ASP were applied to a structured interview. Twenty-six percent of men and 17% of women were diagnosed as ASP, and 24% of men and 9% of women were alcoholic. The authors concluded: (1) unipolar depression (rates: 24% in men, 44% in women) did not appear to increase the risk of alcoholism in women (for men this was difficult to ascertain because the onset of heavy drinking usually occurred before the first depressive episode); (2) controlling for ASP and family history of alcoholism, men had higher risk of developing alcoholism than women; (3) the presence of ASP in a subject (male or female) elevated the risk more than the presence of family history of alcoholism.

The transmission to their children of risk for alcoholism from parents with alcoholism or antisocial personality diagnosis has been explored through follow-up studies and comparative family sibling studies. Several large-scale prospective follow-up studies have linked childhood behaviors of conduct disorder (characterized by impulsivity) and attention deficit disorder (characterized by distractibility) to a significantly higher occurrence of alcoholism in at-risk individuals when they were followed to adulthood (McCord & McCord, 1962; Robins, 1966; Robins, Bates, & O'Neal, 1962; Vaillant & Milofsky, 1982). A study of high-risk children demonstrated the significant power of childhood delinquent acts in predicting subsequent adult alcoholism (Vaillant, 1980). In follow-up of the 397 Boston city-core men in Vaillant's study, 9 of the 16 (56%) who at age 14 were truant and exhibited school behavior problems had DSM III diagnosed alcohol dependence in adulthood, four times the rate of their peers (16%, 60/381). Vaillant points out that most sociopaths who later abuse alcohol do so as part of their antisocial behavior, but most alcoholics do not have sociopathic personality. This study also demonstrated the heterogenous nature of alcoholism. Alcohol dependence occurred in 18% of city-core men. Thirty-four percent of 71 men with several alcohol-abusing relatives were DSM III alcohol dependent compared to only 10% of the 178 men with no alcohol-abusing relatives. Vaillant reported a similar finding in his college risk sample: 9% of the 158 men with one or no alcoholic relatives abused alcohol, whereas 26% of 46 college men with two or more known alcoholic relatives were found to abuse alcohol upon follow-up (Vaillant, 1983).

Tarter, McBride, Buonpane, and Schneider (1977) have shown that hyperkinetic and/or minimal brain damage (HK/MBD) associated childhood behaviors were retrospectively reported more frequently in primary alcoholics compared to secondary alcoholics. However, Hesselbrock et al. (1984) have demonstrated that for both males and females retrospectively self-reported conduct disorder (CD) and attention deficit disorder (ADD) childhood behaviors were significantly more often described by ASP alcoholics as compared to non-ASP adult alcoholics. Analysis of the same sample by three criteria—family history negative, family history positive unilineal, and family history positive—did not demonstrate any difference between CD and ADD childhood behaviors and presence or absence of family history for alcoholism (Hesselbrock, Stabenau, & Hesselbrock, 1985; Stabenau, 1982). Children born to parents with a family history positive for antisocial spectrum disorder (ASP, alcoholism, and hysteria) have been found to have significantly more conduct disorder (August & Stewart, 1983).

Assortative Mating

The frequent occurrence of both alcoholism and antisocial personality diagnosis in family members of both alcoholics and individuals with antisocial personality disorder has suggested that assortative mating has occurred between individuals with alcoholism and ASP.

Rimmer and Winokur (1972) found that 3% of wives and 20% of husbands of alcoholics were also alcoholic. Stabenau and Hesselbrock (1980) reported higher rates of 11% and 35%, respectively. Alcoholic drinking was assessed by family history methods and RDC criteria by Hall, Hesselbrock, and Stabenau (1983a, 1983b). Heavy drinking and probable alcoholic drinking rates were surprisingly similar for the spouses, their same-sex siblings, and the same-sex siblings of their alcoholic marital partners (Table 6.14). Probands' brothers and sisters, however, had almost twice the rate of alcoholic drinking as compared to their spouses' brothers and sisters. These data show a strong assortative mating for heavy and alcoholic drinking style (Hall et al., 1983a, 1983b).

In an analysis of psychopathology in the family for the same sample of proband alcoholics, using FHRDC diagnostic methods, assortative mating between individuals with ASP and alcoholism was highest for husbands with ASP and wives with a history of alcoholism on one or both sides of their family compared to those without alcoholism in the family (Table 6.15) (Stabenau & Hesselbrock,

TABLE 6.14
Assortative Mating: Drinking Style of Alcoholic Probands' Siblings, Spouses, and Spouses' Siblings, in Percent

	Spouses		*Probands' siblings*		*Spouses' siblings*	
Drinking style	*husbands*	*wives*	*brothers*	*sisters*	*brothers*	*sisters*
Abstainer	10	18	16	33	23	38
Social drinker	43	63	43	52	46	51
Heavy drinker and probable alcoholic drinker	16	11	19	7	19	6
Definite alcoholic drinker	31	8	22	8	12	5
TOTAL *N*	74	192	379	388	274	267

Source: Hall et al., 1983b, p. 377. Reprinted by permission.

TABLE 6.15
DSM III Psychiatric Diagnoses in Spouse of Alcoholic Probands by Family History for Alcoholism

Relatives,				*Diagnosis*		
Family History	*n*	*Alcoholism*	*Unipolar*	*Bipolar*	*ASP*	*Drug Abuse*
Wife						
FHN	43	17	5	0	2	2
FHPU	110	6	12	1	1	5
FHPB	35	3	11	0	0	0
Husband						
FHN	11	55	9	0	9	9
FHPU	46	22	2	0	13	3
FHPB	18	31	6	0	17	28

Source: Stabenau & Hesselbrock, 1984b, p. 117. Reprinted by permission.
Note. Data are percentages of relatives with diagnosis.
Based on Family History Research Diagnostic Criteria (FHRDC) from proband information. FHN = Fam. History Neg.; FHPU = Fam. History Pos., one side; FHPB = Fam. History Pos., both sides

1984b). The rate of alcoholism for spouses of alcoholic probands was 8% for wives and 29% for husbands. However, more husbands and wives of alcoholics who were also diagnosed as alcoholic had married probands who had a family history negative for alcoholism. These observations have in part been replicated and suggest models for the interplay of ASP and alcoholism in the expanding network of matings that lead to the observed high frequency of family history positive alcoholism (Merikangas, Weissman, Prusoff, Pauls & Leckman, 1985).

There is strong evidence of assortative mating for height and intelligence in the human population (Crow & Felsenstein, 1968). Assortative mating also appears to exist for personality traits, but to a lesser degree than that observed for physical traits, sociodemographic traits, intelligence, and attitudes and values (Merikangas, 1982). Considerable assortative mating, or the concordance of psychiatric illness between spouses, has been reported, often with increased frequency of morbidity for the disorder in those children at risk (Fischer & Gottesman, 1980).

Population geneticists such as Cavalli-Sforza and Bodmer (1971) believe that "with assortative mating, in principle, gene frequencies do not change" (p. 53); however, "positive assortative mating for a recessive phenotype increases its incidence" (p. 538) and "assortative mating for polygenic traits inflates the genic additive portion of the variance" (p. 543). Since family pedigree segregation figures for both alcoholism and antisocial personality disorder appear to be non-Mendelian, the effect of assortative mating between ASP and alcoholism would be predicted to be additive.

In summary, two genetically determined alcoholisms have been demonstrated: alcoholism associated with antisocial personality disorder and alcoholism in non-ASP individuals who have a family history of alcoholism. Affective disorder, a separate genetically transmitted condition, although often coexisting with alcoholism, does not appear to increase risk for alcoholism as do ASP or family history of alcoholism.

PROSPECTIVE AND MARKER STUDIES OF GENETIC VULNERABILITY TO ALCOHOLISM

Clinical comparisons of alcoholics with non-alcoholic control subjects have demonstrated significant, valid, and reliable differences in current personality profile, perceptual pattern, MMPI profile, and neurocognitive integrity, but most of such findings appear to be an expression of alcohol use and not characteristics of the premorbid state (Barnes, 1979). Family, twin, and adoptive studies have demonstrated strong, genetic, biologic relationships predicting alcoholic outcome in some of the children at risk. The larger question in prospective studies is whether one can identify the specific individual at risk for this disorder.

Allan Gregg once said, "The feature of psychiatry lies in the genetic and the predictive" (Shakow, 1977, p. 6). Research that identifies the mechanisms of genetic and experiental interaction that lead to high risk for alcoholism is essential if causal relationships between genotypes and alcoholic phenotypes are to be firmly established.

This section briefly reviews longitudinal prospective and retrospective studies of proposed internal drive state, metabolic, neurocognitive or other marker differences, and genetic vulnerability toward alcoholism.

Personality and Drive State Assessments

Personality tests of alcoholics have been examined for pre-alcoholic features different from those of non-alcoholics in search of clues for vulnerability for alcoholism. Barnes (1979, 1983) made a comprehensive review of over 250 papers covering objective measures of personality structure or function. The Minnesota Multiphasic Psychologic Inventory (MMPI) has a reputation for reliability and validity for differentiating experimental and control populations. Unfortunately, most MMPI studies of alcoholics appear too contaminated by the effects of alcohol dependence to be useful predictors (Nerviano & Gross, 1983; Skinner & Jackson, 1974). However, a follow-back study by Loper, Kammier, and Hoffman (1973) showed that MMPI Pd and Ma scales were significantly higher for college students who later became alcoholic compared to matched controls. Other studies have shown that the MMPI MacAndrew scale for alcoholism discriminated chronic alcoholics from controls in measuring alcoholic behavior (Apfeldorf, 1978). Barnes (1979) reviewed the contributions of eight other widely used objective measures of personality evaluation in studies comparing alcoholics and controls. He concluded that only two perceptual tests, measurements of field dependence and stimulus intensity modulation, showed promise in exploring the pre-illness characteristics of alcoholics. An index composed of three measures of field dependence has shown that alcoholics are more field dependent than non-alcoholics (i.e., are less able to free themselves from the effects of the field surrounding a novel stimulus in order to accurately detect the novel stimulus) (Witkin, Dyk, Fatterson, Goodenough, & Karp, 1962). Field dependence appears to be a relatively stable characteristic that has been regarded as a predisposing factor to alcoholism. However, the nonspecific nature of these tests is apparent since certain varieties of brain damage are associated with a high degree of field dependence (Bailey, Hustimyer, & Kristofferson, 1961). Field dependence has also been associated with heroin addiction, overeating, and other psychiatric disorders (Barnes, 1979).

The second perceptual test with some potential for trait determination reviewed by Barnes was Stimulus Intensity Modulation. Alcoholics have been found to be stimulus augmentors rather than reducers (Petrie, 1967). Barnes (1979), however, concluded that there were few findings to suggest a precursor alcoholic personality and that the personality pattern noted in clinical alcoholism is most likely a response to the process of addiction and the effects of prolonged alcohol use. Similar conclusions have been reached by Loberg (1980) and Vaillant (1983). In future studies longitudinal measurement of the predictive capacity of MMPI Pd scores, field dependence, and stimulus modulation should involve genetically controlled populations.

Deviant Childhood Behaviors

Several syndromes of deviant childhood behavior (Minimal Brain Damage [MBD]; Attention Deficit Disorder [ADD]; Hyperkinetic [HK] behavior; and Conduct Disorder [CD] have been linked with development of adult antisocial behavior and alcoholism (Cantwell, 1972; Morrison & Stewart, 1974; Robins et al., 1962). Some of the problems in assessing whether any causal relationship exists and, further, whether there are genetic factors in the transmission of vulnerability for these syndromes lie in the diagnostic overlap of the ADD, MBD and HK, CD syndromes. For example, in the United States 30%–40% of children attending child guidance clinics were diagnosed as hyperkinetic (Safer & Allen, 1976), whereas only 1.5% were similarly labeled in the United Kingdom (Rutter, Shaffer, & Shepherd, 1975). The use of DSM III diagnostic criteria for ADD and CD may provide a means for standardized labeling and comparing of deviant childhood behaviors (DSM III, 1980).

This brief review focuses on several concepts and findings. Although there is still behavioral descriptive overlap, diagnostic polarization has moved toward two syndromes. ADD, previously titled MBD, has been suggested as an alteration in CNS function of the child that may be in either a continuum of effects of variants in gross brain damage (Rutter, 1982) a syndrome constituting in part a genetically determined disorder (Wender, 1971) or an interaction of the two factors. Similarly CD, previously described as part of the HK disorder, may be due to a genetically determined impulse disorder principally without identifiable CNS damage features (August & Stewart, 1983) or an interaction of subtle CNS insults and the former.

Clinical reports of ADD have highlighted the predictive quality of CNS integrative defects and pathologic adult outcome (Milman, 1979). However, a follow-up study of grade school children characterized as ADD showed no greater frequency of treatment for alcohol problems in adulthood nor in the frequency of family history of one or both parents being heavy drinkers of alcohol (Howell & Huessy, 1982). On the other hand, 14 adoptee alcoholics more hyperactive than their matched control adoptees were reported to have had at least 10 biologic parents who were alcoholic (Goodwin et al., 1975), and a 5-year follow-up of 23 hyperkinetic teenagers showed that they drank alcohol more frequently than their matched controls (Blouin, Bornstein, & Trites, 1978). In another study, one third of 27 alcoholic adults had demonstrable residual ADD behavior (Wood, Wender, & Reimherr 1983).

The nonspecific nature of the HK syndrome is evident from the study of Eyre, Rounsaville, and Kleber (1982), who found hyperactivity in the childhood of 22% of 157 opiate addicts. Also, Huessy (1984) points out that ADD is overrepresented (8 times as frequent) in adoptees as compared to non-adoptees and that biologic parents of children placed out for adoption have more ADD than control parents.

Family studies by Morrison and Stewart (1974) and Cantwell (1972) suggested alcoholism, antisocial personality, and hysteria were unusually prevalent among adult relatives of hyperactive children (Stewart, DeBlois, & Cummings, 1979). August and Stewart's (1983) study showed that children born to parents with family history positive for antisocial spectrum disorder (ASP, alcoholism, and hysteria) had significantly more conduct disorder (24%) than attention deficit disorder (8%). Children born to family history negative parents had more learning and academic problems, and they and their siblings had more attentional and learning disabilities (ADD 17%) than conduct disorder (0%). The authors concluded that conduct disorder is genetically linked to ASP spectrum disorder, but attention deficit disorder is not.

The relationship between CD childhood symptoms and adult ASP behavior and alcoholism is not entirely clear. However, Hesselbrock et al. (1984) showed that both CD and ADD childhood retrospectively recalled behaviors were significantly more often reported by ASP alcoholics than by non-ASP alcoholics. Hesselbrock, Stabenau, and Hesselbrock (1985) demonstrated a lack of association between parental alcoholism and recall of CD or ADD childhood behaviors.

Thus CD and ADD in alcoholism may be associated with the ASP type of alcoholism and not the family history positive type of alcoholism.

In an effort to assess genetic and environmental factors and the role of ASP and alcoholism in childhood hyperactivity, Cadoret and Gath (1978) studied 96 matched pairs from 143 experimental and 103 control adoptees. Feighner and ICD-9 criteria were used for childhood and adult diagnoses. Thirteen percent (3/24) of hyperactive children had a biologic parent who was antisocial; 25% (4/16) a biologic parent who was alcoholic; and 15.4% (6/39) an antisocial and/or alcoholic parent. There were no significant sex differences for the 10 males and 9 females diagnosed as hyperactive. The authors concluded that there was an association between hyperactivity and adopted-out children and a diagnosis of antisocial personality and/or alcoholism in their biologic parents. The lack of correlation between hyperactivity and measures of medical problems during mothers' pregnancy, labor, delivery, and the neonatal period was felt to be compatible with the hypothesis that genetic factors alone are important (Cadoret & Gath, 1978).

Biologic Mediators and Markers

Advances in the diagnostic assessment of the alcoholic in planning treatment for alcoholism may be made in part through identification of factors that place a person at genetic risk for becoming addicted to alcohol and for the consequences secondary to chronic exposure to alcohol. McClearn (1983), in a review of the significance of animal models in isolating genetic factors of alcohol abuse, noted that "genes can influence voluntary ingestion of ethanol in mice and rats . . . This body of knowledge will not only be pertinent to the question of the extent of the hereditary influence on alcoholism but will also inevitably illuminate the nexus of causal mechanisms and thereby suggest rational therapies and preventive measures" (p. 27).

The data supporting a heterogenous genetic basis for alcoholism imply that predisposition is mediated by separate and possibly interacting biological mechanisms. Many biologic marker systems have been assessed. In several studies associated between blood groups, serum proteins, secretion of AB blood group substance, phenylthiourea sensitivity, color vision defects, and alcoholism have been significant, but most often the associations appear to result from acquired rather than inherited factors (Swinson, 1983). No significant deviation in HLA distribution of 27 HLA antigens in alcoholics as compared to healthy blood donors was found in a recent study (Rosler, Bellaire, Hengesch, Giannitsis, & Jarovici, 1983). Ryback and Eckhardt (1978) have concluded that alcoholism has yet to be satisfactorily defined in terms of a single biochemical or hematological marker without resulting in undue numbers of false negatives as well as clinically embarrassing false positives. They have found that it takes as many as 25 routinely requested laboratory tests to provide 100% correct identification of non-alcoholic patients and an 86% correct identification of alcoholics.

Schuckit and colleagues have reported a number of studies on possible biologic mediators in alcoholism. They have examined a series of young males at risk for alcoholism because they have an alcoholic parent or sibling so diagnosed according to criteria of Woodruf, Goodwin, and Guze, (1974), and a comparison group who do not have a family history positive for alcoholism. Twenty family history positive non-alcoholic at-risk subjects have been found to have significantly elevated blood acetaldehyde concentrations after a moderate dose of alcohol (Schuckit & Rayses, 1979). A replicate study with 15 family history positive subjects and 15 matched controls using a different method for acetaldehyde assessment has been reported by Schuckit with similar results except that acetaldehyde levels were one half those originally reported in the first study (Schuckit, 1984c). However, Behar et al. (1983) found that 11 boys who had either a father or mother alcoholic by Feighner criteria, and who had not drunk more than five sips of alcohol in their life, had no differences in breath or blood acetaldehyde levels nor in objective or subjective measurements of intoxication after ingestion of 0.5 mL/kg alcohol.

Schuckit (1984b) has found that family history positive subjects (twenty-three 21 to 25-year-olds) self-report less intense reaction to alcohol when experiencing similar blood alcohol levels after ingesting 0.75 mL/kg but not 1.1 mL/kg doses of alcohol in a comparison with FHN control subjects, matched on demographic, smoking, drinking history, and height-weight ratios. Schuckit concludes that family history positive subjects may have an "innate differential brain sensitivity to the drug or one acquired during years of normal drinking" (p. 883).

In a separate report Schuckit described reduced body sway in 10 family history positive 21 to 25-year-old subjects compared with 10 matched FHN controls after ethanol loading and during almost identical BAC blood alcohol concentration (Schuckit, 1984c). In an earlier study Lipscomb, Carpenter, and Nathan (1979), however, found that in two substudies of 12 and 21 family history positive subjects (i.e., with an alcoholic biologic relative), there was significantly greater body sway than for family history negative subjects. The effect was independent of alcohol influence and subject's typical drinking pattern.

The risk items from a longitudinal follow-up study of college and city samples of young men had been summarized by a regression analysis. The authors found three factors with high weighting accounted for most of the variance in alcoholism diagnosis of adults. They were (1) family history of alcoholism; (2) a non-Mediterranean ethnicity; and (3) frequency of school behavior problems (Vaillant & Milofsky, 1982). These data support the finding of an earlier follow-up study of children referred to a psychiatric clinic. Forty-five percent of 57 children with juvenile court record became alcoholic in adulthood, significantly more than those with antisocial behavior but no juvenile court record (25%, $n = 57$) or without antisocial behavior (15%, $n = 59$) (Robins et al., 1962).

Neurocognitive Markers and Risk Differences

Neurocognitive and electrophysiologic differences between alcoholics, sociopaths, and control subjects have been reported. The range of diagnostic characteristics used to describe "psychopaths" or "sociopaths" does not necessarily correspond to DSM III or RDC criteria for antisocial personality. Most criticism of the EEG studies lies in the wide heterogeneity of classifications used. However, a review of EEG studies in adult sociopaths confirms the conclusion of earlier reviews that psychopaths show a higher incidence of EEG abnormalities than do controls, but not necessarily higher than the incidence in other psychiatric groups (Syndulko, 1978). One specific EEG pattern has been described. Forty-eight percent of 90 service men with explicit criteria for a diagnosis of three types of psychopathic personality were compared to 20 medically hospitalized controls. They showed 14 and 6 per-second positive spiking, especially during drowsiness and light sleep. None of the controls had this distinct abnormality (Kurland, Yeager, & Arthur, 1963).

Hare (1978) has reviewed autonomic system differences for psychopaths and concluded that "although the psychopaths were poor electrodermal conditioners, they were good cardiovascular ones" (p. 132). Hare and Cox (1978) conclude:

> The psychopath's pattern of heart rate acceleration and small increases in electrodermal activity is hypothesized to reflect the operation of an active, efficient coping process, and the inhibition of fear arousal. As a result, many situations that have great emotional impact for most people would be of little consequence to the psychopath, because he is better able to attenuate aversive inputs than to inhibit anticipatory fear. As indicated elsewhere, however this very efficient "coping" process would be adaptive for survival only when the psychopath could not make use of the premonitory cues and anticipatory fear to facilitate avoidance behavior. To a certain extent, this may help to account for the psychopath's difficulty in avoiding punishment. That is, the cues that would help him to do so are "tuned out" and the mediating effects of anticipatory fear are reduced. (p. 219)

These data suggest that genetically mediated characteristics may shape response to experience for the ASP alcoholic and also may explain natural history and response to treatment differences that have been found.

Neuropsychological deficits observed in alcoholics when compared to control non-drinking subjects have been reviewed in detail (Tarter & Alterman, 1984). These authors suggest that most described deficits are concomitants of alcohol abuse, but the effects of head injury, liver disease, and nutritional deficiency must be evaluated. In addition, long-term effects of alcohol abuse during gestation by mothers of alcoholics must be considered when assessing neurocognitive deficits as vulnerability markers or as genetically determined precursor neuropsychologic (NP) deficits of alcoholism.

The battery of NP tests may be extensive and can be characterized by those in the Halstead Reitan-Indiana Battery. Most reports of alcoholics and of subjects at risk for alcoholism have included the category errors test, which, excepting the verbal instruction, may be called a visual test of abstraction. It particularly measures skills in the testing of hypotheses and adaptivity in maintaining or discarding hypotheses in view of the feedback given from the test itself. The category errors test is highly and indirectly correlated to verbal intelligence.

Lowered verbal IQ and elevated Halstead Reitan Battery Category Error scores have been two reported neurocognitive antecedent "markers" for children at risk for the development of alcoholism. Gabrielli and Mednick (1983), employing the Danish Birth Cohort, demonstrated significantly lowered verbal IQ in 27 male and female children at high risk for alcoholism as compared to 114 children without alcoholic parents (Table 6.16). Knop, Goodwin, Teasdale, Mikkelsen, and Schulsinger (1984), also using the Danish Birth Registry, reported significantly elevated category error scores for 134 males with at least one parent treated for alcoholism compared to 70 control subjects with no known alcoholic parent (Table 6.16). A study of 14 delinquent sons of alcoholic fathers (age 16) showed significantly reduced mean verbal IQ scores by Peabody but not by WISC/WAIS measurement when compared to similarly aged subjects with family history negative for alcoholism (Tarter, Hegedus, Goldstein, Shelly, & Alterman, 1984). None of these studies controlled adequately for possible confounding effects of alcohol use by the experimental or control subjects. It has been demonstrated, however, that social and heavy drinkers as well as alcoholics had category error scores that were directly proportional to the total amount of alcohol consumed and amount per daily occasion (Eckhardt, Parker, Noble, Feldman, & Gottshalk, 1978; Parker & Noble, 1977); One prospective study has examined subjects at risk for alcoholism for possibly separate contributions to neurocognitive performance deriving from (1) the genetic heterogeneity of the alcohol dependence syndrome; and (2) the contribution to deviation in performance caused by early social alcohol use. Hesselbrock and his colleagues (1985) have demonstrated that if NP test results are controlled for age, ounces of absolute alcohol consumed in the 6 months prior to study, and full-scale IQ of the subject, then family history positive subjects have category error scores no different from low-risk control subjects (Table 6.16).

Efforts to compare NP functioning in alcoholics with or without HK/MBD childhood symptoms have demonstrated that high HK/MBD scoring patients performed more poorly than low HK/MBD scoring alcoholics in the Shipley and Ravens tests (DeObaldia, Parsons, & Yohman, 1983). These tests, which measure verbal capacity, led the authors to conclude that such findings support the hypothesis that childhood HK/MBD factors may be a predisposing issue in the early onset, more pervasive type of alcoholism since primary alcoholics also had more HK/MBD symptoms and also performed more poorly on the tests.

TABLE 6.16

Neurocognitive Assessment of Children of Alcoholics and Control Subjects at Differing Risk for Alcoholism

Investigator	*Number*	*Sex*	*Risk Category And Age*		*Halstead Category Error Scale*	*Neurocognitive Test Results* Peabody Verbal IQ	WISC/WAIS Verbal IQ	WISC/WAIS Performance IQ	WISC/WAIS Full Scale IQ	*Trails A&B (Total Time in Seconds)*
Knop et al (1983)	204	M	134 Hisk Risk (HR): parent with hospital diagnosis of alcoholism; 70 low risk (LR): parent unlisted in psychiatric register.	HR	45.7[1]					
			Age 20–22 years	LR	38.5					
Gabrielli et al. (1983)	184	M&F	27 HR: one parent alcoholic; 43 M High Risk (MR): parent with	HR			95.7[2]	106.0[3]	100.6[4]	
			problem controlling alcohol intake; 114 LR: parent not alcoholic or	MR			98.5	106.8	102.7	
			problem drinker. Age 11–13 years.	LR			106.1	110.9	109.1	
Tarter et al. (1984)	39	M	Juvenile delinquents 14 HR: alcoholic fathers, 25 LR: no alcoholic parent.	HR		89.1[5]	90.6[6]	95.9[6]	92.2[6]	114.0[5]
			Age 16	LR		99.3	94.3	98.4	90.5	92.2
Hesselbrock, Stabenau and Hesselbrock (1985)	146	M&F	99 HR: offspring of inpatient alcoholics, age 24 years; 47 LR:	HR		31.9[7]	109.0	111.0	110.4	99.7
			outpatient dental clinic volunteers without parental alcoholism, Age 26 years	LR		29.5	114.2	114.9	115.3	81.8

$p = .04$, [2]$p = .001$, [3]$p = .23$, [4]$p = .004$, [5]$p = .05$, [6]N.S., [7]HR vs LR N.P. Test Scores NS, when corrected for oz. abs. alc. consumed previous 6 months for Full Scale IQ.

Recent studies of children at risk for alcoholism have focused attention on preexisting alterations in central nervous system function prior to extensive social use of alcohol. In brain wave studies of children with a family history for alcoholism, there is a significant decrease in visual (VER) and auditory (AER) evoked responses during tests of cognition (Begleiter, Porjesz, Bihari, & Kissin, 1984; Elmasian, Neville, Woods, Schuckit, & Bloom, 1982). The functional significance of reduced amplitude and increased latency of P3 or P300 in family history positive subjects found in both studies "suggests that they either could not or would not devote as many resources to the task. This interpretation is consistent with the behavioral data indicating that family history positive subjects were slower and less accurate in identifying the targets" (Elmasian et al., 1982, p. 7903). P300 (or P3) changes described for auditory evoked response (AER) in children at risk for schizophrenia include *shorter* latencies but *no* significant differences in amplitude when compared to control groups of children (Itil, Hsu, Saletu, & Mednick, 1974). However, a smaller mean AER P300 amplitude was found for 21 sociopaths as compared to the mean value for 21 normals (Syndulko, 1978).

VER and AER slope, wave form, and frequency bands tend to have higher correlations for MZ as compared to DZ twins (Buffington, Martin, & Becker, 1981), suggesting a degree of genetic control to these brain responses. A study of such possible genetic control of evoked potential wave patterns of alcoholic abusers to visual stimuli has involved 36 male alcoholic abusers who were paired with a female first-degree relative (mother, sister, or daughter). Patterns of visually evoked responses were analyzed. The VER wave forms and slopes of these non-twin relative pairs were less similar than those reported for MZ twins but significantly more similar than those of random pairs (Buffington et al., 1981).

A study of MZ and DZ twins has shown that resting baseline spectral EEG patterns are under genetic control (Propping, 1977). Two studies of spectral brain wave EEG patterns in individuals at risk for alcoholism report conflicting findings. Gabrielli, Mednick, Volavka, Schulsinger, and Itil (1982) found that male but not female 11 to 13-year-old children at risk for alcoholism showed an excess of fast beta EEG activity. Propping, Kruger, and Mark (1981), on the other hand, reported patterns of poorly synchronized EEG wave forms in adult female alcoholics with reduction of alpha frequencies and a preponderance of beta activity and significantly similar patterns in their relatives. But this EEG pattern was not observed for male alcoholics or their relatives. Itil et al. (1974) found that children at risk for schizophrenia had "presence of high frequency beta activity, fewer fast alpha waves, and more very slow low voltage delta activity in computerized EEG" (p. 892). Except for the slow activity, the beta activity appears non-specific in two at-risk populations.

In summary, personality, except for ASP, does not appear to be a genetic precursor trait of alcoholism, but rather, the behavior traits associated with alcoholism appear to be a responsc to chronic alcohol exposure. Childhood conduct disorder and attention deficit disorder associated with adult alcohol outcome are linked to ASP disorder rather than a family history positive for alcoholism. A CNS biologic marker that may be genetically mediated has been a decreased pattern in visually evoked EEG response to cognitive tasks found in male children with a family history positive for alcoholism.

GENE-ENVIRONMENT INTERACTION AND MODELS OF GENETIC TRANSMISSION

This section briefly reviews some of the issues raised in conceptualizing models for the etiology of alcoholism. Optimally, a model should encompass the biologic, psychologic, and sociocultural determinants of alcoholic behavior and, it is hoped, begin to answer the questions, why this condition rather than another? why this person rather than another? and why at this time?

Alcoholism can be described as a progressive condition in individuals who differ in their susceptibility to the disorder, with a special vulnerability in either the social, psychological, or biological area. These differing vulnerabilities can dominate at different stages of the syndrome. Individuals with varying degrees of biologic and genetic risk move into a period of increased vulnerability and, influenced initially by social factors, begin to drink more heavily. Psychological dependence develops in some of these heavy drinkers and alcohol abuse follows. A number of heavy drinkers with genetic vulnerability become alcohol dependent as physical dependence occurs. Some of these individuals develop medical and psychosocial consequences of chronic alcohol misuse and now are labeled chronic alcoholics (Kissin & Begleiter, 1983).

Methods that document the role of psychological and sociocultural factors in alcoholism and their measurement are outside the scope of this review. The fact that these variables are not discussed in detail does not imply their unimportance, but rather allows more concerted focus on the genetic.

Human genetic variation has contributed to the development of a number of discrete psychiatric disorders. Any single disorder is likely to comprise heterogeneous genetic factors, and hence psychiatric diagnosis may not be as important as identifying phenotypes that reflect individual genetic variants (Kidd & Matthysse, 1978). Rieder and Gershon (1978) have proposed strategies to estimate the contribution of specific genetic and environmental factors to identify homogenous subtypes within heterogeneous syndromes such as alcoholism.

Twin and adoption studies already cited provide strong evidence that genes and environment are important for both alcoholism and antisocial personality (Cadoret, Cain, & Crowe, 1983; Cloninger & Reich, 1983). How different pathogenic factors interact to produce phenotypes from genotypes is not as clear. Cloninger Reich, and Guze (1978) list three markedly different mechanisms for genotype-environment interaction: (1) the "additive" model where genetic and environmental factors act independently of one another; (2) the "diathesis-stress" or "interaction" model where a genotype in different environments produces different clinical manifestations, and an environment produces different effects on different genotypes; or (3) the "correlational" model where the genotype of the individual influences the environment encountered by that individual.

Genetic transmission of "familial" alcoholism and "antisocial personality" alcoholism does not follow Mendelian single-gene locus dominant, recessive, or sex-linked patterns of transmission. Cloninger and Reich (1983) have pointed out that single genes code for individual polypeptide chains that are the subunits of enzymes and structural proteins. A number of investigators have concluded that most complex phenotypes depend on a few loci modified by extensive multifactorial (polygenic and environmental) variability (Fraser, 1980; Gottesman & Shields, 1972; Thoday, 1967; Wright, 1968). An example from crossed inbred animal strains, worked out under experimental control and in detail not possible in man, is cortisone-induced cleft palate defects (Fraser, 1980). There are a large number of primary defects leading to cleft palate, but in any particular susceptible strain only a few factors are frequent, and those few usually differ between strains. Thus, only a proportion of susceptible animals develop cleft palate, depending both on multigenic predisposition and/or on the dose of cortisone to which they are exposed during development (Cloninger & Reich, 1983).

Animal studies are necessary in order to advance our understanding of the mechanisms operative during intervening stages of environment and gene interplay. Ginsburg (1977) notes that discordant identical twin human pairs or members of an inbred animal strain having the same "encoded" genotypes do not necessarily have the same "effective" genotypes. Ginsburg (1977) states:

> This is not only a matter of "reaction range" or how the same genes act under various environmental conditions, but also of the "genomic repertoire;" i.e., which of a number of genetically encoded alternatives will be activated during development to interact with environmental conditions in determining the phenotype. These events may be under the control of regulatory genes that interface with developmental events to selectively activate and/or suppress the expression of encoded genetic alternatives. These processes can be readily studied in animals and have powerful implications for the understanding and control of phenotypic variation, particularly in clinical states. (p. 307)

Ginsburg (1977) continues:

> Segregation data following the Mendelian model may involve genes that interface with environmental events during development to determine whether still other (structural) genes become incorporated into the effective genotype. In this way, genetics provides an interface strategy between the environment and the metabolic potential of the cell, the tissue, or the organism. The study of how such genes act

to selectively regulate the phenotypic expression of the genome should provide an effective clue to the means by which appropriate intervention can optimize the potential of the genotype in relation to its latent phenotypic capabilities. (p. 309)

Ginsburg and his co-workers showed that inbred strains of mice with seizure diathesis were sensitive to glutamic acid changes in the early post-partum period. Seizure-prone phenocopies were produced with opposite biochemical and behavioral change that persisted during the lifetime of the mice. Ginsburg (1977) concluded that these data were consistent with the notion that the genotype can be reprogrammed in order to produce change (in either direction) in the phenotype of the treated individual. "Close genetic relationship does not mean genetic uniformity. The phenotype or syndrome is not a unitary one with respect to underlying mechanism, since the symptoms will be consistently alleviated by a particular pharmacological agent in some individuals but not in others" (Ginsburg, 1977, p. 312).

Three different models are used to explain the etiology and expression of alcohol dependence and/or alcoholism in man. These are (1) the unitary trait disease; (2) the multivariate threshold condition; and (3) the spectrum disorder concept. Goodwin has championed the position that we really do not know that alcoholism is not a disease caused by a single key switch mechanism that is critical to all others in the condition (Goodwin, 1983). Reich, Rice, Cloninger, Wette, & James (1979), on the other hand, note that "one of the most outstanding characteristics of common non-Mendelian familial diseases is their extreme variability. Affected individuals often differ in terms of severity, age of onset, and the form that the disease presents. Unaffected individuals may also be heterogenous with respect to the disorder, in that a proportion of them manifest mild or preclinical traits, which signifies that they or their offspring are more liable to develop the illness" (p. 371). Reich and his colleagues utilizing concepts of Curnow and Smith (1975) have proposed a "Multifactorial Model" as appropriate for the detection of phenotypically homogenous groups in alcoholism (Reich, Cloninger, & Guze, 1975). The assumptions of the Multifactorial Model are as follows:

1. All environmental and genetic causes of a trait may be combined into a single continuous variable termed the "liability of an individual";
2. There are one or more "threshold values" of the liability which divide individuals into recognizable phenotypic classes, i.e., if the liability exceeds the threshold the individual is affected, otherwise normal;
3. The distribution of liability in the general population is normal;
4. Genes which are relevant to the etiology of the disorder are each of small effect in relation to the total variation, and act additively;
5. Environmental contributions to the etiology of the disease are due to many events whose effect are additive;

5. Environmental contributions to the etiology of the disease are due to many events whose effect are additive;
6. The variance of the liability to develop the disorder may be made up of different proportions of genotypic and environmental variance in different subforms (Reich et al., 1979) (p. 372)

Cloninger and Reich (1983) summarized the genetic research dilemma and propose steps in resolving it. They state:

> Attempts to circumvent the problem of the heterogeneity are still impeded by the inadequacy of the one gene-one enzyme principle for understanding multidimensional phenotypes. In the absence of homogenous clinical syndromes and/or a detailed understanding of the mechanisms underlying the genotype-phenotype pathway, association studies of a random set of genetic markers or empirically identified biochemical/physiological disturbances are unlikely to be fruitful . . . Therefore a more systematic strategy for evaluating the inheritance of multidimensional phenotypes is needed (Cloninger, Rice, Reich, & McGuffin, 1982). The first stage in a biologically more realistic approach to the heterogeneity problem is to recognize and describe the multiple components that make up a common multidimensional trait. Multivariate methods such as factor analysis or cluster analysis are useful to identify the independent factors or patient subgroups. In the second stage either the inheritance of individual components of a more complex phenotype (Cloninger, Reich, & Wetzel, 1979) or more homogenous subgroups of a heterogenous set of disorders (Matthysse & Kidd, 1981) may be studied. The inheritance of individual component factors is likely to be simpler than the developmentally more complex ultimate phenotype. In the third stage the interaction of individual risk factors in producing the multidimensional phenotype is studied. The inheritance of individual risk factors may be simpler and thereby crucial to understanding the pathophysiological mechanisms underlying the genotype-phenotype pathway. Nevertheless, it is the consequence of the interactions of individual risk factors that is usually most important clinically. (pp. 146–147)

Biologic genetic research on alcoholism requires identification of specific neurobiological and/or developmental factors that allow for a division of a generalized phenotype into subforms that are clinically and etiologically more homogenous. Methods utilizing pairwise comparison of matched probands and controls may not be sensitive enough to detect etiologies that are heterogenous or developmentally complex (Matthysse & Kidd, 1978). Risk factor paradigms have been described for subdividing probands and relatives according to presence or absence of a putative etiologic factor (Rieder & Gershon, 1978) or according to extreme (high and low) values on a quantitive variable (Buchsbaum, Coursey, & Murphy, 1976). Other models testing etiologic causality that involve both biologic and environmental factors employ path analysis strategies (Cloninger, Lewis, Rice, & Reich, 1981). One strategy is to make observations on multiple classes of relatives varying in degree of genetic relationship and extent of shared rearing experiences. A second method is to observe the phenotype under study, the putative environmental factors, and the putative genetic markers. Cloninger, Lewis, Rice, and Reich (1981) believe that "the two research strategies are complementary and may be used to validate one another and increase the precision of parameter estimates" (p. 326). Computer programs have been described that provide maximum likelihood estimates of the parameters from reported correlations (Cloninger, Rice, & Reich, 1979). These investigators also note that current ability to specify relevant factors and quantify their influence is a substantial advance over merely estimating heritabilities (Cloninger, Lewis, Rice, & Reich, 1981).

Two major adoption studies of alcoholism (Cloninger, Bowman, & Sigvardsson, 1981) and antisocial behavior (Cadoret et al., 1983) have demonstrated interaction between identifiable environmental factors and genetic vulnerability factors. Two heritable subtypes of alcoholism were identified by Cloninger et al.: Type I (or milieu-limited) and Type II (or male-limited). Type I alcohol abuse had occurred in 13% of adopted men. Both their biologic fathers and mothers typically exhibited mild alcohol abuse requiring no treatment, and post-natal environment determined both the frequency and severity of alcohol abuse in the susceptible sons. The sons' alcohol abuse was usually mild or isolated but could be severe depending upon the nature of the environmental provocation. With such provocation the relative risk of developing alcohol abuse in congenitally predisposed individuals was increased two-fold; without it the risk was the same as that in the general population. Type II alcohol abuse occurred in 4% of the adopted men. It represented, therefore, about 25% of the alcohol abusers in the study. Their biologic fathers, but not their mothers, had severe alcoholism and criminality requiring some form of treatment. Post-natal environment had no effect on the expected numbers of abusers but may have influenced severity. Alcohol abuse in the susceptible sons was usually recurrent and moderate, although sometimes it was severe. An estimated nine-fold increase in risk of developing alcohol abuse or alcoholism was evident in the affected individuals, regardless of post-natal environment. The frequency of alcoholism in males identified by chronic abuse in this adoption study is similar to that found in males by different clinical methods used for alcoholism diagnosis in the Danish adoption study (Table 6.8). Cadoret et al. (1983) report on three adoption studies of antisocial behavior and find "the regression coefficients demonstrate the consistent importance of environmental variables in all three samples: the presence of a psychiatric disturbance in the adoptive families significantly predicts increased adolescent antisocial behavior in the adoptee in all three studies; and the age at which the adoptee was placed in the adoptive home is a significant predictor in two studies.

In only one study, the Iowa 1980, did the genetic variable appear by itself asan important factor. Finally the interaction of genetic and environmental influences was a significant determinant of adolescent antisocial behavior in all three studies" (p. 307).

Most geneticists consider the fetal alcohol syndrome in man and animals to be principally nongenetic. The genomic encoding theory and studies of alcohol in animal cross-fostering model by Ginsburg, Yanai, and Csze (1975) suggest that the relationship of the fetal alcohol syndrome (FAS) and the transmission of genetic vulnerability toward alcohol dependence should not be dismissed. The probable effects of maternal drinking upon high-risk children during gestation and lactation must be considered, especially when assessing the neurocognitive capacities of these children as adults and whether there is any interaction between in utero alcohol exposure and risk of developing alcohol dependence as an adult (Abel, Jacobson, & Sherwin, 1983; Dexter, Tumbleson, Decker, & Middleton, 1983; Pennington et al., 1983; Pennington, Taylor, Cowann, & Kalmus, 1984; Reyes, Rivera, Saland, & Murray, 1983). Such interaction may have a genetic component.

In summary, three models of gene-environment interaction have suggested alcoholism as a unitary disease with a single genetic on/off switch mechanism; a spectrum disorder in which alcoholism is one possible response to an underlying genetic deficit; and a multivariate condition that is expressed when genetic vulnerability and environmental stress additively interact beyond a threshold. If alcoholism comprises heterogenous phenotypes that reflect independent genetic variants, etiologic biologic research requires isolation of specific neurobiologic and/or developmental factors that identify subforms that are clinically and etiologically more homogeneous.

SUMMARY

1. In most world populations, groups with the highest proportion of drinkers did not have the highest proportion of heavy drinkers; rates of heavy drinking have been four to five times higher for men than women; and alcoholism rates are double for males as compared to females.

2. Alcohol misuse may involve three "diagnostic" stages: abuse including heavy and problematic drinking; alcohol addiction including physical and psychological dependence; and alcoholism including medical and psychosocial consequences of prolonged alcohol misuse in addition to the alcohol dependence syndrome. Rates of abuse are almost twice those for dependence.

3. Some form of genetic influence has been attributed in humans to amount of alcohol consumed; sensitivity of the CNS and body function; and alcohol degradation and elimination. Genetic control over acquisition of tolerance and

development of dependence has been most explicitly obtained from studies of animals selectively bred toward alcohol use.

4. A genetic vulnerability hypothesis for chronic alcoholism has been suggested by twin-pair concordance comparison and controlled adoption cross-fostering studies.

5. Genetic factors influencing the course of alcoholism are male gender associated with highest risk for alcoholism; presence of ASP diagnosis associated with early onset of alcohol dependence; and presence of alcoholism on both sides of the family pedigree associated with greater consequences of prolonged drinking. All three factors contribute, perhaps additively, to the overall risk of alcoholism.

6. Two genetically determined alcoholisms have been demonstrated: alcoholism associated with antisocial personality disorder; and alcoholism in non-ASP individuals who have a family history of alcoholism. Affective disorder, a separate genetically transmitted condition, although often coexisting with alcoholism, does not appear to increase risk for alcoholism as do ASP or family history of alcoholism.

7. Personality, except for ASP, does not appear to be a genetic precursor trait of alcoholism; rather, the personality traits associated with alcoholism appear to be a response to chronic alcohol exposure. Childhood conduct disorder and attention deficit disorder associated with adult alcohol outcome are linked to ASP disorder rather than a family history positive for alcoholism. A CNS biologic marker that may be genetically mediated has been a decreased pattern in visually evoked EEG response to cognitive tasks found in male children with a family history positive for alcoholism.

8. Three models of gene-environment interaction have suggested alcoholism as a unitary disease with a single genetic on/off switch mechanism; a spectrum disorder in which alcoholism is one possible response to an underlying genetic deficit; and a multivariate condition that is expressed when genetic vulnerability and environmental stress additively interact beyond a threshold. If alcoholism comprises heterogenous phenotypes that reflect independent genetic variants, etiologic biologic research requires isolation of specific neurobiologic and/or developmental factors that identify subforms that are clinically and etiologically more homogeneous.

What we know today suggests that genetic vulnerability for chronic alcoholism appears in individuals who indulge in heavy alcohol use and who have antisocial personality traits and/or a biologic alcoholic family member. Genetic factors contribute to metabolism, degradation, and elimination of alcohol but do not explain patterns of heavy drinking. Onset of dependence and complications of chronic alcohol use appear to be under partial control of genetic factors. The future includes the definition of specific at-risk individuals and elucidation of the specific mechanisms under genetic control that may be necessary for such individuals as they move from social to addictive drinking.

ACKNOWLEDGMENTS

The author thanks Drs. Victor Hesselbrock, Ovid Pomerleau, William Shoemaker, and Stuart Sugarman for their review and comments and Carolyn Conti for word processing of this manuscript. This paper was supported in part by NIAAA Center Study Grant No. AA-0510-06.

REFERENCES

Abel, E. L., Jacobson, S., & Sherwin, B. T. (1983). In utero alcohol exposure: Functional and structural brain damage. *Neurobehavioral Toxicology and Teratology, 5,* 363–366.

Amark, C. (1951). A study in alcoholism: Clinical, social-psychiatric and genetic investigations. *Acta Psychiatrica et Neurologica Scandinavica, Suppl. 70,* 253–271.

American Psychiatric Association (1980). Diagnostic and Statistical Manual of Mental Disorders (3rd ed.) (DSM III). Washington, DC: Author.

Andreasen, N. C., Endicott, J., Spitzer, R. L., & Winokur, G. (1977). The family history method using diagnostic criteria. *Archives of General Psychiatry, 34,* 1229–1235.

Apfeldorf, M. (1978). Alcoholism scales of the MMPI: Contributions and future directions. *The International Journal of the Addictions, 13,* 17–53.

Ashley, M. J., Olin, J. S., le Riche, W. H., Kornaczewski, A., Schmidt, W., Corey, P. N., & Rankin, J. G. (1981). The physical disease characteristics of inpatient alcoholics. *Journal of Studies on Alcohol, 42,* 1–13.

August, G. J., & Stewart, M. A. (1983). Familial subtypes of childhood hyperactivity. *Journal of Nervous and Mental Disease, 171,* 362–368.

Bailey, W., Hustinyer, F., & Kristofferson, A. (1961). Alcoholism, brain damage and perceptual dependence. *Quarterly Journal of Studies on Alcohol, 22,* 387–393.

Barnes, G. E. (1979). The alcoholic personality. *Journal of Studies on Alcohol, 40,* 571–634.

Barnes, G. E. (1983). Clinical and prealcoholic personality characteristics. In B. Kissin & H. Begleiter (Eds.), *The pathogenesis of alcoholism; Vol. 6. Psychosocial Factors,* (pp. 113–195). New York: Plenum.

Begleiter, H., Porjesz, B., Bihari, B., & Kissin, B. (1984). Event-related brain potentials in boys at risk for alcoholism. *Science, 225,* 1493–1496.

Behar, D., Berg, C. J., Rapoport, J. L., Nelson, W., Linnoila, M., Cohen, M., Bozevich, C., & Marshall, T. (1983). Behavioral and physiological effects of ethanol in high-risk and control children: A pilot study. *Alcoholism: Clinical and Experimental Research, 7,* 404–410.

Bernadt, M. W., Mumford, J., & Murray, R. M. (1984). A discriminant-function analysis of screening tests for excessive drinking and alcoholism. *Journal of Studies on Alcohol, 45,* 81–86.

Blass, J. P., & Gibson, G. E. (1977). Abnormality of a thiamine-requiring enzyme in patients with Wernicke-Korsakoff Syndrome. *New England Journal of Medicine, 297,* 1367–1370.

Blouin, A. G., Bornstein, R. A., & Trites, R. L. (1978). Teenage alcohol use among hyperactive children: A five year follow-up study. *Journal of Pediatric Psychology, 3,* 188–194.

Bohman, M. (1978). Some genetic aspects of alcoholism and criminality. *Archives of General Psychiatry, 35,* 269–276.

Boyd, J. H., Weissman, M. M., Thompson, W. D., & Myers, J. K. (1983). Different definitions of alcoholism. I: Impact of seven definitions on prevalence rates in a community survey. *American Journal of Psychiatry, 140,* 1309–1313.

Buchsbaum, M., Coursey, R., & Murphy, D. (1976). The biochemical high-risk paradigm: Behavioral and familial correlates of low platelet monamine oxidase activity. *Science, 194,* 339–341.

Buffington, V., Martin, D., & Becker, J. (1981). VER similarity between alcoholic probands and their first degree relatives. *Psychophysiology, 18,* 529–533.

Buss, A. H., & Plomin, R. (1978). *A temperament theory of personality development.* New York: Wiley.

Caddy, G. R. (1977). Toward a multivariate analysis of alcohol abuse. In P. Nathan, A. Marlatt, & T. Loberg (Eds.), *Alcoholism: New directions in behavioral research and treatment* (pp. 71–117). New York: Plenum.

Caddy, G., Goldman, R., & Huebner, R. (1976). Group differences in attitudes toward alcoholism. *Addictive Behaviors, 1,* 281–286.

Cadoret, R. J. (1978). Psychopathology in adopted-away offspring of biologic parents with antisocial behavior. *Archives of General Psychiatry, 35,* 176–184.

Cadoret, R. J. Cain, C. A., & Crowe, R. R. (1983). Evidence for gene-environment interaction in the development of adolescent antisocial behavior. *Behavior Genetics, 13,* 301–310.

Cadoret, R. J., & Gath, A. (1978). Inheritance of alcoholism in adoptees. *British Journal of Psychiatry, 132,* 252–258.

Cahalan, D. (1982). Epidemiology: Alcohol use in American society: In E. L. Gomberg, H. R. White, & J. A. Carpenter (Eds.), *Alcohol, science, and society revisited* (pp. 96–118). Ann Arbor: University of Michigan Press.

Cahalan, D., & Cisin, I. (1968). II: American drinking practices: Summary of findings from a national probability sample. I. Extent of drinking by population subgroups. *Quarterly Journal of Studies on Alcohol, 29,* 130–182.

Cantwell, D. (1972). Psychiatric illness in the families of hyperactive children. *Archives of General Psychiatry, 27,* 414–417.

Cassidy, W., Flanagan, N., Spellman, M., Cohen, M. (1957). Clinical observations in manic depressive disease. *Journal of the American Medical Association, 164,* 1535–1546.

Cavalli-Sforza, L. L., & Bodmer, W. F. (1971). *The genetics of human populations* (pp. 537–550). San Francisco: W. H. Freeman.

Chen, C. (1968). A study of the alcohol-tolerance effect and an introduction of a new behavioral technique. *Psychopharamacologia, 12,* 433–440.

Christiansen, K. (1974). Seriousness of criminality and concordance among Danish twins. In R. Hood (Ed.), *Crime, criminology and public policy* (pp. 63–77). London: Heinemann.

Clark, W. B., & Cahalan, D. (1976). Changes in problem drinking over a four-year span. *Addictive Behaviors, 1,* 251–259.

Clifford, C. A., Fulker, D. W., Gurling, H. M. D., & Murray, R. M. (1981). Preliminary findings from a twin study of alcohol use. *Progress in Clinical and Biological Research, 69,* 47–52.

Cloninger, C. R., & Reich, T. (1983). Genetic heterogeneity in alcoholism and sociopathy. In S. S. Kety, L. P. Rowland, R. L. Sidman, & S. W. Matthysse (Eds.), *Genetics of neurological and psychiatric disorders* (pp. 145–166). New York: Raven Press.

Cloninger, C. R., Bohman, M., Sigvardsson, S. (1981). Inheritance of alcohol abuse. *Archives of General Psychiatry, 38,* 861–868.

Cloninger, C. R., Christiansen, K. O., Reich, T., & Gottesman, I. I. (1978). Implications of sex differences in the prevalences of antisocial personality, alcoholism, and criminality for familial transmission. *Archives of General Psychiatry, 35,* 941–951.

Cloninger, C. R., Lewis, C., Rice, J., & Reich, T. (1981). Strategies for resolution of biological and cultural inheritance. In E. S. Gershon, S. Matthysse, X. O. Breakefield, & R. D. Ciaranello (Eds.), *Genetic research strategies for psychobiology and psychiatry* (pp. 319–332). Pacific Grove, CA: Boxwood Press.

Cloninger, C. R., Reich, T., & Guze, S. B. (1978). Genetic-environmental interactions and antisocial behaviour: In R. D. Hare & D. Schalling (Eds.), *Psychopathic behaviour: Approaches to research* (pp. 225–237). New York: Wiley.

Cloninger, C. R., Reich, T., & Wetzel, R. (1979). Alcoholism and affective disorders: Familial associations and genetic models. In D. W. Goodwin & C. K. Erickson (Eds.), *Alcoholism and affective disorders* (pp. 57–86). New York: SP Medical & Scientific Books.

Cloninger, C., Rice, J., & Reich, T. (1979). Multifactorial inheritance with cultural transmissionn and assortative mating: II. A general model of combined polygenic and cultural inheritance. *American Journal of Human Genetics, 31,* 176.

Cloninger, C., Rice, J., Reich, T., & McGuffin, P. (1982). Genetic analysis of seizure disorders and multidimensional threshold characters. In K. Anderson, W. Hauser, & C. Sing (Eds.), *Genetics and epilepsy* (pp. 291–309). New York: Raven.

Cotton, N. S. (1979). The familial incidence of alcoholism. *Journal of Studies on Alcohol, 40,* 89–116.

Crow, J. F., & Felsenstein, J. (1968). The effect of assortative mating on the genetic composition of a population. *Eugenics Quarterly, 15,* 85–97.

Crowe, R. R. (1974). An adoption study of antisocial personality. *Archives of General Psychiatry, 31,* 785–791.

Curnow, R., & Smith, C. (1975). Multifactorial models for familial diseases in man. *Journal of the Royal Statistical Society, 137,* 134–469.

Dalgard, D., & Kringlen, E. (1976). A Norwegian twin study of criminality. *British Journal of Criminology, 16,* 213–232.

De Obaldia, R., Parsons, O. A. & Yohman, R. (1983). Minimal brain dysfunction symptoms claimed by primary and secondary alcoholics: Relation to cognitive functioning. *International Journal of Neuroscience, 20,* 173–181.

Dexter, J. D., Tumbleson, M. E., Decker, J. D., & Middleton, C. C. (1983). Comparison of the offspring of three serial pregnancies during voluntary alcohol consumption in Sinclair (S-1) miniature swine. *Neurobehavioral Toxicology and Teratology, 5,* 229–231.

Eckardt, M. J., Parker, E. S., Noble, E. P., Feldman, D. J., & Gottschalk, L. A. (1978). Relationship between neuropsychological performance and alcohol consumption in alcoholics. *Biological Psychiatry, 13,* 551–563.

Effron, V., Keller, M., & Guriolo, C. (1974). *Statistics on consumption of alcohol and on alcoholism.* New Brunswick, NJ: Rutgers Center for Alcohol Studies.

Elmasian R., Neville, H., Woods, D., Schuckit, M., & Bloom, F. (1982). Event-related brain potentials are different in individuals at high and low risk for developing alcoholism. *Proceedings of the National Academy of Sciences of the United States of America, 79,* 7900–7903.

Eriksson, C. J. (1980). Elevated blood acetaldehyde levels in alcoholics and their relatives: A reevaluation. *Science, 207,* 1383–1384.

Eriksson, C. J. P., & Peachey, J. E. (1980). Lack of difference in blood acetaldehyde of alcoholics and controls after ethanol ingestion. *Pharmacology Biochemistry & Behavior, 13,* 101–105.

Erwin, V. G., & McClearn, G. E. (1982). Genetic influences on alcohol consumption and actions of alcohol: In M. Galanter (Ed.), *Currents in alcoholism: Vol. 8. Recent advances in research and treatment* (pp. 405–420). New York: Grune & Stratton.

Erwin, V., McClearn, G., & Kuse, A. (1980). Interrelationships of alcohol consumption, actions of alcohol and biochemical traits. *Pharmacology, Biochemistry and Behavior, 13,* Suppl. 297.

Eyre, S. L., Rounsaville, B. J., & Kleber, H. D. (1982). History of childhood hyperactivity in a clinic population of opiate addicts. *Journal of Nervous and Mental Disease, 170,* 522–529.

Feighner, J., Robins, E., Guze, S., Woodruff, R., & Winokur, G. (1972). Diagnostic criteria for use in psychiatric research. *Archives of General Psychiatry, 26,* 57–63.

Fifth Special Report to the U.S. Congress on Alcohol and Health from the Secretary of Health and Human Services (1983). *Genetics and Alcoholism* (pp. 15–24). Rockville, MD: National Institute on Alcoholism and Alcohol.

Filmore, K. (1975). Relationships between specific drinking problems in early adulthood and middle age: An exploratory twenty year follow up study. *Quarterly Journal of Studies on Alcohol, 36,* 882–907.

Filmore, K., & Midanik, L. (1984). Chronicity of drinking problems among men: A longitudinal study. *Journal of Studies on Alcohol, 45,* 228–236.

Fischer, M., & Gottesman, I. (1980). A study of offspring of parents both hospitalized for psychiatric disorders. In L. Robins, P. Clayton, & J. Wing (Eds.), *The social consequences of psychiatric illness*. New York: Brunner/Mazel.

Floderus-Myrhed, B., Pedersen, N., & Rasmuson, I. (1980). Assessment of heritability for personality, based on a short-form of the Eysenck personality inventory: A study of 12,898 twin pairs. *Behavior Genetics, 10,* 153–162.

Frances, R. J., Timm, S., & Bucky, S. (1980). Studies of familial and nonfamilial alcoholism. *Archives of General Psychiatry, 37,* 564–566.

Fraser, F. (1980). Evolution of a palatable multifactorial threshold model. *American Journal of Human Genetics, 32,* 796–831.

Gabrielli, W. F., & Mednick, S. A. (1983). Intellectual performance in children of alcoholics. *Journal of Nervous and Mental Disease, 171,* 444–447.

Gabrielli, W. F., Mednick, S. A., Volavka, J., Schulsinger, F., & Itil, T. M. (1982). Electroencephalograms in children of alcoholic fathers. *Psychophysiology, 19,* 404–407.

Ginsburg, B. E. (1977). Genetic models of behavior disorders. In I. Hanin & E. Usdin (Eds.), *Animal models in psychiatry and neurology* (pp. 307–314). Oxford: Pergamon Press.

Ginsburg, B., Yanai, J., & Csze, P. (1975). A developmental genetic study of the effects of alcohol consumed by parent mice on the behavior and development of their offspring. In M. Chafetz (Ed.), *Research, treatment and prevention.* Proceedings of the Fourth Annual Alcoholism Conference of the National Institute on Alcohol Abuse and Alcoholism (pp. 183–204). Washington, DC: ADAMHA, NIAAA.

Glatt, M. M. (1961). Drinking habits of English (middle class) alcoholics. *Acta Psychiatrica Scandinavica, 37,* 88–113.

Goedde, H. W. (1978). Genetic aspects in the metabolism of drugs and environmental agents. *Annales de Biologie Clinique, 36,* 181–189.

Goedde, H. W., Agarwal, D. P., & Harada, S. (1983). The role of alcohol dehydrogenase and aldehyde dehydrogenase isozymes in alcohol metabolism, alcohol sensitivity, and alcoholism. In M. Rattazzi, J. Scandalios, & E. Whitt (Eds.) *Isozymes: Current topics in biological and medical research: Vol. 8. Cellular localization, metabolism and physiology* (pp. 175–193). New York: Alan R. Liss.

Goodwin, D. W. (1976). *Is alcoholism hereditary?* New York: Oxford University Press.

Goodwin, D. W. (1979). Alcoholism and heredity. *Archives of General Psychiatry, 36,* 57–61.

Goodwin, D. W. (1983). Overview. In M. Galanter (Ed.), *Recent developments in alcoholism* (Vol. 1, p. 7). New York: Plenum Press.

Goodwin, D. W., Schulsinger, F., Hermansen, L., Guze, S. B., Winokur, G. (1973). Alcohol problems in adoptees raised apart from alcoholic biological parents. *Archives of General Psychiatry, 28,* 238–243.

Goodwin, D. W., Schulsinger, F., Hermansen, L., Guze, S. B., Winokur, G. (1975). Alcoholism and the hyperactive child syndrome. *Journal of Nervous and Mental Disease*, 160, 349–383.

Goodwin, D. W., Schulsinger, F., Knop, J., Mednick, S., Guze, S. (1977). Alcoholism and depression in adopted-out daughters of alcoholics. *Archives of General Psychiatry, 34,* 751–755.

Gottesman, I., Shields, J. (1972). *Schizophrenia and genetics, a twin study vantage point.* New York: Academic Press.

Gurling, H. M. D., Clifford, C. A., & Murray, R. M. (1981). Investigations into the genetics of alcohol dependence and into its effects on brain function. In L. Gedda, P. Paris, and W. Nance (Eds.), *Proceedings of the Third International Congress on Twin Studies* (pp. 77–81). New York: Alan Liss.

Guze, S. (1976). *Criminality in psychiatric disorders.* New York: Oxford University Press.

Guze, S., Wolfgram, E., McKinney, J., & Cantwell, D. (1967). Psychiatric illness in the families of convicted criminals. A study of 519 first-degree relatives. *Diseases of the Nervous System, 28,* 651–659.

Haglund, R. M. J., & Schuckit, M. A. (1977). The epidemiology of alcoholism. In N. J. Estes & M. E. Heinemann (Eds.), *Alcoholism: Development, consequences, and interventions* (pp. 28–43). St. Louis: C. V. Mosby.

Hall, R. L., Hesselbrock, V. M., & Stabenau, J. R. (1983a). Familial distribution of alcohol use: I. Assortative mating in the parents of alcoholics. *Behavior Genetics, 13,* 361–372.

Hall, R. L., Hesselbrock, V. M., & Stabenau, J. R. (1983b). Familial distribution of alcohol use: II. Assortative mating of alcoholic probands. *Behavior Genetics, 13,* 373–382.

Harada, S., Agarwal, D., Goedde, H., Takagi, S., & Ishikawa, B. (1982). Possible protective role against alcoholism for aldehyde dehydrogenase isozyme deficiency in Japan. *Lancet, 2,* 827.

Hare, R. D. (1978). Electrodermal and cardiovascular correlates of psychopathy. In R. D. Hare & D. Schalling (Eds.), *Psychopathic behaviour: Approaches to research* (pp. 107–142). New York: Wiley.

Hare, R. D., & Cox, D. N. (1978). Psychophysiological research on psychopathy. In W. H. Reid (Ed.), *The psychopath: A comprehensive study of antisocial disorders and behaviors* (pp. 209–222). New York: Brunner/Mazel.

Hesselbrock, M., Babor, T. F., Hesselbrock, V., Meyer, R. E., & Workman, K. (1983). "Never believe an alcoholic"? On the validity of self-report measures of alcohol dependence and related constructs. *International Journal of the Addictions, 18,* 593–609.

Hesselbrock, M., Hesselbrock, V., Babor, T. F., Stabenau, J. R., Meyer, R. E., & Weidenman, M. (1984). Antisocial behavior, psychopathology and problem drinking in the natural history of alcoholism. In D. W. Goodwin, K. T. Van Dusen, & S. A. Mednick (Eds.), *Longitudinal research in alcoholism* (pp. 197–214). Boston: Kluwer-Nijhoff.

Hesselbrock, V., Hesselbrock, M., & Stabenau, J. (1985). Alcoholism in men patients subtyped by family history and antisocial personality. *Journal of Studies on Alcohol, 46,* 59–64.

Hesselbrock, V., Hesselbrock, M., & Workman-Daniels, K. (in press). The effect of major depression and antisocial personality on alcoholism: Course and motivational patterns. *Journal of Studies on Alcohol.*

Hesselbrock, V. M., Stabenau, J. R., & Hesselbrock, M. N. (1985). Minimal brain dysfunction and neuropsychological test performance in offspring of alcoholics. In M. Galanter (Ed.), *Recent developments in alcoholism* (Vol. 3, pp. 65–81). New York: Plenum Press.

Hesselbrock, V. M., Stabenau, J. R., Hesselbrock, M. N., Meyer, R. M., & Babor, T. F. (1982). The nature of alcoholism in patients with different family histories for alcoholism. *Progress in Neuro-Psychopharmacology and Biological Psychiatry, 6,* 607–614.

Hesselbrock, V., Stabenau, J., Hesselbrock, M., Mirkin, P., & Meyer, R. (1982). A comparison of two interview schedules (SADS-Lifetime and NIMH-DIS). *Archives of General Psychiatry, 39,* 674–677.

Horvath, T. (1975). Clinical spectrum and epidemiologic features of alcohol dementia. In J. G. Rankin (Ed.), *Alcohol drugs and brain damage.* Toronto: Addiction Research Foundation.

Howell, D. C., & Huessy, H. R. (1982, November). *The relationship between adult alcoholism and childhood behavior problems.* Paper presented at the 110th Annual Meeting of the American Public Health Association, Montreal, Canada.

Hrubec, Z., & Omenn, G. S. (1981). Evidence of genetic predisposition to alcoholic cirrhosis and psychosis: Twin concordances for alcoholism and its biological end points by zygosity among male veterans. *Alcoholism: Clinical and Experimental Research, 5,* 207–215.

Huessy, H. R. (1984). Genetics of alcoholism. *Hospital and Community Psychiatry, 35,* 620–621.

Itil, T. M., Hsu, W., Saletu, B., & Mednick, S. (1974). Computer EEG and auditory evoked potential investigations in children at high risk for schizophrenia. *American Journal of Psychiatry, 131,* 892–900.

Jacobson, G. R. (1983). Detection, assessment, and diagnosis of alcoholism: Current techniques. In M. Galanter (Ed.), *Recent developments in alcoholism* (Vol. 1, pp. 377–413). New York: Plenum Press.

Jellinek, E. (1946). Phases in the drinking history of alcoholics: Analysis of a survey conducted by the official organ of the A.A. *Quarterly Journal of the Studies on Alcohol, 7,* 1–88.

Jellinek, E. (1960). *The disease concept of alcoholism.* New Haven: Hillhouse Press.

Kaij, L. (1960). *Alcoholism in twins: Studies on the etiology and sequels of abuse of alcohol.* Stockholm: Almqvist & Wiksell.

Kaij, L., & McNeil, T. F. (1979). Genetic aspects of alcoholism. *Advances in Biological Psychiatry, 3,* 54–65.

Kidd, K. K., & Matthysse, S. (1978). Research designs for the study of gene-environment interactions in psychiatric disorders. *Archives of General Psychiatry, 35,* 925–932.

Kissin, B., & Begleiter, H. (Eds.). (1983). Preface. *The pathogenesis of alcoholism: Vol. 6. Psychosocial factors* (pp. vii-ix). New York: Plenum Press.

Klatsky, A. L., Siegelaub, A. B., Landy, C., & Friedman, G. D. (1983). Racial patterns of alcoholic beverage use. *Alcoholism: Clinical and Experimental Research, 7,* 372–377.

Knop, J., Goodwin, D., Teasdale, T. W., Mikkelsen, U., & Schulsinger, F. (1983). A Danish prospective study of young males at high risk for alcoholism. In D. W. Goodwin, K. T. Van Dusen, & S. A. Mednick (Eds.), *Longitudinal research in alcoholism* (pp. 107–124). Boston: Kluwer-Nijhoff.

Korsten, M., Matsuzaki, S., Feinman, L., & Lieber, C. (1975). High blood acetaldehyde levels after ethanol administration: Differences between alcoholic and non-alcoholic subjects. *New England Journal of Medicine, 292,* 386–389.

Kurland, H., Yeager, G., & Arthur, R. (1963). Psychophysiological aspects of severe behavior disorders. *Archives of General Psychiatry, 8,* 599–604.

Lanskey, D., Nathan, P., & Lawson, D. (1978). Blood alcohol level discrimination by alcoholics: The role of internal and external cues. *Journal of Consulting and Clinical Psychology, 46,* 953–960.

LeBlanc, A., Kalant, H., & Gibbins, R. (1973). Behavioral augmentation of tolerance to ethanol in the rat. *Psychopharmacologia, 30,* 117–122.

Lee, K., Moller, L., Hardt, F., Haubek, A., & Jensen, E. (1979). Alcohol-induced brain damage and liver damage in young males. *Lancet, 8146,* 759–761.

Lewis, C. E., Helzer, J., Cloninger, C. R., Croughan, J., & Whitman, B. Y. (1982). Psychiatric diagnostic predispositions to alcoholism. *Comprehensive Psychiatry, 23,* 451–461.

Lewis, C. E., Rice, J., & Helzer, J. E. (1983). Diagnostic interactions: Alcoholism and antisocial personality. *Journal of Nervous and Mental Disease, 171,* 105–113.

Lindros, K. O., Stowell, A., Pikkarainen, P., & Salaspuro, M. (1980). Elevated blood acetaldehyde in alcoholics with accelerated ethanol elimination. *Pharmacology Biochemistry & Behavior, 13,* 119–124.

Lipscomb, T. R., Carpenter, J. A., & Nathan, P. E. (1979). Statis ataxia: A predictor of alcoholism? *British Journal of Addiction, 74,* 289–294.

Lipscomb, T. R., & Nathan, P. E. (1980). Blood alcohol level discrimination: The effects of family history of alcoholism, drinking pattern, and tolerance. *Archives of General Psychiatry, 37,* 571–576.

Loberg, T. (1980). Neuropsychological deficits in alcoholism: Lack of personality (MMPI) correlates. In H. Begleiter (Ed.), *Biological effects of alcohol* (pp. 797–808). New York: Plenum Press.

Loper, R., Kammeier, M., Hoffman, H. (1973). MMPI characteristics of college freshman males who later became alcoholics. *Journal of Abnormal Psychology, 82,* 159–162.

Ludwig, A. M. (1983). Why do alcoholics drink? In B. Kissin & H. Begleiter (Eds.), *Pathogenesis of alcoholism: Psychosocial factors* (Vol. 6, pp. 197–214). New York: Plenum Press.

Mandell, W. (1983). Types and phases of alcohol dependence illness. In M. Galanter (Ed.), *Recent developments in alcoholism* (Vol. 1, pp. 415–447). New York: Plenum Press.

Matthysse, S., & Kidd, K. (1978). The value of dual mating data in estimating genetic parameters. *Annals of Human Genetics, 41,* 477–480.

Matthysse, S., & Kidd, K. (1981). Pattern recognition in genetic analysis. In E. Gershon, S. Matthysse, X. Breakfield, & R. Ciaranello (Eds.), *Genetic research strategies in psychobiology and psychiatry* (pp. 333–340). Pacific Grove, CA: Boxwood Press.

McClearn, G. E. (1983). Genetic factors in alcohol abuse: Animal models. In B. Kissin & H. Begleiter (Eds.), *The pathogenesis of alcoholism: Biologic factors* (Vol. 7, pp. 1–30). New York: Plenum Press.

McCord, W., & McCord, J. (1962). A longitudinal study of the personality of alcoholics. In D. J. Pittman & C. R. Snyder (Eds.), *Society, culture, and drinking patterns* (pp. 413–430). London: Feffer & Simons.

McKenna, T., & Pickens, R. (1981). Alcoholic children of alcoholics. *Journal of Studies on Alcohol, 42,* 1021–1029.

Mello, N. K. (1983). A behavioral analysis of the reinforcing properties of alcohol and other drugs in man. In B. Kissin & H. Begleiter (Eds.), *The pathogenesis of alcoholism: Biological factors* (Vol. 7, pp. 133–198). New York: Plenum Press.

Merikangas, K. R. (1982). Assortative mating for psychiatric disorders and psychological traits. *Archives of General Psychiatry, 39,* 1173–1180.

Merikangas, K. R., Leckman, J. F., Prusoff, B. A., Pauls, D. L., Weissman, M. M. (1985). Familial transmission of depression and alcoholism. *Archives of General Psychiatry, 42,* 367–372.

Merikangas, K. R., Weissman, M. M. Prusoff, B. A., Pauls, D. L., & Leckman, J. F. (1985). Psychiatric disorders in the offspring of probands with depression and alcoholism. *Journal of Studies on Alcohol, 46,* 199–204.

Meyer, R. E., Babor, T. F., & Mirkin, P. M. (1983). Typologies in alcoholism: An overview. *The International Journal of the Addictions, 18,* 235–249.

Milman, D. H. (1979). Minimal brain dysfunction in childhood: Outcome in late adolescence and early adult years. *Journal of Clinical Psychiatry, 40,* 371–380.

Mirin, S. M., Weiss, R. D., Sollogub, A., & Michael, J. (1984). Psychopathology in the families of drug abusers. In S. M. Mirin (Ed.), *Substance abuse and psychopathology* (pp. 80–106). Washington, DC: American Psychiatric Press.

Morrison, J., & Stewart, M. (1974). Bilateral inheritance as evidence for polygenicity in the hyperactive child syndrome. *Journal of Nervous and Mental Diseases, 158,* 226–228.

Murray, R. M., & Gurling, H. (1982). Alcoholism: Polygenic influence on a multifactorial disorder. *British Journal of Hospital Medicine, 27,* 328–331.

Murray, R. M., & Gurling, H. M. D. (1982b). Genetic contributions to normal and abnormal drinking. In M. Sandler (Ed.), *The psychopharmacology of alcoholism* (pp. 1–41). New York: Raven Press.

Murray, R. M., & Stabenau, J. R. (1982). Genetic factors in alcoholism predisposition. In E. M. Pattison & E. Kaufman (Eds.), *Encyclopedic handbook of alcoholism* (pp. 135–144). New York: Gardner Press.

Myers, R. D., & Ewing, J. A. (1980). Aversive factors in alcohol drinking in humans and animals. *Pharmacology Biochemistry & Behavior, 13, Suppl. 1,* 269–277.

Nerviano, V. J., & Gross, H. W. (1983). Personality types of alcoholics on objective inventories. *Journal of Studies on Alcohol, 44,* 837–851.

Omenn, G. (1975). Alcoholism, a pharmacogenetic disorder. *Modern Problems of Pharmacopsychiatry, 10,* 12–22.

Park, P. (1973). Developmental ordering of experiences in alcoholism. *Quarterly Journal of Studies on Alcohol, 34,* 473–488.

Parker, E. S., & Noble, E. P. (1977). Alcohol consumption and cognitive functioning in social drinkers. *Journal of Studies on Alcohol, 38,* 1224–1232.

Partanen, J., Bruun, K., & Markkanen, T. (1966). *Inheritance of drinking behavior.* Helsinki: Finnish Foundation of Alcohol Studies.

Penick, E. C., Powell, B. J., Othmer, E., Bingham, S. F., Rice, A. S., & Liese, B. S. (1984). Subtyping alcoholics by coexisting psychiatric syndromes: Course, family history, outcome. In D. W. Goodwin, K. T. Van Dusen, & S. A. Mednick (Eds.), *Longitudinal research in alcoholism* (pp. 167–196). Boston: Kluwer-Nijhoff.

Penick, E. C., Read, M. R., Crowley, P. A., & Powell, B. J. (1978). Differentiation of alcoholics by family history. *Journal of Studies on Alcohol, 39,* 1944–1948.

Pennington, S., Boyd, J., Kalmus, G., & Wilson, R. (1983). The molecular mechanism of fetal alcohol syndrome (FAS): I. Ethanol-induced growth suppression. *Neurobehavioral Toxicology and Teratology, 5,* 259–262.

Pennington, S., Taylor, W., Cowen, D., & Kalmus, G. (1984). A single dose of ethanol suppresses rat embryo development in vivo. *Alcoholism: Clinical and Experimental Research, 8,* 236.

Petrie, A. (1967). *Individuality in pain and suffering.* Chicago: University of Chicago Press.

Powell, B. J., Penick, E. C., Othmer, E., Bingham, S. F., & Rice, A. S. (1982). Prevalence of additional psychiatric syndromes among male alcoholics. *Journal of Clinical Psychiatry, 43,* 404–407.

Propping, P. (1977). Genetic control of ethanol action on the central nervous system. *Human Genetics, 35,* 309–334.

Propping, P., Kruger, J., & Mark, N. (1981). Genetic disposition to alcoholism. An EEG study in alcoholics and their relatives. *Human Genetics, 59,* 51–59.

Reich, T., Cloninger, C. R., & Guze, S. B. (1975). The multifactorial model of disease transmission: I. Description of the model and its use in psychiatry. *British Journal of Psychiatry, 127,* 1–10.

Reich, T., Rice, J., Cloninger, C. R., Wette, R., & James, J. (1979). The use of multiple thresholds and segregation analysis in analyzing the phenotypic heterogeneity of multifactorial traits. *Annals of Human Genetics, 42,* 371–390.

Reyes, E., Rivera, J. N., Saland, L. C., & Murray, H. M. (1983). Effects on maternal administration of alcohol on fetal brain development. *Neurobehavioral Toxicology and Teratology, 5,* 263–267.

Rieder, R. O., & Gershon, E. S. (1978). Genetic strategies in biological psychiatry. *Archives of General Psychiatry, 35,* 866–873.

Rimmer, J., & Winokur, G. (1972). The spouses of alcoholics: An example of assortative mating. *Diseases of the Nervous System, 33,* 509–511.

Robins, L. N. (1966). *Deviant children grown up: A sociological and psychiatric study of sociopathic personality.* Baltimore: Williams & Wilkins.

Robins, L. N. (1982). The diagnosis of alcoholism after DSM III. In E. Pattison & E. Kaufman (Eds.), *Encyclopedic handbook of alcoholism* (pp. 40–54). New York: Gardner Press.

Robins, L., Bates, W., & O'Neal, P. (1962). Adult drinking patterns of former problem children. In D. J. Pittman & C. R. Snyder (Eds.), *Society, culture, and drinking patterns* (pp. 395–412). London: Feffer & Simons.

Robins, L., Helzer, J., Croughan, J., Williams, J., & Spitzer, R. (1979). *The National Institute of Mental Health Diagnostic Interview* (DIS, Version II). Washington, DC: National Institute of Mental Health.

Robins, L., Helzer, J., Weissman, M., Orvaschel, H., Gruenberg, E., Burke, J., & Reiger, D. (1984). Lifetime prevalence of specific psychiatric disorders in three sites. *Archives of General Psychiatry, 41,* 949–958.

Rosler, M., Bellaire, W., Hengesch, G., Giannitsis, D., & Jarovici, A. (1983). Genetic markers in alcoholism: No association with HLA. *Archives of Psychiatry and Neurological Sciences, 233,* 327–331.

Rutter, M. (1982). Syndromes attributed to "minimal brain dysfunction" in childhood. *American Journal of Psychiatry, 139,* 21–33.

Rutter, M., Shaffer, D., & Shepherd, M. (1975). *A multiaxial classification of child psychiatric disorders.* Geneva: World Health Organization.

Ryback, K., & Eckhardt, M. (1978). Toward a biochemical definition of alcoholism. In I. Hanin & E. Usdin (Eds.), *Biological markers in psychiatry and neurology*. New York: Pergamon Press.

Safer, D., & Allen, R. (1976). *Hyperactive children: Diagnosis and management*. Baltimore: University Park Press.

Schaeffer, K. W., Parsons, O. A., & Yohman, J. R. (1984). Neuropsychological differences between male familial and nonfamilial alcoholics and nonalcoholics. *Alcoholism: Clinical and Experimental Research, 8,* 347–351.

Schuckit, M. A. (1980). Biological markers: Metabolism and acute reactions to alcohol in sons of alcoholics. *Pharmacology Biochemistry & Behavior, 13,* 9–16.

Schuckit, M. A. (1984a). Relationship between the course of primary alcoholism in men and family history. *Journal of Studies on Alcohol, 45,* 334–338.

Schuckit, M. A. (1984b). Subjective responses to alcohol in sons of alcoholics and control subjects. *Archives of General Psychiatry, 41,* 879–884.

Schuckit, M. A. (1984c). Prospective markers for alcoholism. In D. W. Goodwin, K. T. Van Dusen, & S. A. Mednick (Eds.), *Longitudinal research in alcoholism* (pp. 147–163). Boston: Kluwer-Nijhoff.

Schuckit, M. A., Goodwin, D. A., & Winokur, G. (1972). A study of alcoholism in half siblings. *American Journal of Psychiatry, 128,* 1132–1136.

Schuckit, M. A., & Morrissey, E. R. (1979). Psychiatric problems in women admitted to an alcoholic detoxification center. *American Journal of Psychiatry, 136,* 611–617.

Schuckit, M. A., Pitts, F. N., Reich, T., King, L. J., & Winokur, G. (1969). Alcoholism: I. Two types of alcoholism in women. *Archives of General Psychiatry, 20,* 301–306.

Schuckit, M. A., & Rayses, V. (1979). Ethanol ingestion: Differences in blood acetaldehyde concentrations in relatives of alcoholics and controls. *Science, 203,* 54–55.

Schuckit, M. A., Rimmer, J., Reich, T., & Winokur, G. (1970). Alcoholism: antisocial traits in male alcoholics. *British Journal of Psychiatry, 117,* 575–76.

Shakow, D. (1977). Schizophrenia selected papers. *Psychological Issues, Monograph 38.* New York: International Universities Press.

Skinner, H. A., & Jackson, D. N. (1974). Alcoholic personality types: Identification and correlates. *Journal of Abnormal Psychology, 83,* 658–666.

Spitzer, R., Endicott, J., & Robins, E. (1978). Research diagnostic criteria: Rationale and reliability. *Archives of General Psychiatry, 35,* 773–782.

Stabenau, J. (1982). The social implications of antisocial personality as an "early onset" type of alcoholism. *Report of the Fourth World Congress for the Prevention of Alcoholism and Drug Dependency* (pp. 403–405). Berrien Springs, MI: University Printers.

Stabenau, J. (1983). Typing alcoholism: Factors of genetic importance. *Seventh World Congress of Psychiatry* (p. 356). Vienna: World Psychiatric Association.

Stabenau, J. (1984). Implications of family history of alcoholism, antisocial personality, and sex differences in alcohol dependence. *American Journal of Psychiatry, 141,* 1178–1182.

Stabenau, J. (1985). Basic research on heredity and alcohol: Implications for clinical application. *Social Biology, 32,* 297–324.

Stabenau, J., Dolinsky, Z., & Fischer, B. (1986). Alcohol consumption: Effect of gender and psychopathology. *Alcoholism: Clinical and Experimental Research, 10,* 355–356.

Stabenau, J., & Hesselbrock, V. (1980). Assortative mating, family pedigree and alcoholism. *Substance and Alcohol Actions/Misuse, 1,* 375–382.

Stabenau, J., & Hesselbrock, V. (1984a). Genetic precursors and consequences in alcoholism. In A. Hemmi & K. Tuhkanen (Eds.), *World Psychiatric Association Regional Symposium* (p. 63). Helsinki: Kyriiri oy.

Stabenau, J., & Hesselbrock, V. (1984b). Psychopathology in alcoholics and their families and vulnerability to alcoholism: A review and new findings. In S. Mirin (Ed.), *Substance abuse and psychopathology* (pp. 108–132). Washington, DC: American Psychiatric Press.

Stewart, M. A., DeBlois, C. S., & Cummings, C. (1979). Psychiatric disorder in the parents of hyperactive boys and those with conduct disorder. *Journal of Child Psychology and Psychiatry and Allied Disciplines, 21,* 283–292.

Straus, R. (1983). Types of alcohol dependence: In B. Kissin & H. Begleiter (Eds.), *The pathogenesis of alcoholism: Psychosocial factors* (Vol. 6, pp. 1–16). New York: Plenum Press.

Suwaki, H., & O'Hara, H. (1985). Alcohol-induced facial flushing and drinking behavior in Japanese men. *Journal of Studies on Alcohol, 46,* 196–198.

Swinson, R. P. (1983). Genetic markers and alcoholism. In M. Galanter (Ed.), *Recent developments in alcoholism* (Vol. 1, pp. 9–24). New York: Plenum Press.

Syndulko, K. (1978). Electrocortical investigations of sociopathy: In R. Hare & D. Schalling (Eds.), *Psychopathic behavior approaches to research.* New York: Wiley.

Tabikoff, B., Melchoir, C., Hoffman, P. (1984). Factors in ethanol tolerance. *Science, 224,* 523–524.

Tarter, R. E., & Alterman, A. I. (1984). Neuropsychological deficits in alcoholics: Etiological considerations. *Journal of Studies on Alcohol, 45,* 1–9.

Tarter, R. E., Hegedus, A. M., Goldstein, G., Shelly, C., & Alterman, A. I. (1984). Adolescent sons of alcoholics: Neuropsychological and personality characteristics. *Alcoholism: Clinical and Experimental Research, 8,* 216–222.

Tarter, R. E., McBride, H., Buonpane, N., & Schneider, D. U. (1977). Differentiation of alcoholics. *Archives of General Psychiatry, 34,* 761–768.

Taylor, J. R., & Helzer, J. E. (1983). The natural history of alcoholism. In B. Kissin & H. Begleiter (Eds.), *The pathogenesis of alcoholism: Psychosocial factors* (pp. 17–65). New York: Plenum Press.

Thacker, S. B., Veech, R. L., Vernon, A. A., & Rutstein, D. D. (1984). Genetic and biochemical factors relevant to alcoholism. *Alcoholism: Clinical and Experimental Research, 8,* 375–383.

Thoday, J. (1967). New insights into continuous variation. In J. Crow & J. Need (Eds.), *Proceedings of the Third International Congress of Human Genetics* (pp. 339–350). Baltimore: Johns Hopkins University Press.

Trice, H. M., & Wahl, J. R. (1958). A rank order analysis of the symptoms of alcoholism. *Quarterly Journal of Studies on Alcohol, 19,* 638–648.

Utne, H. E., Hanse, F. V., Winkler, K., & Schulsinger, F. (1977). Alcohol elimination rates in adoptees with and without alcoholic parents. *Journal of Studies on Alcohol, 38,* 1219–1223.

Vaillant, G. E. (1980). Natural history of alcoholism: I. A preliminary report. In S. B. Sells, R. Crandall, M. Roff, J. S. Strauss, & W. Pollin (Eds.), *Human functioning in longitudinal perspective* (pp. 147–156). Baltimore: Williams & Wilkins.

Vaillant, G. E. (1983). *The natural history of alcoholism.* Cambridge, MA: Harvard University Press.

Vaillant, G. E., & Milofsky, E. S. (1982). The etiology of alcoholism: A prospective viewpoint. *American Psychologist, 37,* 494–503.

Virkkunen, M. (1979). Alcoholism and antisocial personality. *Acta Psychiatrica Scandinavica, 59,* 493–501.

Vogel, F., & Motulsky, A. G. (1979). *Human genetics: Problems and approaches.* New York: Springer-Verlag.

VonWartburg, J. (1971). Metabolism of alcohol in normals and alcoholics: Enzymes. In B. Kissin & H. Begleiter (Eds.), *The biology of alcoholism: Vol. 1. Biochemistry* (pp. 13–102). New York: Plenum Press.

VonWartburg, J. P. (1980). Acetaldehyde. In M. Sandler (Ed.), *Psychopharmacology of alcohol* (pp. 137–147). New York: Raven Press.

VonWartburg, J., & Buhler, R. (1984). Alcoholism and aldehydism: New biomedical concepts. *Laboratory Investigation, 50,* 5–15.

Weissman, M. M., Myers, J. K., & Harding, P. S. (1980). Prevalence and psychiatric heterogeneity of alcoholism in a United States urban community. *Journal of Studies on Alcohol, 41*, 672–681.

Wender, P. (1971). *Minimal brain dysfunction in children*. New York: Wiley-Interscience.

West, J. R., & Hodges-Savola, C. A. (1983). Permanent hippocampal mossy fiber hyperdevelopment following prenatal ethanol exposure. *Neurobehavioral Toxicology and Teratology, 5,* 139–150.

Whalley, L. J. (1980). Social and biological variables in alcoholism: A selective review.. In M. Sandler (Ed.), *Psychopharmacology of alcohol* (pp. 1–15). New York: Raven Press.

Wilson, J. R., Erwin, V. G., DeFries, J. C., Petersen, D. R., & Cole-Harding, S. (1984). Ethanol dependence in mice: Direct and correlated responses to ten generations of selective breeding. *Behavior Genetics, 14,* 235–257.

Wilson, J. R., Erwin, V. G., McClearn, G. E., Plomin, R., Johnson, R. C., Ahern, F. M., & Cole, R. E. (1984). Effects of ethanol: II. Behavioral sensitivity and acute behavioral tolerance. *Alcoholism: Clinical and Experimental Research, 8,* 366–374.

Winokur, G. (1974). The division of depressive illness into depression spectrum disease and pure depressive disease. *International Pharmacopsychiatry, 9,* 5–13.

Winokur, G., Reich, T., Rimmer, J., & Pitts, F. N. (1970). Alcoholism: III. Diagnosis and familial psychiatric illness in 259 alcoholic probands. *Archives of General Psychiatry, 23,* 104–111.

Witkin, H., Dyk, R., Fatterson, H., Goodenough, D., & Karp, S. (1962). *Psychological differentiation*. New York: Wiley.

Wood, D. R., Wender, P. H., & Reimherr, F. W. (1983). The prevalence of attention deficit disorder, residual type, of minimal brain dysfunction, in a population of male alcoholic patients. *American Journal of Psychiatry, 140,* 95–98.

Woodruff, R., Goodwin, D., & Guze, S. (1974). *Psychiatric diagnosis*. New York: Oxford University Press.

Wright, S. (1968). *Evolution and the genetics of populations* (Vol. 1). Chicago: University of Chicago Press.

Author Index*

*Full bibliographic references can be found at the end of each chapter.

P

Q

R

S

Subject Index